America's Environmental Report Card

America's Environmental Report Card
Are We Making the Grade?

Second Edition

Harvey Blatt

The MIT Press
Cambridge, Massachusetts
London, England

This book was set in Sabon by Toppan Best-set Premedia Limited.

Library of Congress Cataloging-in-Publication Data

Blatt, Harvey.
America's environmental report card : are we making the grade? / Harvey Blatt.—2nd ed.
   p. cm.
Includes bibliographical references and index.
ISBN 978-0-262-51591-7 (pbk. : alk. paper)
1. United States—Environmental conditions. 2. Pollution—United States.
I. Title.
GE150.B58   2011
363.700973—dc22

2010026307

Contents

Preface

Every nation has environmental problems, a result of the fact that humans are part of the natural environment but are sometimes unhappy about the limitations their surroundings place on them. (No doubt other animals are sometimes unhappy about this as well, but their options for changing things are rather limited.) We may want to live on a mountainside, but the types of rocks and their orientation on the slope make construction risky. The frequent landslides and resulting destruction of houses in the developed parts of hilly California are an example of this. Or we may want to grow a variety of warm-weather thirsty crops on our acreage but cannot because of the cold climate, inadequate rainfall, and high cost of water.

People in many nations have similar environmental problems, typically involving water, soil, and air. Solutions to the problems in one country can often be applied in other countries as well. As a transplanted American, I am aware of many of the environmental problems in the United States: water shortages and pollution, air pollution, soil degradation, and contaminated food, to name a few. I grew up in and around New York City and later lived in Detroit, Michigan; Columbus, Ohio; Ithaca, New York; Austin, Texas; Los Angeles, California; Houston, Texas; and central Oklahoma. I now live in Jerusalem, Israel. Major cities, small towns, and rural areas are part of my background and have allowed me to experience many of the environmental problems that concern Americans.

I believe the problems have gotten worse during my lifetime, and this is the reason I wrote the first edition of this book. However, many things seem to be changing in the United States concerning environmental problems, a circumstance resulting from the marked change in national policy with the election of President Barack Obama. His view of the nation's environmental problems and the need to solve them differs significantly from the view of his predecessor. The next few years may witness noticeable improvements in the way the nation deals with the environment.

This new edition of *America's Environmental Report Card* differs from the earlier one in several respects. Every chapter has been completely rewritten and updated with environmental information that has appeared since the first edition was written in 2003 and 2004. There is a new chapter on the deteriorating infrastructure in the United States— our water and sewer lines, power lines, roads, bridges, railroads, dams, airports, and levees. The chapter on energy in the first edition has been split into two— Fossil Fuels and Alternative Energy— in response to the emphasis on the development of alternative energy by President Obama. The new concluding chapter emphasizes the cultural shift that is necessary if the United States and the rest of the world are to survive, and how the world's spiritual heritage may be the key to effecting the change.

Many Americans—indeed, many of the world's citizens—do not see the environment as something that is integral to their daily lives. It tends to be considered an outside issue, often associated with scientists and academics rather than something that is central to their lives. On an everyday basis, most of us do not link what is happening in the environment with our daily activities. The environment is viewed as something "out there," akin to the quest for world peace or the rings around Saturn rather than something that is with us and affecting us every day of our lives. Yet we cannot live without the water we drink, the soil in which our crops grow, and the air we breathe.

The concept of sustainability is not part of most people's lives. The fact that the planet's resources are limited and that these resources must sustain not just the generations today but also those generations to come, and that those generations have the right to be able to use those same resources and to meet what they feel are their needs is not something that affects our daily actions. Our responsibility to future generations takes a back seat to what we perceive as our current needs. But nearly all of what we think of as needs, at least in Western societies, are actually wants and could easily be done without with no great loss to our happiness. We need to treat the earth as if we intended to stay.

My objective in this book is to inform students, policymakers, politicians, and natural resource managers about the environmental problems we have created and to suggest ways of solving them that are within reach. I would appreciate comments from readers about omissions or errors they believe I have committed.

Introduction

Don't blow it. Good planets are hard to find.

Since publication of the first edition of this book in 2005, the volume of material on the environment has continued to mushroom. Chief among this increase have been the twin issues of energy sources and climate change. These two issues are intimately intertwined because the coal used to generate almost half of America's electric power, 80 percent of China's, and 56 percent of India's, together with the gasoline used to fuel the ever-expanding number of automobiles, generate most of the carbon dioxide that most climate scientists believe is the main anthropogenic cause of climate change. And the amount of the gas in the air continues to increase despite international agreements to decrease it.

At the end of 2009, polling revealed that Americans have lost faith in both scientific assessments of climate change and their government's response to it. Forty percent of Americans say they have little or no trust in what scientists say about the environment. Politicians fare even worse on the topic.

Nevertheless, recent data indicate that previous estimates of the rate of glacial melting and sea level rise may be gross underestimates. Antarctic and Arctic ice and mountain glaciers are retreating faster than anyone predicted, meaning that our low-lying coastal areas will be in increased danger of flooding sooner than expected. In addition, the strength of Atlantic hurricanes is increasing. Because of the catastrophe of Hurricanes Katrina and Rita in 2005, some environmentalists have questioned whether New Orleans can be saved, and at what cost, and whether it is worth doing.

Water supplies and purity continue to be matters of concern. Despite conservation efforts, water shortages in the United States are spreading. Water managers in thirty-six states anticipate local or regional water shortages by 2020. Water costs increased 27 percent between 2002 and

2007 in the United States and 58 percent in Canada. Most irrigation of fruits and vegetables in Florida and California is now accomplished with reclaimed wastewater, and these two states have most of the nation's desalination plants, which are increasing in importance as sources of water.

Recent data indicate that Americans share their bodies with thousands of industrial chemicals and that these toxins may be a cause of the rapid increase in Alzheimer's disease, dementias, autism, and perhaps other maladies. Naturally occurring genetic changes occur too slowly to account for the rapid increases, so they more likely result from environmental factors. Equally frightening is the recognition of endocrine disrupters among the industrial chemicals in our bodies and their effects on the development of both male and female sexual organs and reproductive abilities.

A recent addition to environmental problems in the United States is the recognition at the federal level of our crumbling infrastructure and the urgent need to repair and upgrade it. Most of the nation's underground water mains are quite old and break regularly at unknowable times and places. With water shortages spreading, we cannot afford to allow these breaks to cause the routine average loss of 15 percent of the water that cities pump to our citizens.

The nation's electrical transmission lines are inadequate to carry the current from our growing alternative energy sources and are handicapping the transition to wind, solar, and geothermal energy. A major refurbishing of the electrical grid is sorely needed.

Less than 1 percent of the increasing amount of food we import is inspected for cleanliness and purity. Inspections are equally inadequate in the domestic food industry, and outbreaks of illness caused by salmonella and *Escherichia coli* are increasing. Pesticide residues on the food we eat are a persistent problem, perhaps equal in importance to the industrial chemicals in the water we drink.

The reduction in air pollution is the one bright spot in the litany of environmental problems. The air we breathe is much cleaner than it was a few decades ago and appears ready to improve still more with the advent of hybrid cars, electric cars, and biofuels. Industrial emissions are still excessive, but better enforcement of existing pollution laws and perhaps the introduction of new ones are priorities of the Obama administration. There is reason to be hopeful about continued improvement in air quality in the United States.

In 2009, for the first time in twenty-five years of polling, economic concerns replaced environmental concerns as the major worry of the

American public, putting us in the same boat as developing nations such as China and India. This will very likely be reversed in all countries as their economies improve—in the United States in the next few years, but not in China and India for many decades. The first concerns of governments in developing countries are food, shelter, and clothing for their impoverished populations, and these cannot be provided overnight.

A secondary reason for the decline in environmental concern among citizens and legislators is the language used in discussions about the environment. "Cap-and-trade" means little to most people and would be easier to sell as a "pollution reduction refund." The term "energy efficiency" makes people think of shivering in the dark. It would be accepted better if the need were couched as "saving money for a more prosperous future." Perhaps the word "environment" should be replaced with wording like "purity of the air we all breathe and the water our children drink."

Polluting industries play this word spin all the time. Coal companies promise "clean green coal," an oxymoron if there ever was one. The natural gas industry, whose product is much less polluting than coal but hardly comparable to alternative energy sources, refers to "clean fuel green fuel." Frank Luntz, a communications consultant and pollster, suggested in 2002 that environmentalists refer to themselves as "conservationists" and emphasize "common sense" over scientific argument.

Advertising techniques are always used to manipulate public opinion. Few people want to save a "jungle," but much support can be gotten to save a "tropical rain forest." Saving a "swamp" is a harder sell than "protecting a mangrove forest." "Saving alligators" does not have the appeal of "filtering our water supply" or "protecting the coast from hurricane flooding." Words have the power to lead and mislead.

The quality of our environment is in our hands. We have caused its deterioration, and we can restore it. But it will require a major effort extending over many decades. The environment did not deteriorate overnight and cannot be restored overnight. We must be willing to put lots of money into the effort, but we can manage it. Compared to other industrialized nations, the United States has the lowest tax burden, lowest water bills, lowest gasoline cost at the pump, and lowest many other things. All Americans like and are proud of this accumulation of lowests. There is no evidence that the cost of environmental improvement will cause us to lose our collection of lowests. Other nations are also spending money to improve their environment.

The United States should be leading humanity to a better, more sustainable world, and not be seen as recalcitrant humans dragged kicking and screaming to a better environment. Americans not only can do it but must. We are the richest nation in the world and can afford it. Or more accurately, we can't afford not to.

1

Water: No Cholesterol, Fat Free, Zero Sugar

You never miss the water till the well runs dry.
—Rowland Howard, *You Never Miss the Water*, 1876

The amount of water used in the United States is staggering. In 2005, it was 410 million gallons *per day*, not including the 15 to 20 percent lost to leaky pipes. Total consumption has varied by only 3 percent since 1990. Per capita use peaked in 1970 at 1,815 gallons but has since declined continuously to 1,363 in 2005, a result of conservation by industry, agriculture, and home owners (table 1.1). Power plants use about half of the 410 million gallons, agriculture 31 percent, homes and businesses use 11 percent, and the remaining 8 percent includes use by mining, livestock, aquaculture, and individual domestic wells.[1]

But despite conservation efforts, water shortages are spreading, and experts believe we are moving into an era of water scarcity throughout the United States. We are used to hearing of shortages in the arid and semiarid Southwest, but there are now problems in the Midcontinental grain belt, South Carolina, New York City, southern Florida, and other areas most Americans think of as water rich. In 2003, the General Accounting Office published a survey that found that water managers in thirty-six states anticipate water shortages locally, regionally, or state-wide within the next ten years. There already is a tristate water war among Alabama, Florida, and Georgia.[2]

Unfortunately, the gravity of the situation has not yet set in for most Americans, who tend to view water shortages as temporary—the result of short-term droughts, poor water management by local authorities, or an unusually light snowfall in mountain areas. The erroneous nature of this view is reflected in the fact that between 2002 and 2007, municipal water use rates in the United States increased by 27 percent. People in other nations have seen even larger increases: 32 percent in the United

Table 1.1
U. S. water withdrawals per day, 1975–2000

	Total (billion gal.)	Per capita (gallons)	Irrigation %	Public supply %	Rural %	Industrial %	Steam electric utilities %
1975	420	1,972	33.3	6.9	1.7	10.7	47.6
1980	440	1,953	34.1	7.7	1.3	10.2	47.7
1985	399	1,650	34.3	9.5	2.0	7.8	46.9
1990	408	1,620	33.6	10.0	1.9	7.4	47.8
1995	402	1,500	33.3	10.0	2.2	7.2	47.3
2000	408	1,430	33.6	10.5	2.3	5.6	48.0

Source: "Estimated Use of Water in the United States," U.S. Geological Survey Circular 1268, 2005.

Kingdom, 45 percent in Australia, 50 percent in South Africa, and 58 percent in Canada.[3]

The water problems foreseen decades ago by hydrologists threaten farm productivity, limit population and economic growth, increase business expenses, and drive up prices. Nearly every product uses water in some phase of its production. Reclaimed sewer water is now in wide use for agricultural and other nondrinking purposes. Desalination plants are springing up around the country.

About 30 percent of the water American families consume is used outdoors for watering lawns and gardens, washing cars, maintaining swimming pools, and cleaning sidewalks and driveways.[4] Clearly, nearly all these uses are unnecessary. They remain from the days when the nation had a lower population, fewer houses with large lawns, fewer cars to wash, and fewer swimming pools, and Americans were more willing to expend energy by using a broom on driveways and sidewalks.

The lack of water is imposing limits on how the United States grows. Freshwater scarcity is a new risk to local economies and regional development plans across the country. In 2002, California put into effect a state law that requires developers to prove that new projects have a plan for providing water for at least twenty years before local water authorities can approve their projects. Builders in the humid Southeast are facing limits to planting gardens and lawns for new houses.

The Water Future

According to Peter Gleick in 2008, president of the Pacific Institute, a think-tank specializing in water issues, "The business-as-usual future is a bad one. We know that in five years we'll be in trouble, but it doesn't have to be that way. If there were more education and awareness about water issues, if we started to really think about the natural limits about where humans and ecosystems have to work together to deal with water, and if we were to start to think about efficient use of water, then we could reduce the severity of the problems enormously. I'm just not sure we're going to."[5] It seems that no one has looked at the subject from the point of view of what is sustainable. There does not seem to be anyone in state or federal governments thinking about the long-range big picture that would put the clamps on large-scale development. Politicians rarely want to tell their constituents that they must curb their insatiable appetites for anything.

Where Does Our Water Come From?

The water our lives depend on originates in the world's oceans, from where it evaporates and is carried by air currents over land surfaces. The chief proximate sources are large river systems such as the Mississippi and Ohio in the East and Midwest, and the Colorado and Rio Grande in the West; large lakes such as the five Great Lakes along the Canadian border; and underground aquifers such as the Ogallala in the Midcontinent from north Texas to South Dakota. The water in each of these sources is either decreasing or experiencing increased pollution from the artificial chemicals we inject into it—or both. The Colorado and Rio Grande no longer reach the sea year round because a growing share of their waters are claimed for various uses.

The Colorado River

The Colorado River, with an annual flow of 5 trillion gallons of water, is perhaps the best example of the unsustainable overuse of river water in the United States. A common misconception of water use in the basin and in the West in general is that rapidly growing urban areas are the main users of the region's limited water. In fact, 85 to 90 percent of the water is used in agriculture, mainly to grow food for cattle.[6] Only 10 to 15 percent of the water is used directly by the 25 million people served by the river who live in Los Angeles, Phoenix, and other communities. How much of the water is used to keep swimming pools filled and lawns watered in this dry climate is unknown. But clearly the river's water is oversubscribed, because the river's channel is dry at its entrance into the Gulf of California (figure 1.1). Five trillion gallons of water per year is not enough to satisfy both the needs and wants of 25 million people.

The shortage of water in the Colorado River was recognized many decades ago, and there have been many lawsuits by those who felt slighted by their legislated allocations. The problem was most severe in years when annual rainfall was less than average, so to alleviate this problem, the federal government built many dams and reservoirs along the river to store water and smooth out yearly variations. But lawsuits persisted.

Finally, after years of wrangling and facing the worst drought in a century, and with the prediction that climate change will probably make the Southwest drier in the future, federal officials in 2007 forged a new pact with the states on how to allocate water if the river runs short. The pact puts in place new measures to encourage conservation and manage

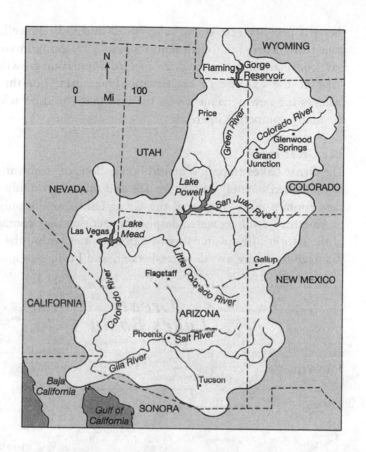

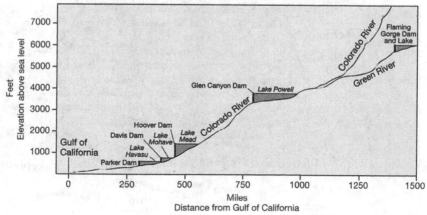

Figure 1.1
Drainage area served by the Colorado River and the dams constructed to minimize the effect of yearly variations in precipitation.

the two primary reservoirs, Lake Mead and Lake Powell, which have gone from nearly full to half-empty since 1999.[7] Some environmentalists have complained that managing expected population growth was emphasized at the expense of conservation measures, but the government believes the new agreement was the best that could be achieved among the many competing interests.

The Great Lakes

The Great Lakes contain 6 quadrillion gallons of freshwater, 20 percent of the world's supply (see figure 1.2 and table 1.2). Only the polar ice caps contain more. However, the Great Lakes supply only 4.2 percent of America's drinking water, despite the fact that they contain 90 percent of the nation's freshwater supply. Communities within the Great Lakes' drainage basin are awash in freshwater, and businesses and residents in

Figure 1.2
Drainage basin of the Great Lakes. (Atlas of Canada)

Table 1.2
Numerical information about the Great Lakes

	Lake Ontario	Lake Erie	Lake Huron	Lake Michigan	Lake Superior
Surface area (sq. miles)	7,540	9,940	23,010	22,400	31,820
Water volume (cu. mil.)	393	116	849	1,180	2,900
Elevation (feet)	246	571	577	577	609
Average depth (feet)	283	62	195	279	483

the area want to keep it that way. In October 2008 their desires were codified when President George W. Bush signed the Great Lakes–St. Lawrence River Basin Water Compact that had previously been approved by the eight states bordering the lakes and the adjacent Canadian provinces. Ken Kilbert, director of the Legal Institute of the Great Lakes, stated that the document was "the best legal step so far to protect the most important resource in our area from diminishment."[8]

Water withdrawals from the Great Lakes total 43 million gallons per day, with almost two-thirds withdrawn on the U.S. side. Nearly all of the water is returned to the basin through runoff and discharge. Only 5 percent is made unavailable by evapotranspiration or incorporation into manufactured products.[9] Considering that the water volume in the five lakes totals 5,438 cubic miles and climate change is forecast to increase precipitation in the area of the Great Lakes, there is not a looming problem with water supply for those with access.

The compact protects against most new or increased diversions of water outside the Great Lakes Basin. Diversions refer to the transfer of water from the Great Lakes to areas outside the Great Lakes watershed. The compact also promotes conservation and efficiency programs that enforce better use of water within the basin, 72 percent of which is used in power plants and is recycled. Public water systems use 13 percent, industry consumes 10 percent, and other uses total 5 percent.[10]

Many politicians believe they see water wars on the horizon, and there is no way for the Great Lakes states to prevent the federal government from taking the water if it wants to do so. Probably the Great Lakes Compact will not be the final word on distribution of the water in the lakes. The balance of political power in Washington has been tilting south and west for decades, and agricultural interests in the nation's midwestern breadbasket will increasingly covet the water in the lakes as water levels in the Ogallala aquifer they depend on continue to drop.

The Ogallala Aquifer

Twenty percent of America's water use comes from underground aquifers, the largest of which by far is the series of sandstones and conglomerate called the Ogallala Formation (figure 1.3).[11] It extends over an area of about 174,000 square miles in parts of eight states, from Wyoming and South Dakota in the north to New Mexico and Texas in the south. About 27 percent of the irrigated land in the United States overlies this aquifer system, which yields about 30 percent of the nation's groundwater used for irrigation. Water from the Ogallala aquifer serves an area that produces 25 percent of U.S. food grain exports and 40 percent of wheat, flour, and cotton exports. In addition, it provides drinking water to 82 percent of the people who live within the aquifer boundary.[12]

Ogallala water irrigates more than 14 million acres of farmland, areas with only 16 to 20 inches of rainfall—not enough for the abundance of corn, wheat, and soybeans American farmers have come to expect. The aquifer averages 200 feet thick and holds more than 70 quadrillion gallons of water (70,000,000,000,000,000 gallons) in its pores. The water accumulated undisturbed from rainfall over millions of years, but for the past eighty-five years, the water has been withdrawn from thousands of wells at a rate that is eight times the current replenishment rate from the low annual rainfall.[13] Farmers are pumping more groundwater. In 1950, 30 percent of irrigation water came from aquifers; in 2005, 62 percent did.[14] Water levels have declined 30 to 60 feet in large areas of Texas, and many farmers in the High Plains are now turning away from irrigated agriculture. Wells must be deepened, and the costs of the deepening and increased pumping have caused some agricultural areas to be abandoned. If overpumping of the Ogallala continues, the aquifer may be effectively dry within a few decades, with disastrous effects on the economy of a large area of the United States.

Prospects for the Future

Our ability to irrigate at low cost is coming to an end, not only in the Midcontinent but in other areas as well. As noted earlier, the Great Lakes will come under increasing pressure from states in the Midwest and Southwest up to 1,500 miles away to share the enormous volume of water currently under the control of the eight states bordering the lakes.

The cost of transporting water is determined largely by how far it has to be carried and how high it has to be lifted. The elevations of the three largest Great Lakes are between 577 feet and 609 feet, but the elevations of the area served by the Ogallala range from about 2,000 feet to 3,600

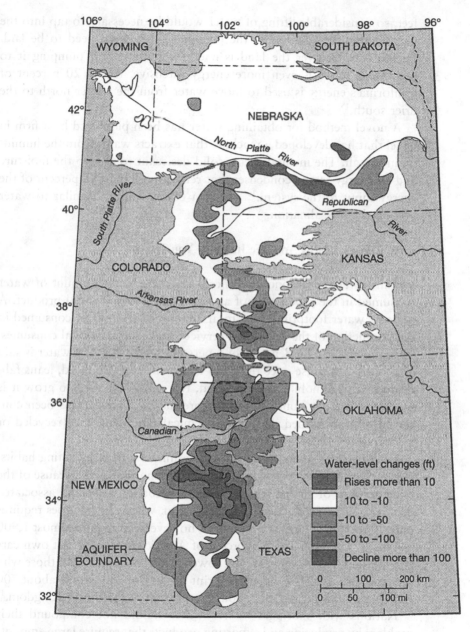

Figure 1.3
Changes in water level in the Ogallala aquifer between 1850 and 1980. The declines have continued to the present day. (U.S. Geological Survey).

feet, so considerable lifting of water would be necessary to tap into the Great Lakes, in addition to the pipelines that would need to be laid. Pumping water over the land is energy intensive, and pumping it to higher elevations is even more energy intensive. About 20 percent of California's energy is used to move water from the wetter north to the drier south.[15]

A novel method for obtaining water has been pioneered by a firm in Israel that has developed a machine that extracts water from the humidity in the air. The method uses a solid desiccant to absorb the moisture and an energy-saving condenser that reuses more than 85 percent of the energy input to the system.[16] The cost of the water is similar to water produced by desalination.

Virtual Water: Now You See It, Now You Don't

Virtual water is an economic concept referring to the amount of water consumed in the production of an agricultural or industrial product. A person's water footprint is the total amount of freshwater consumed in the production of the goods and services that that individual consumes. Virtual water is a hidden part of a person's water use. The water is said to be virtual because once the grain is grown, beef produced, jeans fabricated, or automobile manufactured, the real water used to grow it is no longer actually contained in the product as water. It has been consumed or transformed into other chemicals and cannot be recycled or recovered (table 1.3).

Each person's water footprint is determined largely by eating habits. Vegetarians have a lower water footprint than omnivores because of the large amount of virtual water needed to produce meat and associated dairy products. Producing a pound of corn, wheat, or potatoes requires only 30 to 160 gallons of water; beef, however, can require almost 1,900 gallons (figure 1.4). The 10 percent of Americans who do not own cars and families with fewer cars have lower water footprints than those who are more affluent. The water footprint of the United States is about 700 gallons per year per person, about double that in the United Kingdom.[17]

Nations with shortages of freshwater should not compound their problem by producing and exporting products that require large amounts of water in their production. For example, in Israel, a nation that is mostly arid to semiarid and where water shortages are common, the export of oranges has been discouraged since the 1980s because it is a relatively thirsty crop and it makes no sense to send Israel's water to

Table 1.3
Virtual water in various food and manufactured products

Amount of product	Water consumed (gallons)
FOOD	
1 cup coffee	37
1 pound corn	108
1 pound wheat	156
1 pound rice	185
1 quart milk	208
1 pound soybeans	363
1 pound broken rice	407
1 pound poultry	542
1 dozen large eggs	592
1 pound pork	608
1 pound beef	1,800
MANUFACTURED	
Diaper	215
Cotton shirt	300
Bed sheet	2,584
Jeans	2,875
Passenger car	106,000
Average house	1,590,000

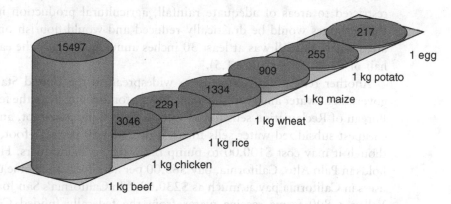

Figure 1.4
Amounts of virtual water in various foodstuffs, expressed in liters per kilogram. (Office for Economic Co-operation and Development)

more water-rich countries. The United States exports huge amounts of virtual water in its agricultural products and automobiles. The United States and the European Union countries export to the Middle East and North Africa as much water as flows down the Nile into Egypt for agriculture each year. The volume is more than 40 billion tons, embedded in 40 million tons of grain.

How Do We Use It?

Water is used in three main areas: agriculture, industry, and homes. Usage grew three times faster than America's population during the twentieth century. The increase was due largely to the expansion of agriculture, by far the biggest consumptive user of water in the United States.

Agriculture

Farming drinks 34 percent of the nation's water, most of it from groundwater. The profligate use of groundwater is the reason a large part of America's most productive cropland can be located in areas with relatively low annual rainfall. Much of the midcontinental grain belt averages less than 25 inches of rain per year; the San Joaquin Valley in California produces half of the nation's fruits and vegetables but receives only 8 to 12 inches of rainfall in an average year. If farming were restricted to areas of adequate rainfall, agricultural production in the United States would be drastically reduced and would flourish only in areas where rainfall was at least 30 inches annually, roughly the eastern half of the country (figure 1.5).

Another reason agriculture is so widespread in the United States is government water and crop subsidies. Water for farming from the federal Bureau of Reclamation sells for $10.00 to $15.00 per acre-foot, and the cheapest subsidized water sells for as little as $3.50 per acre-foot, even though it may cost $100.00 to pump the water to the farmers. Households in Palo Alto, California, pay $65.00 per acre-foot, and some urban users in California pay as much as $230.00.[18] In California's San Joaquin Valley, 6,800 farms receive water from the federally funded Central Valley Project, built in 1936 for $3.6 billion. The Environmental Working Group reported in 2005 that in 2002, farms received $538 million in combined water and crop subsidies, $416 million of which was for water. In 2002, the average price for irrigation water from the Central Valley Project was less than 2 percent of what residents of Los Angeles

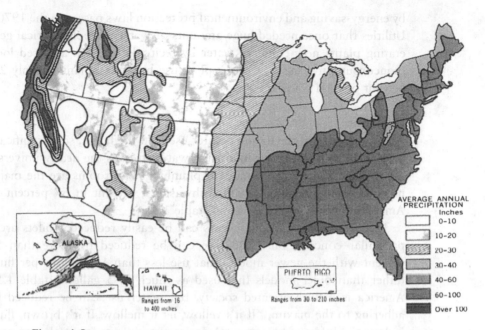

Figure 1.5
Average annual precipitation in the United States. (U.S. Water Resources council, 1968, *The Nation's Water Resources*, 1968)

paid for drinking water, one-tenth the estimated cost of replacement water supplies, and about one-eighth of what the public paid to buy its own water back to restore the San Francisco Bay and delta.[19]

Increased agricultural efficiency in water use has contributed significantly to water conservation efforts in the United States. Irrigation methods have been improved by decreasing the use of flood irrigation and up-into-the-air water sprinklers in favor of techniques such as having drip tubes extending vertically from the sprinkler arm immediately above the plants. Another efficient method is drip irrigation, pioneered in Israel, in which water is released in measured amounts from tubes on the ground directly above the plant roots. Losses of water to evaporation and runoff are nearly eliminated by these methods.

Industry

Nearly half (48 percent) of the 408 billion gallons used in the United States goes to power plants, which have greatly reduced their water requirements from the past. They have made the biggest reduction in water use in recent decades, a result of water-saving technology driven by energy-saving and environmental protection laws passed in the 1970s. Utilities that once needed huge amounts of water to cool electrical generating plants now conserve water by recirculating it in a closed loop (nonconsumptive use).[20] Consumptive use by power plants is only 2.5 percent of total water use.[21]

Home Use

About one-fifth of the nation's water use is in the home, so a significant part of the reason for our increasing water stress is the nearly universal access Americans have to modern plumbing. Bathrooms are the major users of water in homes, with dishwashers, present in 57 percent of American homes, ranking second (table 1.4).

The use of water in our homes can be easily reduced. Toilets are a particular concern, but water use can be reduced by more than 50 percent with the newer models that use less than 1.3 gallons per flush rather than older models that used as much as 5 gallons (table 1.5). America is a flush-oriented society, but water use can be reduced by adhering to the maxim, "If it's yellow, let it mellow. If it's brown, flush it down." Unfortunately, most Americans appear to want closure after toilet use, and this can be provided by flushing. No-flush urinals have been available for many years, but consumers have resisted them.

The silent toilet bowl leaks in American bathrooms are only slightly less scandalous than the breaks in city water mains that lose 15 to 20 percent of the water piped through them (chapter 2). It has been esti-

Table 1.4
Allocation of water indoors in the typical American home

Use	Gallons per capita, daily	Percentage
Toilets	18.5	26.7
Clothes washers	15.0	21.7
Showers	11.6	16.8
Faucets	10.9	15.7
Leaks	9.5	13.7
Baths	1.2	1.7
Dishwashers	1.0	1.4
Other uses	1.6	2.2
Total	69.3	100.0

Source: American Water Works Association, "Water Use Statistics," 2010 Available at http://google.co.il/#hl=en&sourceehp&q=Allocation+of+Water+indoor +in+the+typical+American+home.

Table 1.5
Daily indoor water use

Use	Gallons per capita	Percentage
Toilets	8.2	18.0
Clothes Washers	10.0	22.1
Showers	8.8	19.5
Faucets	10.8	23.9
Leaks	4.0	8.8
Baths	1.2	2.7
Dishwashers	0.7	1.5
Other uses	1.6	3.4

mated that 20 percent of all toilets leak, and this accounts for 14 percent of home water use.

Clothes washers are present in 81 percent of American homes. Newer models use half the water of older models, but washing machines are durable and are not replaced often, so the change to newer models will be slow. Rapid change can be instituted in our showering habits, not by showering less frequently but by running the water briefly, only before and after soaping, instead of the common American habit of standing under the full flow for perhaps five minutes to relax a stressed body.

Conservation is the most cost-effective solution to water scarcity.

Forty million acres of America are covered in lawns, our largest irrigated crop, and one that can be accurately described as ecological genocide. Home lawn and landscape irrigation consumes an average of more than 8 billion gallons of water daily, equivalent to 14 billion six-packs of beer.[22] According to the Environmental Protection Agency (EPA), one-third of all residential water use in the United States is devoted to irrigation—almost none of it necessary. Many cities and some states in the Southeast and Southwest report that 50 percent of their residential water use is outdoors, primarily for lawns. In 2008, satellite data revealed that lawns (99.96 percent) and golf courses (0.04 percent) in the United States cover nearly 50,000 square miles, or 32 million acres, an area roughly the size of New York State.

Probably the largest manicured and watered lawn in the United States surrounds the White House. It extends over 18 acres and is under the jurisdiction of the National Parks Service. In order to encourage better uses of America's lawn areas, President Obama in 2009 authorized the cultivation of an organic vegetable garden fertilized with compost for his family over 1,100 square feet (0.01 acres) of the lawn, the first garden at the White House since a Victory Garden in 1943. At the height of the Victory Garden movement during World War II, gardens were supplying 40 percent of the nation's fruits and vegetables. Someday the pampered front lawn that today's Americans admire so much may be considered an ugly vestige of an ignorant time.

What Do We Pay for It?

The average American household spends, on average, only $523 per year on water and sewer charges, in contrast to an average of $707 per year on soft drinks and other noncarbonated refreshment beverages.[23] Compared with other developed countries, the United States has the lowest burden for water and wastewater bills when measured as a percentage of household income. Where water is concerned, price does not indicate value to Americans.

Many studies have shown that water demand is responsive to price changes. An attack on the consumers' wallets is the surest way to get their attention, and to encourage consumers to conserve water, prices need to be increased. State utility commissions must allow utilities to use a rate structure that reflects a consumer's water usage. Consider these examples of price structures for water use:

• Most of the 60,000 water systems in the United States charge uniform rates; consumers pay the same rate per gallon no matter how much they use each month. One-third of municipalities do the opposite: the more water you use, the less you pay. Only one-fifth of utilities charge higher rates for those who use more. In Israel, where water shortages are common, a system of block rates or tiered pricing is used: the per-unit charge for water increases as the amount used increases. The first block of water (gallons) is relatively cheap, recognizing that everyone needs a basic amount of water for sanitation, cooking, and cleaning. But the price increases rapidly for each succeeding block; those who take fifteen-minute showers, fill swimming pools, wash their cars using a running hose, and regularly water large lawns have exceptionally large water bills. According to an EPA study in 2000, only 9 percent of utilities in the United States use block rates.[24]
• Utilities can charge seasonal rates, in which prices rise or fall depending on water demands and weather conditions. Water should be more expensive when demand is high. Only 2 percent of American water companies charge more during summer months.

• A corollary to seasonal pricing is time-of-day pricing, in which prices are higher during a utility's peak demand periods.

• A relatively new method for encouraging lower water use is a digital water meter.[25] Its heart is an electronic device called a water manager. The water user buys a smart card at a local convenience store that, like a long-distance telephone card, is programmed for a certain number of credits. At home, the purchaser punches the card's code into a small keyboard and pushes the LOAD key. The water manager automatically sends a signal to the water company to supply water. When the user runs out of credits, he or she pushes the LOAN key, and the utility gives the user a bridge loan until he or she purchases another card. Studies in the United Kingdom revealed that households using the water manager reduced water use by 21 percent.[26] The UK Environment Agency said that a shift to widespread metering is essential for the long-term sustainability of water resources.

More than 100 studies of the relationship between residential water use and pricing indicate that a 10 percent increase in price lowers use by 2 to 4 percent. In industry, a 10 percent increase in price lowers demand by 5 to 8 percent. In economic terms, water demand is said to be inelastic, meaning that when price increases, consumption decreases at a smaller amount than the increase in price.[27]

Reusing Dirty Water

Through the natural water cycle, the earth has recycled and reused water for billions of years. However, when used in discussions of water availability to consumers, *recycling* generally refers to projects that use technology to speed up natural processes. The number of such projects is increasing dramatically in the United States because of increasing pressure on freshwater resources. Recycled water can satisfy water demands for irrigating crops, cooling water in power plants, mixing concrete in construction work, watering a lawn, mopping a floor, or flushing a toilet. Hundreds of American cities now use recycled water for nondrinking purposes. Most irrigation of fruits and vegetables in California and Florida is accomplished with recycled wastewater. In Israel, about one-third of water needs is met by reclaimed and recycled municipal wastewater, or sewage water. Water reuse and recycling is second only to conservation as a means of boosting water supplies.

Water recycling is a three-step filtration process. When water enters the treatment facility, solids are settled out, and the wastewater is sucked up into thousands of tiny straws less than three-hundredths the thickness of a human hair, which help separate out bacteria. This is followed by reverse osmosis, a process where intense pressure is used to force the water molecules through a sheet of plastic. Dissolved salts cannot pass through the membrane. Biological processes may also be used to remove contaminants. Microorganisms consume the organic matter as food. After the bugs do their work, chlorine, ultraviolet light, hydrogen peroxide, and radiation may used to kill the organisms before the water is released from the purification plant into streams and the ocean. The entire process ensures that not even the tiniest bacterium, virus, chemical, or hormone can survive. According to California's Department of Health Services, water from such a modern plant is purer than expensive mountain spring water but is piped into streams and the ocean because current state regulations do not permit the water to be fed directly into homes.

Instead of being fed into streams after leaving the purification plant, the water may be injected underground to replenish depleted groundwater supplies that supply drinking water to millions of humans above ground. Underground injection adds another step, and perhaps an unnecessary one, to the decontamination process. A new half-billion-dollar purification plant in Orange County, California, processes 70 million gallons of sewage per day that is pumped underground but will eventually stream out of faucets in people's homes.

Only about a dozen water agencies in the United States recycle treated sewage to replenish drinking water supplies, but none steers the water directly into household taps. The concept of toilet-to-tap drinking water is hard for many people to swallow. Many Americans have a psychological barrier to imbibing water that at one stage had fecal matter floating in it. But with education, and as water shortages become more severe, their fecophobia will be overcome.

Israeli scientists have developed a system that instantly purifies contaminated water, removing organic, biological, and chemical contaminants.[28] The technology has been miniaturized to fit into the top of a cork that can be plugged into virtually any size bottle, container, or tap. One cork can purify 250 gallons of water before being replaced, and, according to the developers, it costs no more than a large coffee and pastry at an upscale coffee shop. The device is ideal for hikers, soldiers in the field, or victims of disasters and can prevent the deaths of the 1.6 million children under the age of five who die each year in the undeveloped world from drinking untreated water. Impure water is the major killer of people in the Third World.

Taking Out the Salt

The ocean holds 97 percent of the earth's water, but its salinity renders it unusable for drinking and for most other uses. It contains about 35,000 parts per million (ppm) of dissolved materials (3.5 percent). The EPA's guideline for drinking water recommends a maximum of 500 ppm; most drinking water in the United States contains 100 ppm or less. Expensive technology is required to make seawater potable, but the benefit is incalculable: an unending supply of freshwater. As one water specialist has said, "When you're running out of water, you don't care about what the energy bill is." The world's largest desalination plant opened in Ashkelon, Israel, on the Mediterranean Sea coast in 2005; it supplies 5 to 6 percent of the nation's demand and 13 percent of domestic consumer needs.[29]

Freshwater produced by desalination of seawater costs two to three times more than water obtained by conventional water treatment, but water is so cheap in the United States that doubling or even tripling the cost is something most Americans can easily bear. At present, desalination's contribution to the total U.S. water supply is negligible. There were about 250 desalination plants in the United States in 2005 and every state has at least one, but they have the capacity to provide less than 0.4 percent of the water used in the United States, and most of this water is used by industries, not municipalities.[30] Florida has nearly half of the plants, with Texas and California in second and third place.

Most of the existing plants are designed to handle brackish water (1,000–10,000 ppm) rather than seawater, and cost about half as much money to build. Brackish water is present at depths of less than 500 feet over about half of the conterminous United States (figure 1.6) and is a large potential source of water that has not been tapped.

The Bureau of Reclamation forecasts that by 2020, desalination technologies will contribute significantly to ensuring a safe, sustainable, affordable, and adequate water supply for the nation.[31] The ability to make ocean water potable guarantees an inexhaustible water supply, albeit at prices higher than Americans are used to paying. Desalination is an inevitable part of America's water future.

Poisoning Our Water

Americans are making a two-pronged attack on their water supply. Not only do they use it extravagantly and wastefully, but they pour harmful chemicals in it as well.[32] In 2007, 232 million pounds of toxic chemicals

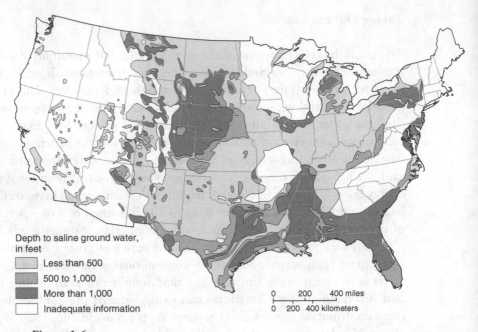

Depth to saline ground water,
in feet

▨ Less than 500
▨ 500 to 1,000
▨ More than 1,000
☐ Inadequate information

0   200   400 miles
0   200 400 kilometers

Figure 1.6
Depth to saline groundwater in the United States. (U.S. Geological Survey. Hydrologic
Investigations. Atlas HA-199)

were dumped into 1,900 waterways. Indiana and Virginia were the
leading dumpers. The top three waterways in the nation for the most
total toxic chemicals discharged in 2007 were the Ohio River, New River
(which flows through North Carolina, Virginia, and West Virginia), and
the Mississippi River. The Ohio River also was number one for toxic
chemicals that are cancer causing and chemicals that cause reproductive
disorders.

In a Gallup Poll in 2007, pollution of drinking water, rivers, lakes,
and reservoirs was named by Americans as their greatest environmental
concern (60–68 percent of Democrats, 41–46 percent of Republicans).[33]
Large numbers of industrial chemicals are present in our blood, although
in very small amounts. Whether they affect our health and longevity is
uncertain, but there are reasons to be concerned. Basic toxicity data are
not publicly available for about three-quarters of the 3,000 chemicals
produced in the highest volume each year, excluding pesticides. And 1.2
trillion gallons of untreated industrial waste, sewage, and storm water
are discharged into U.S. waters annually.[34] To this noxious cocktail is
added runoff from the animal manure in the monstrous livestock feedlots
that increasingly cover the landscape (chapter 5).

The effect of these feedlots on water purity was brought home in 2009 to residents of Brown County, Wisconsin.[35] One cow produces as much waste as 18 people. The 41,000 dairy cows in the county produce more than 260 million gallons of manure each year. In measured amount, that waste acts as a fertilizer, but the cows produce far more manure than the land can absorb. Because the amounts are excessive, bacteria and chemicals flow into the ground and contaminate residents' tap water. In the town of Morrison in Brown County, more than 100 wells were polluted by agricultural runoff within a few months. As parasites and bacteria seeped into drinking water, residents suffered from diarrhea, stomach illnesses, and severe ear infections. A resident in a town a few miles away commented that "sometimes it smells like a barn coming out of the faucet." At an elementary school a few miles from a large dairy, signs above drinking fountains warn that the water may be dangerous for infants.

Rivers and aquifers are not the only casualty of pollutants. In any given year, about 25 percent of beaches in the U.S. are under advisories or are closed at least once because of water pollution.[36]

Pharmaceutical companies are among the industries that are major sources of drug pollution. Wastewater treatment plants downstream of pharmaceutical factories have exceptionally high levels of antibiotic drugs, opiates, barbiturates, and tranquilizers.

Another pollution source is the storm water that runs off lawns, streets, and driveways. It contains motor oil, fertilizers, and pesticides that will eventually end up in the nation's waterways. Impermeable surfaces like concrete prevent storm waters from soaking into the ground, which can trap potential pollutants.

Many organizations, federal, state, and private, have examined our surface waters and groundwater, and their results are consistent and scary: half of America's rivers and lakes are too polluted to safely swim in. In 2001 the U.S. Geological Survey examined 139 streams in thirty states and looked for ninety-five industrial chemicals. At least one was present in each of the 139 streams, and a mixture of seven or more were present in half the streams.[37] Groundwaters were less contaminated but also commonly contained multiple pollutants.

In 2002 the H. John Heinz III Center for Science, Economics, and the Environment published the results of a five-year study of the nation's streams and groundwaters. It determined that 13 percent of the streams and 26 percent of the groundwaters were seriously polluted.[38]

A study published in 2004 reported on the results testing thousands of rivers, aquifers, wells, fish, and sediments across the country over a

ten-year period. More than 400 scientists analyzed 11 million samples
for more than 600 chemicals. They detected pesticides in 94 percent of
all water samples (and in 90 percent of fish samples).[39] Many of the
pesticides have not been used for decades, but they continue to persist
in the environment. Persistent toxins do not break down and go away;
they keep polluting the water they are in. Clearly, ending the application
of chemical pollutants into the environment does not immediately end
their presence in our water.

In a study published in November 2009, the EPA reported that
mercury, a pollutant released primarily from coal-fired power plants, was
present in all fish samples it collected from 500 lakes and reservoirs. At
half the lakes and reservoirs, mercury concentrations exceeded levels the
EPA deems safe for people eating average amounts of fish (figure 1.7).
And a person does not need to eat much fish for a seafood meal to raise
mercury levels. In an experiment in 2006, David Duncan of National
Geographic ate some halibut and swordfish in San Francisco and the
next day had his blood drawn and tested for mercury content. The level
of mercury had more than doubled from an earlier blood test—from 5
micrograms per liter of blood to 12 micrograms. There is no way to

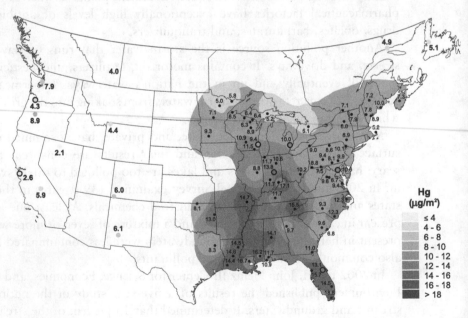

Figure 1.7
Wet deposition of mercury from the atmosphere, 2006. (Environmental Protection Agency)

know whether Duncan suffered permanent damage from the higher mercury level, but children have suffered losses in IQ at concentrations of only 5.8 milligrams.

Mercury was not the only contaminant in the lakes and reservoirs analyzed by the EPA. Polychlorinated biphenyls (PCBs), banned in the late 1970s but still present in the environment, were present in 17 percent of the water bodies. PCBs have been linked to cancer and other health effects.

A ray of hope surfaced in 2009 with the EPA's announcement in its annual Toxics Release Inventory that water pollution decreased by 5 percent between 2006 and 2007. However, releases of PCBs into the environment increased by 40 percent due to disposal of supplies manufactured before the substances were banned in 1979. Mercury releases, mostly due to mining, increased by 38 percent. Dioxin releases increased by 11 percent, and lead releases increased by 1 percent. Releases of all persistent, bioaccumulative, and toxic chemicals or metals increased by 1 percent. These increases will likely be reflected in water analyses over the next few years.

That chemicals in combination can be more deadly than either chemical alone and in lower concentrations was recently demonstrated in a study of salmon by federal scientists.[40] Five of the most common pesticides used in California and the Pacific Northwest acted in deadly synergy by suppressing an enzyme that affects the nervous system of salmon. Some fish died immediately. Exposures to a single chemical, however, did no harm. As expected, harmful effects on the salmon were observed at lower pesticide levels when chemicals were applied in combinations. Earlier studies had found that three of the pesticides can be lethal to salmon and can inhibit their growth by impairing their ability to smell prey, impair their ability to swim, and make it difficult to spawn and avoid predators.

More than 2,300 chemicals that can cause cancer have been detected in U.S. drinking water. Although the amounts are usually small and considered safe by the EPA, the surgeon general has stated, "No level of exposure to a chemical carcinogen should be considered toxicologically insignificant to humans."[41]

Seventy years ago in the United States, one person in fifty could expect to get cancer in his or her lifetime. Today one in three people and one in two males can expect to get cancer. The risk that a fifty-year-old white woman will develop breast cancer soared to 12 percent from 1 percent in 1975. Studies reveal that 90 percent of breast cancer cases are not caused by "bad" genes. Synthetic chemicals are a likely cause. The sharp increase in cases of autism since 1990 is increasingly thought to have environmental rather than genetic causes.[42] Between 1979 and 1997 cases of Alzheimer's disease and other dementias more than tripled in men and rose by nearly 90 percent in women in England and Wales, with similar results in other countries.[43]

Some of the enormous increase in cancer and Alzheimer's occurrence can probably be attributed to increased longevity. In 1940 the average American lived sixty-four years; in 2006 it was seventy-eight years, and the immune system of humans is known to deteriorate with age. But it seems unlikely that a fourteen-year increased life span could by itself be responsible for the enormous increases in catastrophic bodily diseases that have occurred. It is much more likely that increased body burdens of industrial chemicals are largely responsible for the sharp increases in some cancers and brain diseases. People are exposed to carcinogenic chemicals in pesticides, deodorants, shampoos, hair dyes, makeup, foods, cleaning products, sunscreens, electronics, furniture, walls, paints, carpeting, and a host of other common commercial products. A study in the United Kingdom found that the average woman applies 515 chemicals to her face each day in makeup, perfumes, lotions, mascara, and other beauty products. Pollution is built into the modern world.

Television personality Bill Moyers discovered that his blood contains eighty-four synthetic chemicals. Tests commissioned by the Environmental Working Group found that the blood or urine of all of the subjects they studied was contaminated with an average of thirty-five consumer product ingredients, including flame retardants, plasticizers, and stainproof coatings. These mixtures of compounds, found in furniture, cosmetics, fabrics, and other consumer goods, have never been tested for safety.[44] In another study, the group tested umbilical cord blood collected by the American Red Cross. This blood of unborn babies contained an average of 287 different industrial chemicals and pollutants per sample. Most of these compounds detected are believed to cause cancer or birth defects, or are neurotoxins.[45]

In 2005 the Centers for Disease Control and Prevention published the results of a study of blood and urine in Americans.[46] In testing volunteers for 148 industrial chemicals and harmful pollutant elements, they found mercury, pesticides, hydrocarbons, dioxins, PCBs, phthalates (plasticizers), DDT (banned since 1973 in the United States but still used in countries from where we import food), insect repellent, and other harmful chemicals. There can be no question that our bodies are heavily contaminated with the products we have manufactured that make our everyday lives more comfortable.

A study in 2009 found up to forty-eight toxic chemicals in blood and urine samples of five prominent female environmental activists from various parts of the country.[47] The chemicals found are present in everyday consumer products. Each of the women's samples contained fire retardants, Teflon chemicals, fragrances, bisphenol A (BPA), and perchlorate. Flame retardants are found in foam furniture, televisions, and computers. Teflon is used in nonstick coatings and grease-resistant food packaging. BPA is a plastics chemical; perchlorate, an ingredient in rocket fuel, can contaminate tap water and food. Fragrances have been associated with hormone disruption in animal studies. A physician with the Environmental Working Group noted that animal studies show that the chemicals can be potent at very low levels of exposure. Although the rising number of chronic diseases has many roots, increased exposure to chemicals is one likely cause.

In an incredibly detailed blood test in 2009, David Duncan, author of *Experimental Man*, underwent several hundred scientific and medical tests costing $25,000, in which he was tested for 320 chemical toxins.[48] The tests revealed he had 185 of these known toxins in his body. There are about 80,000 industrial chemicals in existence, so testing for "only" 320, much less than 1 percent of them, barely scratches the surface of our probable bodily pollution. The average person's bloodstream may well contain thousands or tens of thousands of industrial chemicals. It is noteworthy that the body is known to hide its poisons in its fat, cells, and other areas of the body to keep them out of the bloodstream, so even an analysis for all 80,000 industrial chemicals might not uncover all of the ones in the body.

Not only is our drinking water much less than pure, 40 percent of our rivers and 46 percent of our lakes are too polluted for fishing, swimming, or aquatic life.[49] Two-thirds of U.S. estuaries and bays are either moderately or severely degraded from eutrophication (nitrogen and phosphorous pollution). Chesapeake Bay and the Gulf of Mexico nearshore waters have become notorious for the level of their pollution.[50]

Even the reservoir that holds 90 percent of America's fresh drinking water is polluted.[51] More than a century of industrial dumping has spread pollution throughout the Great Lakes. Fish caught from this largest source of drinking water are often unsafe to eat.

Americans are often told by their government that no nation has better-quality drinking water than the United States. This is certainly true when only short-term effects are considered. Water-borne diseases are uncommon, and parasites and disease-causing microorganisms have been largely eliminated from the water that pours from taps. However, the statement about water purity fails to consider the long-term effects of chemical pollutants in the water. Certainly the small amounts of pesticides and other industrial chemicals in the water are not lethal in the short run, but their effect during a lifetime of ingesting them cannot be benign. And it is not necessary. There is no necessity for agriculture to use pesticides that end up in rivers (chapter 5) or for industry to pour its poisonous liquid waste into rivers and inject them underground into aquifers. Industry does it because it is an inexpensive way to dispose of stuff they do not want, and thanks to decades of lobbying of our elected representatives, it is perfectly legal.

The Hudson River

One of the most notorious examples of river pollution is the PCB contamination in the Hudson River, an important source of drinking water for a high percentage of the people in New York State.[52] In 1947, General Electric started using PCBs in one of its manufacturing plants on the eastern shore of the river. It was not illegal at the time, although major health and safety problems with PCBs had been detected eleven years earlier. The chemicals are suspected human carcinogens and increase the risk of birth defects in children born to women who eat fish from the polluted Hudson River. They cause damage to the nervous system, immune system, and reproductive system in adults. GE legally dumped more than 1 million pounds of the chemicals into the Hudson River over a thirty-year period.

In 1974 the EPA established that there were high levels of PCBs in Hudson River fish and set the safety threshold at 5 ppm PCBs in fish for human consumption. Two years later, Congress passed the Toxic Substance Control Act banning the manufacture of PCBs and prohibiting their use except in totally enclosed systems, and the public was warned about the dangers of eating fish from contaminated parts of the Hudson River. All commercial fisheries were closed. It was determined that GE had caused the pollution. In 1983, 193 miles of the upper Hudson River were added to the Superfund National Priority List. A year later the EPA reduced the acceptable safety limit for PCBs from 5 ppm to 2 ppm.

In 1993, sediment in the river adjacent to a GE plant was found to contain 20,000 ppm of PCBs. Blood tests of Hudson Valley residents in 1996 revealed elevated levels of PCBs in non–fish eaters, who presumably ingested the chemicals in drinking water. Tree swallows and bald eagles in the area were found to have 55 to 71 ppm of PCBs in their body fat, qualifying them as hazardous waste.

Under Superfund law, polluters are responsible for cleaning up the messes that they make. GE spent millions of dollars on an ultimately unsuccessful campaign to persuade the federal government not to implement a dredging and cleanup plan to rid the river of PCBs. The thirty-year economic and environmental struggle between GE and the government lasted until 2009, when GE finally began dredging the river bottom sediment. Two-and-a-half million cubic yards of toxic sludge will be dredged and transported to a landfill in Texas. The project is expected to cost $750 million and take at least six years. GE is still fighting to reduce the amount of dredging it must do.

Two-and-a-half million cubic yards of Hudson River toxic sludge will be wrapped in heavy plastic, like a burrito, loaded into open railcars, and shipped to the Texas landfill in trains at least eighty cars long. By the third year of the EPA-approved plan, two to three trains a week will arrive at the dump site. At the landfill, excavators on platforms will rip open the bags and transfer the sludge to 110-ton mining trucks. The trucks will haul and deposit the sludge into a pit 75 feet deep into red clay and lined with two layers of heavy polyethylene. Then it will be covered with 3 feet of clay.

Chesapeake Bay

Maryland's Chesapeake Bay has had pollution problems for a hundred years with no cleansing solution in sight. Nitrogen and phosphorus runoff from widespread agriculture in the bay's watershed is the cause. Oyster harvests declined from 53,000 tons in 1880 to 10,000 tons in 1980 to 100 tons in 2003. Oysters cleanse the water by filtering up to 5 quarts of water per hour, a task they can no longer perform adequately. In 1880 there were enough oysters to filter all the water in the bay in three days; by 1988 it took more than a year. The bivalve population has been decimated, and a dead zone now covers up to a third of the bay. A *dead zone* is a volume of water that lacks enough oxygen for aerobic animal life to exist.

Fish contain high levels of mercury, and there are algal blooms and voracious bacteria that threaten the health of people who fish, boat, and swim in the estuary.[53] Health authorities advise against swimming until two days after a significant rain because the rain can sweep in animal manure and human waste from older sewage systems and leaky septic tanks. A spokesman for Maryland's health department warned that people should not let cuts or open wounds contact the water.

Other Dead Zones

Pesticides and artificial fertilizers are used in massive amounts on America's farms, and most of it washes off the farms, into local streams, and eventually into the Mississippi River. In addition to making about half the streams and rivers in the watershed unsafe for drinking, swimming, or recreational contact, the pollution has created a dead zone in the nearshore Gulf of Mexico near the mouth of the river. It covers about 8,000 square miles, the size of Massachusetts, and has been growing since measurements began in 1985.

The hypoxia—very low levels of dissolved oxygen—is caused mainly by excess nitrogen from fertilizer used on crops, with corn using the most. The dissolved nitrogen flows into the Gulf and spurs the growth of excess algae. The algae cause an oversupply of organic matter that decays on the Gulf floor, depleting the water of oxygen. There are no fish, shrimp, or crabs in the dead zone and little marine life of any kind. Fish that survive in areas with slightly higher levels of oxygen have reproductive problems.

Numerous dead zones can be found around the coastline of the United States and at 405 locations worldwide. In most cases, the cause is the same: runoff of artificial fertilizer into nearshore waters.

A piece of hopeful news arrived in 2007 when North America's first full-scale commercial water treatment facility capable of removing phosphorous began operating in Edmonton, Canada. It is also possible to remove nitrogen but the process is not yet widespread.

Endocrine Disrupters

Among the many well-publicized concerns about specific pollutants such as lead and mercury in water is a group of chemicals that affect our sexual characteristics.[54] Apprehension is growing among scientists that the cause of these maladies may be a class of chemicals called endocrine disrupters, widely used in agriculture, industry, and consumer products. Some also enter the water supply when estrogens in human urine pass through sewer systems and then through water treatment plants.

These chemicals interfere with the endocrine system in our bodies, a system that regulates many functions such as growth, development and maturation, and the way various organs operate.[55] There are 966 known or suspected endocrine-disrupting chemicals in existence, and often they are found in the environment. They are ubiquitous in modern life, found in plastic bottles, cosmetics, some toys, hair conditioners, and fragrances. At least forty chemical compounds used in pesticides, and many in prescription medications, are known to be endocrine disrupters. Among the harmful effects these chemicals are known to cause are sexual and reproductive anomalies, which have been documented in studies of rats, toads, mice, fish, dogs, panthers, reptiles, polar bears, and birds. In the Potomac River watershed near the nation's capital, more than 80 percent of male smallmouth bass are producing eggs.

In 2008, research showed that humans are affected as well.[56] Among the most common endocrine disrupters are chemicals called phthalates, which suppress male hormones and sometimes mimic female hormones. Boys born to women exposed to widespread chemicals in pregnancy have smaller and imperfect penises and feminized genitals. They also have a shorter distance between their anus and genitalia, a classic sign of feminization. A study in Holland showed that boys whose mothers had been exposed to PCBs grew up wanting to play with dolls and tea sets rather than with traditionally male toys. As expressed by Gwynne Lyons, an advisor to the British government on the health effects of chemicals, "The basic male tool kit is under threat." If terrorists were putting phthalates in our drinking water, we would be galvanized to defend ourselves, but we seem less concerned when we are poisoning ourselves.

Sperm counts are dropping precipitously. Studies in more than twenty countries have shown that they have dropped by 60 percent over fifty years.[57]

There is also some evidence that endometriosis, a gynecological disorder, is linked to exposure to endocrine disrupters. Researchers also suspect that the disrupters can cause early puberty in girls.

Women in communities heavily polluted with gender benders in Canada, Russia, and Italy have given birth to twice as many girls as boys, which may offer a clue to the reason for a mysterious shift in sex ratios worldwide. Normally 106 boys are born for every 100 girls, but the ratio is slipping. It has been calculated that over the years, 250,000 babies who would have been boys have been born as girls instead in the United States and Japan alone.

In June 2009, the Endocrine Society, an organization of scientists specializing in this field, issued a landmark fifty-page warning to Americans.[58] "We present evidence that endocrine disrupters have effects on male and female reproduction, breast development and cancer, prostate cancer, neuroendocrinology, thyroid metabolism and obesity, and cardiovascular endocrinology." The effects of endocrine disruption can be subtle. For example, a number of animal studies have linked early puberty to exposure to pesticides, PCBs, and other chemicals. It is well known that women with more lifetime menstrual cycles are at greater risk for breast cancer, because they are exposed to more estrogen. A woman who began menstruating before age twelve has a 30 percent greater risk of breast cancer than one who began at age fifteen or later. American girls in 1800 had their first period, on average, at about age seventeen. By 1900 that had dropped to fourteen. Now it is twelve, and endocrine disruption is probably at least partly responsible.

In the United States, the EPA has shown little interest in studying endocrine-disrupting chemicals, and no legislation pending in the U.S. Congress addresses these.

As a Native American chief once said, "Only when the last tree has been felled, the last river poisoned and the last fish caught, man will know, that he cannot eat money."

Ocean Pollution

The world ocean is not immune from the onslaught of water pollution, although it may be one of the ugliest when oil pollution is involved. Contrary to the impression one gets from media reports, oil spills from beached tankers are only a minor part of the problem.[59] Large oil spills contribute only 5 percent to the ocean's oil pollution, and by 2015, when all oil tankers in U.S. waters will be required to be double-hulled, large oil spills may be a thing of the past because the United States is the biggest importer of oil. There are 706 million gallons of oil pollution per year, most of which can, in principle, be controlled (table 1.6).

Control in principle does not necessarily mean control in practice, as illustrated by the BP disaster in the Gulf of Mexico in April 2010, where a blowout preventer failed on a well, causing the worst oil spill that has ever occurred in U. S. waters. Petroleum from 18,000 feet below the sea floor spewed into Gulf waters for three months, perhaps 200 million gallons. The cause of the blowout was human error. There were multiple warning signs, and safety procedures were not followed.

All major oil companies have intensive safety programs and processes to prevent spills. But every human enterprise has a failure potential, as the 2010 Gulf disaster clearly demonstrates. Since 1946 50,000 oil wells have been drilled in the Gulf, and 3,858 of them are currently producing 11 percent of America's domestic supply. The BP disaster is the first

Table 1.6
Sources of oceanic oil pollution

Source	Percentage
Storm drains in cities	51.4
Routine maintenance of ships	19.4
Power plants and motor vehicles	13.0
Natural seeps	8.8
Large oil spills	5.2
Offshore drilling	2.1

Source: OceanLink, n.d. "World Oil Pollution: Causes, Prevention and Clean-Up."

major spill, nevertheless, and this safety record is incomparably better than other commercial and noncommercial activities in the United States such as driving cars, flying, slaughtering animals for human consumption, or accidents at home.

Probably the easiest source to control is the largest: the oil that drains into the ocean from America's cities. Its origin is from people dumping motor oil down storm drains after driveway oil changes, supplemented by road and urban street runoff. Motor oil produced by car owners' oil changes should be brought to garages for appropriate disposal.

The contribution from power plants and automobiles will decrease as renewable and nonpolluting energy sources gradually replace fossil fuels.

Water as a Human Right

The U.S. Constitution says that Americans are entitled to life, liberty, and the pursuit of happiness. Implied in these entitlements are access to adequate food and water. Thus, freshwater is a legal entitlement rather than a commodity or service provided on a charitable basis. Few people would quarrel with this. Access to fresh, clean water is a joint responsibility of federal, state, and local governments. Of course, the term *access* does not mean that individuals are not responsible for their own welfare, only that governments must be concerned about the essential needs of their citizens.

Until the past few decades, access to clean water has been taken for granted by nearly everyone in the United States, a situation made possible by our geographical location and the circumstance that all of the country's major rivers and their tributaries have their headwaters within national borders—the Mississippi, Ohio, Missouri, Colorado, Columbia, and Rio Grande. In addition, entirely contained within our borders is the world's largest underground water resource, the Ogallala aquifer. The amount of water used in the United States is staggering: on a per capita basis, we use far more water than any other nation does.

But all good things must come to an end. Many areas of the United States have begun to experience water problems related to population distribution (too many people in southern Arizona); inadequate or deteriorating infrastructure (old and very leaky underground water pipes); profligate use on lawns (average 10,000 gallons per year per lawn in suburbia), flower gardens, and golf courses (753 billion gallons per year); and willful pollution of both surface and subsurface water supplies (agricultural runoff and injection of pollutants into the subsurface). It almost seems as though Americans have a death wish as far as water is concerned.

What are the responsibilities of governments and individuals to ensure water supplies and avoid a water catastrophe? Some things are the responsibility of governments at various levels. Only governments have the resources to rebuild and upgrade the infrastructure and the legislative ability to stop the injection of pollutants into the subsurface, which poisons our aquifers. At the federal level, ensuring that water is clean is the responsibility of the EPA, and President Obama's EPA head, Lisa Jackson, has begun cracking down on public and private polluters. In September 2009, an investigation found that companies and other workplaces had violated the Clean Water Act more than 500,000 times in the past five years alone, but fewer than 3 percent of polluters had ever been fined or otherwise punished. The water provided to more than 49 million people has contained illegal concentrations of chemicals such as arsenic or radioactive substances like uranium.[60] Jackson has ordered an assessment of the agency's shortcomings, promised stronger enforcement, and added new chemicals to the long list of contaminants.

As with most other environmental laws, responsibility is shared. Washington sets the health standards, but the states write and enforce the permits, which tell polluters what can and cannot be discharged into the water. The EPA has the authority to crack down on polluters if a state fails to enforce the laws. However, some consistent polluters are unregulated, such as large animal-feeding operations (chapter 5). Power plant emissions into the air are regulated, but the toxics they discharge into the water, such as cadmium, lead, and arsenic, are not (chapter 6).

The EPA needs additional money and staff to accomplish its legislatively required goals.

Individuals can design the cared-for property around their houses to reduce water use. Agricultural use, the main consumer of water, can be reduced by eliminating water wasters such as above-ground irrigation sprayers and adopting water-saving devices such as drip irrigation. Population distribution will adjust automatically. If water is not available, people will relocate to areas where it is.

A right to water cannot imply a right to an unlimited amount of water. Resource limitations, ecological constraints, and economic and political factors limit water availability and human use. Given such constraints, how much water is necessary to satisfy this right? Enough solely to sustain life? Enough to grow all food sufficient to sustain a life? Enough to sustain a certain economic standard of living? International discussions among experts in water use lead to the conclusion that a human right to water should apply only to basic needs for drinking, cooking, and fundamental domestic uses such as sanitation and bathing. Water for swimming pools, golf courses, flower gardens, and so on cannot be accepted as a human right. Not pricing water correctly is at the root of many problems with water.

The United Nations and many private and governmental organizations have determined that each person needs a minimum of about 12 gallons of water per day for drinking, cooking, sanitation, and personal and household hygiene. Amounts above this are not necessary, only desirable. And of course we all desire it. But it is becoming apparent that some limitations must exist if we are to live harmoniously with our fellow citizens. Comprehensive discussions about water management among America's political leaders are sorely lacking.

2

Infrastructure: Pipes, Wires, Roads, Bridges, Railroads, Dams, Airports, and Levees

There are no Republican bridges. There are no Democratic drinking water purification facilities. We all use these systems.

—Casey Dinges, American Society of Civil Engineers

If you knew the condition of what's under the city, you probably wouldn't walk on the sidewalk.

—Fred Graf, senior research engineer for PECO, Philadelphia's electric and natural gas utility

Infrastructure refers to the basic constructed features that undergird our civilization: water and sewer mains, gas and liquid transmission lines, electrical grid, highways, bridges, railroads, dams, airports, and levees. Collectively they are called *public works*, although they may be developed and operated by either the private sector or government. When proposed by the federal government, they often are referred to as "pork barrel" projects or "earmarks"—projects proposed by members of Congress to enhance their state's industry, attractiveness, or culture and improve the legislator's chances of being reelected. Although the beneficiaries of these projects are mostly or entirely the legislator's constituency, they are paid for by all the nation's taxpayers.

One of the more recent and infamous pork barrel projects was the Big Dig in Boston, Massachusetts, a project to take a preexisting 3.5-mile elevated highway and relocate it underground, which cost the nation's taxpayers $14.6 billion, an astonishing $4.2 billion per mile. Other interesting expenditures in 2007 were appropriations of $7.4 billion for the National First Ladies' Library in Canton, Ohio, and financing for a project to improve "rural domestic preparedness" in Kentucky. In 2009, $3.8 billion was appropriated for the Old Tiger Stadium Conservancy in Michigan and $1.9 billion for the Pleasure Beach water taxi service project in Connecticut. One proposed project in recent years that failed to be approved was federal funding to build a memorial in North Dakota to the late band leader Lawrence Welk. Each year the group Citizens Against Government Waste publishes the *Congressional Pig Book Summary* listing the pork projects deemed most egregious.

The Real Needs

It is easy to laugh at or condemn outrageous examples of wasteful federal spending such as these, but not all expenditures classed as pork should be criticized. The United States is in dire need of infrastructure maintenance and upgrading, and the vast reach and large amounts of money involved require that a substantial part of the cost be borne by the federal government. Government spending on infrastructure fell sharply after construction of the interstate highway system many decades ago and has risen only slightly since then[1] (figure 2.1). Leadership from Washington is essential, and President Obama plans to devote a large amount of federal money to infrastructure projects, the first president to do so in many decades.

Strolling down city sidewalks, we implicitly assume that we are treading on solid earth. In reality, just beneath our feet lies a vast and dark world inhabited by thousands of miles of water mains, gas mains, sewers, immense labyrinths of communications cables and fiber optics, electrical conduits, storage tanks, tunnels, abandoned subway stations, and assorted other things.

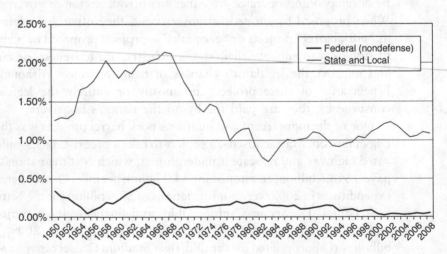

Figure 2.1
Government spending for public projects. (http://sharedprosperity.org/bp217.html)

Table 2.1
Infrastructure report card

Category:	Grade
Aviation	D
Bridges	C
Dams	D
Drinking Water	D–
Energy (National Power Grid)	D+
Hazardous Wasters	D
Inland Waterways	D–
Levees	D–
Public Parks and Recreation	C–
Rail	C–
Roads	D–
Schools	D
Solid Waste	C+
Transit	D
Wastewater	D–

Source: American Society of Civil Engineers, *Infrastructure Report Card for 2009* (2009).

Every few years, the American Society of Civil Engineers (ASCE) releases its Infrastructure Report Card that evaluates the condition of the nation's infrastructure (table 2.1). Between 2005 and 2009, Aviation declined from D+ to D, Roads from D to D–, and Transit from D+ to D. The only improvement was for Energy, the national power grid, which improved from D to D+. This is not a group of grades any American child would want to bring home to parents, but the parents are us. These are our grades.

Our infrastructure is poorly maintained, unable to meet current and future demands, and, in some cases, it is wholly unsafe. In its 2005 report card, ASCE said it would take $1.6 trillion in repairs over a five-year period to bring all categories to passing grades. Today that number stands at $2.2 trillion.[2] This estimate of $440 billion per year for five years to upgrade our infrastructure may be compared to 2009 federal spending of $600 billion for defense and $420 billion for Medicare in a total budget of $3 trillion.

Infrastructure modernization can be an important part of stimulating the economy and lowering the rate of unemployment. Moreover, infrastructure problems that are ignored do not go away; they only get worse.

Crumbling infrastructure is, in fact, as much a domestic threat to America as terrorism is. If ignored, either one could have catastrophic effects on our civilization.

Much of the infrastructure we use today was proposed and built during the administration of President Franklin Delano Roosevelt in the 1930s to reduce the 25 percent unemployment rate during the Great Depression. His Works Progress Administration built 78,000 bridges and viaducts and improved 46,000 more. It also constructed 572,000 miles of rural roads and 67,000 miles of urban streets.[3] These types of massive public works programs have faded into history. Unless a dramatic catastrophe occurs, such as a major bridge collapse that causes the death of many people, sewage backing up into people's homes, or an electrical failure that results in a blackout of most of the Northeast, Americans tend to ignore infrastructure and its need for continual maintenance, repair, and upgrading.

Pipes for Water

The bulk of the water and sewer lines beneath American streets were installed in three phases: at the end of the nineteenth century, in the 1920s, and just after World War II as adjustments to periods of population growth in cities and expansion into suburbs. Hence, most of the water mains in American cities are very old; many were built more than one hundred years ago.[4]

Americans use 10 trillion gallons of water every year that reaches our homes through 880,000 miles of water mains. Depending on the age of the pipe, the pressure in the pipe, and other conditions, the pipes may be made of concrete (12 percent), PVC plastic (39 percent), cast iron (48 percent), or clay (1 percent).[5] Each of these materials may deteriorate with age. Concrete is composed of calcium carbonate (artificial limestone) that dissolves over the decades in contact with water, PVC plastic may crush because of freezing and thawing of the soil around it or because of a new building erected overhead, cast iron may rust, and clay is not a durable material.

Because of their age, many of the water mains in American cities leak. The largest single consumer of water in most cities is not a family or a factory but leaky water pipes. Millions of gallons of water are lost every day in every major city in the United States. There are 250,000 to 300,000 breaks in water mains each year—one break for about every 3 miles of pipe.[6] Fifteen to 20 percent of drinking water in the United States is lost before it reaches the tap, a loss totaling tens of billions of gallons a year.[7] New York City loses 15 percent of its water to leaks in its 6,000 miles of water mains, and ruptures occur five hundred or six hundred times a year. Buffalo loses 40 percent to leaks and Boston 36 percent. Pittsburgh loses 12 million gallons of water each year, 15 percent of its water production, because of breaks in water mains. St. Louis's water system predates the Civil War. In 2007, Suburban Sanitary Commission crews in the nation's capital repaired 2,129 breaks and leaks; 2008 was better, in that only 1,700 pipes broke. Because of damage from Hurricane Katrina in 2005, the water pipes in New Orleans leak 50 million gallons each day, 18 billion gallons per year.[8] According to the American Water Works Association, a "huge wave" of water mains laid fifty to one hundred or more years ago is approaching the end of useful life, and "we can expect to see significant increases in break rates and repair costs over the coming decades."[9]

The Environmental Protection Agency (EPA) has projected that unless cities invest more to repair and replace their water and sewer systems, nearly half of the water system pipes in the United States will be in poor, very poor, or "life-elapsed" status by 2020. Currently about 4,400 miles of cast-iron pipe are replaced each year, and an estimated 13,200 miles of new pipe are installed—2 percent of this part of our infrastructure.[10]

Many cities can recite horrifying stories resulting from inadequate or broken water or sewage pipes. In Prince George County, Maryland, on January 18, 2009, a 42-inch pipe installed in 1965 broke and sent a river almost 75 feet wide flowing from the yards of some homes into the basements of others.[11] In Pittsburgh in 2000, a 5-foot portion broke off from a 20-inch diameter water main, spilling 20 million gallons and disrupting service for many hours to 8,000 homes and several hospitals.[12] In West New York, New Jersey, in 2007, a seventy or eighty-year-old 2-foot-wide pipe ruptured, stopping water service for 200,000 people and sending water rushing down the street like rapids in a raging river. It took two and a half days to restore service.[13] In December 2008, a forty-four-year-old water main five and a half feet in diameter broke shortly before 8:00 A.M. in a hilly area of Bethesda, Maryland, sending 150,000 gallons of water a minute rushing down a highway, much to the distress of cars in its path. In the same year, public outcry had forced cancellation of a proposed monthly fee of twenty dollars to speed up replacement of the area's pipes, which are currently being replaced at an inadequate rate of 27 miles per year.

The effects of inadequate delivery of water to needed locations go beyond drinking water. In some cities, firefighters deal with water pressure in hydrants inadequate to supply the amount of water needed to fight house fires. In the nation's capital in 2007, a 100-year-old 6-inch water main was unable to supply water quickly enough to put out a fire in a four-story condominium building. Thirteen percent of the area's underground water mains have only 6-inch diameters, and, of course, many of these aged pipes leak.

The seriousness of leakage from water pipes is compounded by the fact that a break in a water pipe can be a gateway for harmful bacteria. Estimates are that microbial pathogens in public drinking water supplies sicken hundreds of thousands of people each year, in part because of breaks in water mains.[14]

Making accurate estimates of the amount of money needed to modernize the nation's water and wastewater infrastructure is difficult because of the lack of a national database to gauge the age and condition of existing infrastructure, as well as uncertainty surrounding future regulatory requirements and technological breakthroughs. Studies by government and utilities agree that cities and towns will need to spend $250 billion to $500 billion more over the next twenty years to maintain the drinking water and wastewater systems required for modern living.[15]

Currently local governments and ratepayers provide 90 percent of the costs to build, operate, and maintain public water and sewer systems. Federal funding for drinking water and wastewater treatment has declined 24 percent since 2001.[16] Americans can easily afford to pay for maintaining and improving their water distribution systems. We pay about $2.50 for 1,000 gallons of water, less than half the amount paid for tap water in other developed countries. If we continue to ignore our water and sewer systems (figure 2.2), the nation can expect increased threats to public health, environmental degradation, and severe economic losses.

Pipes for Sewers

The EPA estimates that 1.2 million miles of sewers snake underground across the United States. Another EPA report in 2001 found that each year in the United States, sewers back up in basements an estimated 400,000 times, and municipal sanitary sewers overflow on 40,000 occasions.[17] The main pipes under the street are the city's responsibility; pipes from the street to a home, called service lines, are the responsibility of

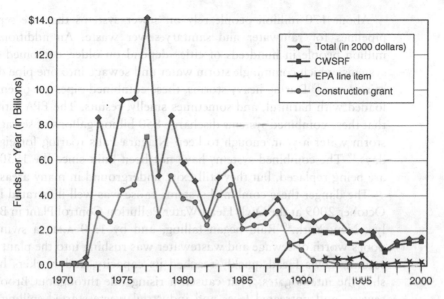

Figure 2.2
Federal funding for wastewater infrastructure, 1970–2000. *Note:* CWSRF = Clean Water State Revolving Fund. (Source: EPA, Report to Congress: Impacts and Control of CSOs and SSOs, Washington, D.C.: EPA, 2004.)

the homeowner. But it is not always clear who is responsible for sewage bubbling up in a basement, and, as with other parts of our underground infrastructure, local officials and voters have tended to treat sewers as an out-of-sight, out-of-mind problem.

Older municipal sewer systems are powered almost entirely by gravity rather than by large pumps, which means that when they are working properly, they move millions of gallons of sewage daily across considerable distances with only a minimum expenditure of energy, a feat of efficiency virtually unparalleled in the annals of engineering. New York was one of the first cities to build a large sewer system, starting construction in 1849, the year President James Polk annexed California; the sewer system operates almost entirely by gravity.

Most of the time, backups in sewers are caused by tree roots, although many of the items produced and discarded in our industrial civilization can also cause problems. If a pipe has even a pinhole leak, which often occurs at joints, trees near the pipes will find it, and tiny root tendrils will begin slowly working their way into the opening and grow, eventually forming dense root balls through which solids cannot pass. No way has been found to stop this.

About 170 million people rely on sewer systems that use separate pipelines for rainwater and sanitary-sewer waste. An additional 40 million people in hundreds of cities depend on older, combined sewer systems, which commingle storm water and sewage into one pipe during rainstorms.[18] During heavy storms, these combined pipes are often overloaded with harmful, and sometimes smelly, results. The EPA estimates that these combined systems discharge 850 billion gallons of sewage and storm water a year, enough to keep Niagara Falls roaring for eighteen days.[19] The combined systems have not been built since the 1950s and are being replaced, but they still exist underground in many areas.

The danger these combined systems cause was well illustrated in late October 2009 at the Owls Head Water Pollution Control Plant in Brooklyn, New York.[20] Rain began falling, and by 1:00 A.M., a swimming pool's worth of sewage and wastewater was rushing into the plant every second. Owls Head quickly reached its capacity, and workers had to shut the intake gates. That caused a rising tide throughout Brooklyn's sewers, and untreated feces and industrial waste started spilling from emergency release valves into Upper New York Bay and Gowanus Canal. According to one of the plant's engineers, "It happens anytime you get a hard rainfall. Sometimes all it takes is 20 minutes of rain, and you've got overflows across Brooklyn."

Between 2007 and 2009, more than 9,400 of the nation's 25,000 sewage systems (38 percent) reported violating the law by dumping untreated or partly treated human waste, chemicals, and other hazardous materials into rivers and lakes and elsewhere, according to data from state environmental agencies and the EPA. Around New York City, samples collected at dozens of beaches or piers have detected the types of bacteria and other pollutants tied to sewage overflows from the city's 7,400 miles of sewer pipes.

In 2001, a report from Johns Hopkins researchers linked sewer overflows with diseases such as giardiasis, hepatitis A, and cryptosporidiosis.[21] The study found that 68 percent of waterborne disease outbreaks over a forty-seven-year period were preceded by high precipitation events, which are increasing in frequency because of global warming (chapter 9).

New York is not alone with this problem. Researchers estimate that as many as 4 million people become sick annually in California alone from swimming in waters containing pollution from untreated sewage. Today sewage systems are the nation's most frequent violators of the Clean Water Act. Among the guilty since 2006 are San Diego, Houston, Phoenix, San Antonio, Philadelphia, San Jose, and San Francisco.

The experience of Charles Moore at Seal Beach in southern California in 2000 illustrates the problem. A sewage spill had occurred in an inland city upriver, sending virulent microbes downriver and into the shallow surf offshore. No warning signs were posted. The bacteria infected a not-quite-healed sore on his elbow and caused a potentially fatal skin infection. He was hospitalized for a week with fluid oozing from an arm swollen to double its size.

In March 2006, the Hawaii Department of Health had to shut down Honolulu's Waikiki beach for about a week following a major sewer spill. The microbe-contaminated waters were blamed for several illnesses and implicated in the death of a man who became infected with flesh-eating bacteria. In addition to the medical problems, there were severe economic repercussions statewide. Waikiki accounts for nearly half of Hawaii's tourist business and expenditures, which were about $5 billion in 2002.

The effect of heavy rains is well illustrated in California. The state's 2006 annual quality report gave Huntington Beach all A's in dry weather but C's or lower in wet weather. The same disparity occurred in nearly half of the 356 California beaches included in the study. Between 627,000 and 1.5 million cases of beach-related gastroenteritis occur annually in Los Angeles and Orange counties.

The National Resources Defense Council counted more than 20,000 days of beach closure and advisories in 2005 in the twenty-nine ocean and Great Lakes states. The organization attributed this increase of 5 percent over 2004 to burgeoning coastal development and a year of heavy rainfall, which caused extra sewage runoff. Probably improved monitoring and additional aging of sewer pipes also contributed.

The American Society of Civil Engineers said in 1999 that the nation's 500,000-plus miles of sewer lines were thirty-three years old on average.[22] Sewers in Newark, New Jersey, which date back to 1852, are so old that they are listed in the National Register of Historic Places. The EPA estimates that as much as $400 billion in extra spending is needed over the next decade to fix the nation's sewer infrastructure. The federal government has not yet addressed this problem (figure 2.2).

As an example of the cost of sewer maintenance to residents in a city, I can cite Portland, Oregon, which has some of the highest water and sewer rates in the country. The average residential sewer bill in the city has risen from about $14 a month in the early 1990s when the city began mandated improvements to $45 a month in 2007.

Derivatives of the contents of toilets are not the only things that pour from sewage treatment plants into unwanted places. Medicine residues are significantly more abundant in sewage downstream from public treatment plants that handle waste from drug manufacturers.[23] Apparently some of the drug plant's product is going down the drain. The effect on human health is not entirely known, but certainly it is not beneficial. Among the chemicals that survive treatment in the sewage plant are endocrine disrupters, which affect glands and hormones that regulate many bodily functions.[24]

Sewage plants are important parts of America's infrastructure, and in 2006 Democrats in Congress proposed a plan to protect sewage plants from terrorists. Not all Republicans believed this was a serious problem, even though Senator Lincoln Chafee of Rhode Island pointed out that "they're going to blow up our poop."

Pipes for Gas and Hazardous Liquids

Buried beneath the United States and offshore waters are more than 2 million miles of natural gas steel pipelines, operated by more than 3,000 companies, that transfer this energy source across the nation. Natural gas supplies about one-quarter of our energy needs so that the maintenance of this part of the infrastructure is clearly of great importance. The location, construction, and operation of these pipelines are regulated by either federal or state agencies, depending on whether the pipe crosses state lines. More than seventy-five full-time federal pipeline inspectors conduct inspections from five regional offices throughout the United States. However, the majority of inspections are made by state inspectors who work for state regulatory agencies.

Since 2005 there have been hundreds of gas pipeline incidents that have killed 68 people and injured 280 others. The most recent was an explosion and fire that occurred in the evening of September 9, 2010, in San Bruno, California, when a 54-year-old natural gas pipeline exploded. Eight people were killed, 30 more were injured, 38 houses were leveled, and many more dwellings were damaged. The cause is unclear—perhaps corrosion, a bad weld, or a crack in the pipe.

Pipeline companies routinely inspect their pipes for corrosion and defects using highly sophisticated pieces of equipment known as pigs: robotic devices that are propelled down pipelines to evaluate the interior of the pipe. Pigs can test pipe thickness and roundness, check for signs of corrosion, and detect minute leaks and any other defect along the interior of the pipeline that may impede the flow of gas or pose a potential safety risk for the operation of the pipeline. According to the Department of Transportation, pipelines are the safest method of transporting petroleum and natural gas. The leading cause of serious incidents such as fatalities and injuries related to the pipelines occurs during excavation rather than because of breaks in the pipe. About forty pipeline incidents occur each year that result in fatalities or hospitalization, mostly related to excavations around the service lines between the large city main under the street and the private home receiving the gas.

Wires for Electricity

In 1940, 10 percent of energy consumption in the United States was used to produce electricity. In 1970, that percentage was 25 percent. Today it is 40 percent, showing the growing importance of electricity as a source of energy supply. According to the Electric Power Research Institute, the nation's appetite for electricity has grown by 25 percent since 1990, and is growing at approximately 2.5 percent per year. The current electricity demand in the United States is roughly four trillion watts (four terawatts), and current projections place demand at between eleven and eighteen terawatts by 2050.

Electricity has the unique ability to convey both energy (lighting, batteries) and information (television, cell phones) in an ever-increasing array of products, services, and applications throughout society. There are more than 3,000 electric utilities, three-quarters of them stockholder owned, with the rest owned by federal agencies, rural electric cooperatives, and municipalities. The average price paid in April 2008 was eleven cents per kilowatt-hour, although prices vary from state to state depending on local regulations, generation costs, and customer mix.[25]

The power delivery system is largely based on technology developed in the 1950s or earlier and installed as many as fifty years ago. The strain on this aging system is beginning to show, particularly as consumers ask it to do things it was not designed to do. The equipment was deployed before Frank Sinatra was in his prime (1950s and 1960s), before a man walked on the moon (1969), before the Internet (1969), and before cell phones (1973).

Since 1998 the frequency of power outages and the number of citizens affected has increased dramatically.[26] New Yorkers recall the August 2003 blackout that left 50 million people in the Northeast United States and Canada without power. (Contrary to popular mythology, there was not a surge in births nine months after the blackout.) That disaster was only the tip of the iceberg. The amount of reserve electricity generation capacity available for emergency shortages in New England fell below the 15 percent safety margin in 2009 and will be below this safety margin in the Midwest and mid-Atlantic regions by 2012, requiring some combination of new power plants, more transmission lines, and electricity conservation by consumers and business.[27] Compounding the problem is the fact that almost half of existing power plants are fueled by coal, and these plants are increasingly criticized because coal is the most polluting of the fossil fuels whose use dominates our energy supply.

About 79 percent of the nation's electricity transmission lines are above ground, and 21 percent are buried. Burying power lines eliminates potential problems such as icing of power lines in northern climates and lines being knocked down by falling tree limbs on windy days. But buried power lines are more susceptible to damage from floods and mudslides and cannot necessarily be repaired quickly. The limiting factor for burying power lines is the cost, which can be more than ten dollars a foot. Burying existing overhead power lines costs ten to fifteen times more than the cost of stringing them from poles. The national trend in cities is to bury lines when the city's infrastructure is being upgraded. Now that billions from President Obama's 2009 stimulus package will be going into upgrading infrastructure, the electric grids may become smarter and more commonly underground.

The NERC says that the electric transmission system is being used closer to its limits more of the time than in the past, and the demand for power in summer is expected to increase by 18 percent by 2017. The power blackout in the Northeast in 2003 was a harbinger of what the future holds if remedial measures are not taken soon. About 2,000 miles of high-voltage lines were built in 2006, a 1 percent increase in existing lines.

America operates about 157,000 miles of high-voltage (more than 230 kilovolts) electric transmission lines.[28] Since 1982, growth in peak demand for electricity—driven by population growth, bigger houses, bigger televisions, more air conditioners, and more computers—has exceeded transmission growth by almost 25 percent every year. Yet annual investment in new transmission facilities has decreased by about 30 percent.[29] The resulting increased grid congestion increases transmission and distribution losses, which have doubled to 10 percent since 1970. To correct deficiencies in the existing system and enable the smart power system of the future will require an additional $8 billion to $10 billion annually. That additional investment will lead to an average increase of 3 to 5 percent in consumers' electric bills. The investment will pay for itself many times over by increasing the nation's economic growth rate and reducing the cost to consumers for power disturbances. Economic output (gross domestic product per capita) is strongly correlated with the consumption of electric power.[30]

The need for alternative sources of energy to supplement or replace fossil fuels has been well publicized, but the need for new electric transmission lines has not. Many transmission lines and the connections between them are too small for the amount of power that companies would like to squeeze through them. The difficulty is most acute for long-distance transmission but shows up at times even over distances of a few hundred miles. The power grid is fragmented, with about 200,000 miles of power lines supporting 830 gigawatts, and divided among 500 owners. Built gradually over time, the grid exists as a patchwork of remedies rather than a singular coordinated unit. Big transmission upgrades often involve many companies, many state governments, and numerous permits, and every addition to the grid provokes fights with property owners. These barriers have caused electrical generation to grow four times faster than the ability of the grid to transmit it. Modernizing the electric infrastructure is an urgent national problem. The aging system cannot indefinitely sustain the nation's demand for electricity, which is growing at 2 percent per year.

The electrical grid in the United States is divided into 140 balancing areas, some overseeing as little as 100 megawatts and others more than 100 gigawatts, a thousand times more. Grid operators balance supply and demand in each area and oversee the flow of power through their individual systems. An area may have too much supply at a time when a neighboring area is running short. But the oversupplied area may not be able to assist the undersupplied area because of transmission constraints that exist between the two regions.

What is required is a smart grid—an electrical transmission network with these features:[31]

• Intelligent. Capable of sensing system overloads and rerouting power to prevent or minimize a potential outage; of working autonomously when conditions require resolution faster than humans can respond, and cooperatively in aligning the goals of utilities, consumers, and regulators.

• Efficient. Capable of meeting increased consumer demand without adding infrastructure.

• Accommodating. Accepting energy from any fuel source and capable of integrating any better ideas and technologies as they are market proven and ready to come online.

• Motivating. Enabling real-time communication between consumer and utility so consumers can tailor their energy consumption based on individual preferences, such as price or environmental concerns.
• Resilient. Increasingly resistant to attack and natural disasters as it becomes more decentralized and reinforced.

The Department of Energy in 2009 allocated $620 million to thirty-two smart grid programs across the country, and another $1 billion is being added from private investment money.

Currently, almost 300,000 megawatts of wind projects, more than enough to meet 20 percent of our electricity needs, are waiting in line to connect to the grid because there is inadequate transmission capacity to carry the electricity they would produce.[32] Waiting in a queue are 70,000 megawatts in the Upper Midwest, 50,000 megawatts in Texas, 40,000 megawatts in the Lower Midwest, 40,000 megawatts in the Great Lakes/Mid-Atlantic, and 13,000 megawatts in California. Many of the windiest sites have not been built for this reason. Achieving the 20 percent figure or even approaching it would require moving large amounts of power over long distances, from the windy, lightly populated plains in the middle of the country to the coasts where a large part of the population lives. Currently this cannot be done. The way the national grid exists now would require half the country to move to the Midcontinent to make use of the wind power there. This situation is analogous to being able to manufacture a needed product but lacking the rail or highway network to get it to the consumers.

Concern about inadequate transmission is shared by the solar, geothermal, and hydropower industries as well. As of January 2009 in California, more than 13,000 megawatts of large solar power plants were waiting to connect to the grid. The hot deserts in the Southwest where the solar power is best generated have few power lines. There is a dire need to construct transmission lines to connect the scattered sources of power supply in the United States to the populated areas that need it. Just as the nation needs to be connected by a maze of superhighways for an effective transportation network, we also need to construct a "green power superhighway."

The key to a cost-effective plan is the use of high-voltage transmission lines in place of the low-voltage lines in use today (table 2.2). One 765 kilovolt line can carry as much power as six 345 kilovolt lines, reducing the amount of land needed by a factor of four. Given their efficiency, very high voltage lines will significantly reduce congestion and transmission losses, which will reduce power costs. A 765 kilovolt grid could

Table 2.2
Relationships among transmission voltage and construction cost

Transmission voltage	Cost per mile ($)	Capacity (MW)	Cost per unit of capacity ($/MW-Mile)
230	$2.077 million	500	$5,460
345	$2.539 million	967	$2,850
500	$4.328 million	2,040	$1,450
765	$6.579 million	5,000	$1,320

Source: Green Power Superhighways, American Wind Energy Association/Solar Energy Industries Association (2009)

cut peak load losses by 10 gigawatts. New high-voltage transmission lines for renewable energy sources could be to energy what President Eisenhower's interstate highway system was to transportation.

While conventional power plants can be constructed close to where their electricity is needed, renewable energy sources are commonly not near population sources, and the power must be transmitted long distances. For example, solar power is most efficiently generated in southwestern United States, but the power may be needed in Chicago, 1,500 miles away. To solve this problem, supporters of renewables favor construction of a new generation of high-voltage direct current (DC) supergrids linking regions rich in wind, solar, or other renewable energy sources with populated areas thousands of miles away. Over distances greater than 600 miles, alternating current (AC) transmission lines become increasingly inefficient, losing more than 10 percent of the energy pumped into them, and they become more expensive than DC, whereas a high-voltage DC line would lose less than 3 percent. Conversion of DC to the AC that consumers need adds only an extra 0.6 percent loss.[33]

The greatest challenge facing electric distribution with the existing grid is responding to rapidly changing customer needs for electricity. Increased use of information technologies, computers, and consumer electronics has lowered the tolerance for outages, fluctuations in voltages and frequency levels, and other quality disturbances.

Crumbling Highways

The 4 million miles of roads in the United States range in quality from one-lane loose gravel to six-lane asphalted interstate highways. Paved roads cover about 60,000 square miles, about 2 percent of the nation's total area. Almost all are surfaced with asphalt, an increasingly expensive petroleum product. Ninety percent of travel in the United States occurs on highways, and three-quarters of all domestic goods are shipped by road.[34]

As with other parts of our infrastructure, the roads are deteriorating, and federal and state funds are inadequate to maintain them. The federal highway trust fund is the source of federal funds for roads and bridges, the monies coming almost entirely from federal taxes on gasoline and diesel fuel, which have not increased since 1993; they remain at 18.4 cents a gallon for gasoline and 24.4 cents a gallon for diesel fuel. Since 1993, the trust fund's purchasing power has decreased by 30 percent because of inflation and skyrocketing construction costs.

The fund has now run dry and in fact had an $8.3 billion deficit on September 30, 2008.[35] Congress had to quickly appropriate $8 billion in 2008 and $7 billion more in 2009 so the fund could pay its bills and remain solvent. The last funding bill passed by Congress in 2005 was 24 percent less than the Federal Highway Administration said it needed. States are similarly strapped. Most state gasoline taxes are based on volume (number of gallons sold), not price, so that state revenues for road maintenance have not increased even as the price of gasoline has risen sharply.

Our national highway system is fifty years old in many places; when it was constructed during the Eisenhower administration, it carried 65 million cars and trucks. Today that number has nearly quadrupled to 246 million.[36] The number of miles driven jumped 41 percent between 1990 and 2007, from 2.1 trillion to 3 trillion, two-thirds of it on urban roads. According to the American Association of State Highway and Transportation Officials, one-third of the nation's major roads are in "poor or mediocre condition," and 36 percent of major urban highways are congested (not a surprise to motorists).[37] Motorists spent 4.2 billion hours a year stuck in traffic in 2007, wasting 2.8 billion gallons of gas.

According to a study by the Road Information program and the American Association of State Highway and Transportation Officials, the continued increase in urban traffic puts significant stress on roads as transportation funding either falls or fails to keep pace with the rate of road degradation. The study also found that poor roads cost drivers a considerable amount of money in extra operating expenses, such as accelerated vehicle depreciation, additional repair costs, increased fuel consumption, and tire wear, or an average per driver of $458 a year. The stimulus package passed in 2009 will provide $27 billion for highway projects, but the report's authors said this amount can only be a beginning to more investment. The president of the Northern Virginia Transportation Alliance, a group that favors more state investment in road and rail projects, said that a 10 cent increase in the gas tax would cost each Virginia motorist an average of $60 a year, much less than the poor roads are costing them.

The rate of road deterioration depends not only on the number of vehicles it services, but also on the weight of the vehicles. This means that heavy trucks are disproportionately responsible for highway damage. The average American or Japanese car weighs about 3,000 pounds; trailer trucks can weigh more than 100,000 pounds. According to the Department of Transportation (DOT), combination trucks weighing 80,000 to 100,000 pounds pay just 50 percent of the cost of the damage they cause to the highway system. Trucks weighing more than 100,000 pounds pay only 40 percent.[38] The DOT has pointed out that this violates a tenet of highway taxation, dating back to the creation of the Highway Trust Fund, that "different vehicle classes should be charged in proportion to their contribution to highway investment requirements." Drivers of automobiles are subsidizing the trucking industry.

Collapsing Bridges

A 2000 report by the Federal Highway Administration (FHA) indicated that an average of about 2,500 new and well-constructed bridges are built each year. But there are 594,709 bridges longer than twenty feet in the United States. Their average age is forty-three years, seven years shy of the maximum age for which most are designed. Twenty percent are more than fifty years old. The FHA reported in 2006 that 25.8 percent were "structurally deficient" (12.4 percent) or "functionally obsolete" (13.4 percent).[39] The percentage of deficient bridges was highest in the District of Columbia (52.2 percent) and lowest in South Dakota (4.4 percent). "Structurally deficient" generally means the bridge cannot carry the traffic it was designed to accommodate. Functionally obsolete bridges also have major design problems that diminish their load-carrying capacity.

In its most recent report card on the nation's infrastructure, in 2009, the American Society of Civil Engineers gave the nation's bridges a grade of C. About one-quarter of our bridges are in trouble. A $17 billion annual investment is needed to substantially improve current bridge conditions. Total bridge expenditures by all levels of government for system preservation and expansion are $10.5 billion.

Many of the inadequate bridges were built between the 1930s and 1960s, when designers did not fully understand the effects of metal fatigue or other structural challenges. In addition to the dangers of design inadequacies, the useful safe life of bridges has been shortened because they carry far more vehicles per day than when they were built decades ago. Compounding the problem is the fact that modern inspection techniques were not available until recently. Modern bridge inspections use ultrasound, wireless sensors, X-ray technology, and other techniques to detect cracks, corrosion, and other defects.

The Interstate 35W bridge in Minneapolis that collapsed into the Mississippi River in August 2007, killing 13 and injuring 145, was built in 1967 and was one of 760 identically built bridges around the country, 35 percent of which were found to be structurally deficient by the Federal Highway Administration in 2006. An examination of bridges with a similar but not identical design revealed that 58 percent were deficient.[40]

Seventy-three percent of road traffic in the United States and 90 percent of truck traffic travels over state-owned bridges.[41] States tend to ignore bridge and road maintenance because if they wait until they are in poor condition, they become eligible for federal funds (federal funds cannot be used for routine maintenance). As noted earlier, the federal Highway Trust Fund is now bankrupt.

Resurrecting Railroads

The first railroad in the United States was chartered in 1826 in Massachusetts and was 3 miles in length. It carried its first passengers two years later. By 1840, 2,800 miles of track had been laid, a number that increased rapidly each decade, to peak at 254,000 in 1916. Since that time, the number of miles of track has decreased continuously, in part because of a concerted campaign by American carmakers to acquire rail lines and close them, and in part because of a decision by Congress in the 1950s to build the world's most extensive interstate highway network. Today there are only 161,000 miles of railroad track, about the same amount that existed in the late 1890s.

Despite the loss of track, freight railroad traffic and freight volumes have increased dramatically during the past forty years. Railroads today are primarily high-volume freight haulers, accounting for more than 40 percent by weight of the nation's freight, carried in 1.3 million freight cars. Almost half the freight is coal (figure 2.3).

Within the past few years, the rising cost of gasoline for eighteen-wheel trucks has sparked an increased commercial interest in railroads

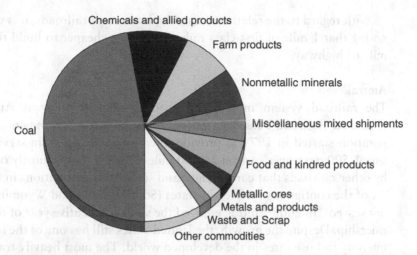

Figure 2.3
Classes of products carried by freight railroads in the United States. (Association of American Railroads, 2008)

as haulers of freight. A train can haul a ton of freight 423 miles on 1 gallon of diesel fuel, about a three-to-one fuel efficiency advantage over eighteen-wheelers, and the railroad industry is increasingly touting itself as an ecofriendly alternative to the trucks that have dominated freight transport for decades. To handle their new green status, the railroads spent nearly $10 billion in 2008 to add track, build switchyards and terminals, and open tunnels. According to the DOT, freight rail tonnage will rise nearly 90 percent by 2035.[42] Employment in the industry has increased accordingly. In 2002, the major railroads laid off 4,700 workers; in 2006 they hired more than 5,000.

The rail network is privately owned, was largely built almost one hundred years ago, and includes over 76,000 railroad bridges, mostly built before 1920. Each freight car is carrying more weight today over old bridges. The condition of the bridges is considered proprietary and is known only to the executives of the railroad companies. About 41 percent of the bridges are made of wood, 36 percent of steel, and 23 percent of masonry.[43] Bridges and tunnels are likely to cost more to repair—and as much as $100 million to replace—than other components of railroad infrastructure networks, such as tracks and signals. As a result, railroads are more likely to invest in other components sooner and to consider extensive bridge or tunnel repair or replacement as one of their last investment options.

With regard to the relative costs of highways and railroads, it is worth noting that 1 mile of first-class railroad track is cheaper to build than 1 mile of highway.

Amtrak

The railroad system most familiar to passengers today is Amtrak (National Railroad Passenger Corporation), a quasi-governmental corporation started in 1971 to provide intercity passenger train service. It uses 1,500 passenger cars on 21,000 miles of track that is mostly owned by other railroads that carry freight and serves 500 destinations in forty-six of the contiguous forty-eight states (South Dakota and Wyoming are not served). In 2008 Amtrak enjoyed the sixth consecutive year of record ridership. Despite the growth, the United States still has one of the lowest intercity rail use rates in the developed world. The most heavily traveled routes are in the Northeast Corridor between Boston and Washington, D.C.. Amtrak is not required by law to operate a national route system.

Although Amtrak served 28.7 million passengers in 2008, rail travel is a bit player in passenger transportation, accounting for only 0.1 percent of U.S. intercity passenger miles (5.7 million out of 5.3 billion total, of which private automobile travel makes up the vast majority). The 25 million passengers served by Amtrak stands in sharp contrast to the 700 million served by commercial airlines. The American love affair with the automobile and the interstate highway system funded by federal funds since the 1950s was the death knell for nationwide passenger rail service.

Amtrak does not pay its own way and needs federal subsidies to stay in business. Each budget cycle, many lawmakers complain about this continuing subsidy, saying that a transportation system that cannot pay its own way does not deserve to exist. These congressional representatives do not seem to, or do not want to, recognize the subsidies regularly given to the air travel and automotive industries. If Amtrak were similarly treated, it would be highly profitable and convenient, and a much greater cross-section of the population would use it.

Many decades ago, in the days before the airlines and the industries associated with car travel increased their lobbying efforts, the railroads received enormous federal subsidies. Today Amtrak receives about $2 billion each budget cycle. In contrast, the Federal Aviation Administration, essentially a large subsidy for the airlines, has a yearly federal budget of at least $15 billion. Embedded federal subsidies for travel by road (construction, repair, expansion) are about $47 billion annually.[44]

The Obama administration has resurrected an emphasis on passenger rail transport. High-speed rail has emerged as the cornerstone of the president's desire to remake the nation's transportation agenda by switching the government's financial focus away from building highways and roads to building mass transport.[45]

This makes good environmental sense. According to the National Association of Railroad Passengers, airplanes use energy at a rate 20 percent higher per passenger mile than passenger trains. Cars are even worse, at 27 percent more per passenger mile.[46] The president's proposed budget for 2010 gives an additional $2 billion to the DOT, most of it to be used for rail and aviation improvements. Amtrak will receive $1.3 billion. Nearly half of the $48 billion in stimulus money for transportation projects will be spent on rail, buses, and other nonhighway projects.

The need for improvement in the nation's passenger rail system was dramatically illustrated in June 2009 when two trains carrying hundreds of passengers rear-ended in Washington State, killing nine and injuring more than seventy. Federal safety officials had warned the train company that the cars could be unsafe in crashes and called for them to be replaced, or at least strengthened. But nothing was done because of financial constraints.

When a northbound train belonging to the Chicago Transit Authority derailed in July 2006, injuring more than 150 people, the National Transportation Safety Board reported that the line had been placed into service fifty-five years earlier and that many parts of the track had never been replaced. The report described corroded rails and fasteners, and rotten wood on the ties, and questioned why the problems had not been identified and repaired before the derailment.

Although federal financing for capital improvements to transit systems has been rising, the share going to the largest systems has been shrinking because they have had to compete with new, smaller systems. The nation's seven largest systems, which carry 80 percent of the nation's rail riders, have received only 23 percent of federal financing eligible for bringing systems into a state of good repair, according to the transit administration.

The need to reduce energy consumption is clear to all Americans today. A Harris poll in December 2005 revealed that the public believes that the federal government should set transportation policy and that trains should have an increased share of public transportation, with an emphasis on commuter trains rather than long-distance trains.[47] Nearly two-thirds want freight trains to carry an increased share of the nation's freight. Trucks ranked below air freight as second choice.

Unsafe Dams

Most people who think of dams envision the giant federal dams, which are less than 5 percent of all dams in the United States: Hoover Dam in Nevada, Grand Coulee in Washington, or one of the thirty-four dams along the Tennessee River and its tributaries. But there are 85,000 dams more than six feet high in the United States, and tens of thousands of smaller dams. In the 1950s and 1960s, a dam went up in the United States every six minutes. They were built for a variety of reasons: flood control, hydroelectric power, water supply and irrigation, navigation, tourism, and recreation in the lakes that form behind the dam. But dams, like the other infrastructure, require maintenance and replacement as they age. Although a rock-filled dam in Syria built around 1300 B.C.E. is still in use, dams in general have a much shorter life expectancy, both because they weaken with age and because they fill with sediment carried into the reservoir behind the dam. Human agriculture and construction have greatly increased erosion rates, and 13 percent of the river's sediment load is trapped in reservoirs, significantly reducing their useful life span. [48] According to the American Society of Civil Engineers, the average life expectancy of a dam is fifty years, and the average age of the 85,000 dams in the United States in 2009 was fifty-one years. By 2020, 85 percent will be older than fifty years (figure 2.4). [49]

Antiquated dams become increasingly vulnerable to such events as seismic shifts from below, water pressures that scour them from behind, invasive species that choke intake and outflow pipes, and the effects of a fast-changing climate that bring flooding on scales for which the dams were not designed. A study in 2009 by the American Society of Civil Engineers reported that the number of high-hazard dams—dams whose failure would cause loss of human life—was 1,819. [50] The cost to repair them was estimated at more than $10 billion. As downstream land development increases, so will the number of dams with high hazard potential. States reported that 3,300 dams were unsafe. [51] A 2003 study by the Association of Dam Safety Officials placed the cost of bringing U.S. dams into safety compliance and to remove obsolete dams at $36.2 billion. [52]

The urgency of the problem was emphasized by two storms in New England in 2005 and 2007 that caused the overtopping or breaching of

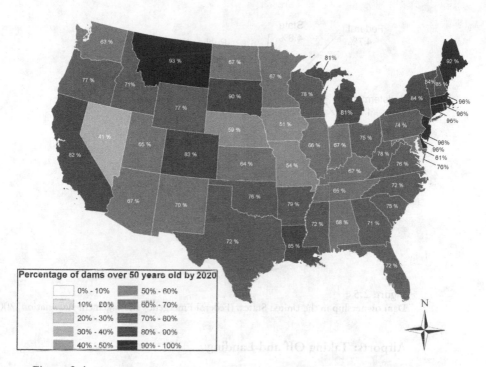

Figure 2.4
Percentage of dams more than fifty years old by 2020, EOS (American Geophysical Union), December 1, 2009.

more than 400 dams in three states.[53] Clearly many of the nation's dams are disasters waiting to happen. About 500 have been torn down in the past few decades for structural or environmental reasons, such as the blocking of sediment to downstream areas and blocking of fish passage. Dam removal averages about one-third the cost of repair to structurally deficient dams.[54] In the past decade or so, the rate of dam removal has exceeded the rate of dam construction.[55]

More than half of the dams are privately owned, which is one reason dams are among the nation's most dangerous structures (figure 2.5).[56] Many private owners cannot afford to repair aging dams; some resist spending the necessary money by tying up governmental repair demands in court or campaigning to weaken state dam safety laws. Thousands of dams have been abandoned by their owners, and over time, title to them has become obscure. Fifteen percent of the listings in America's National Inventory of Dams have no known owner.

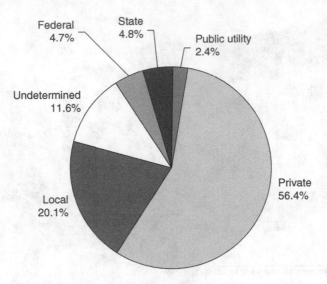

Figure 2.5
Dam ownership in the United States. (Federal Emergency Management Association, 2005)

Airports: Taking Off and Landing

There are 510 airports in the United States with commercial service; in 2005 they handled more than 660 million passengers. The busiest airports were at Atlanta, Chicago, and Dallas. Today's largest aircraft require 12,000 feet of runway to take off and 7,500 feet to land.

Although travel on domestic airlines declined in 2008 because of the poor economic situation, demand for air travel is certain to increase in the coming years as it did regularly before the recession. One factor adding to the pressure on infrastructure at airports is the increasing size of some commercial aircraft, with passenger capacities of 800 or more. Another is the tendency of airlines to replace larger aircraft with many smaller ones to maximize profit margins by flying with fewer empty seats. Furthermore, the tremendous growth in regional and commuter carriers and budget carriers is increasing the number of aircraft operations at America's busiest airports. The number of aircraft handled by air traffic control is expected to increase by nearly 30 percent between 2004 and 2015.[57]

The National Transportation Safety Board has recently emphasized the need for more air traffic controllers at major airports. Because of a new labor contract imposed by the Federal Aviation Administration (FAA) in 2006, record numbers of controllers have retired or resigned.

More than 1,200 did so in 2008. The number of fully trained professional controllers at the nation's airports is now the lowest since 1992. Fifty-two percent of controllers at twenty-five airports work six-day weeks, a schedule not conducive to improved safety. Also, the Federal Aviation Administration's Office of Runway Safety has not produced a national runway safety plan since 2002, went two years without a permanent director, and has had a 45 percent staff cut since 2003. Clearly safety considerations at America's airports are not getting the attention they deserve. It may require a major airport disaster and many deaths to change the way things are operating.

Federal Aviation Administration facilities include 420 air traffic control centers. A report from the DOT inspector general in 2008 said that FAA facilities have an expected useful life of twenty-five to thirty years. But 59 percent of FAA facilities are more than thirty years old. The average age of its control towers is twenty-nine years old, and its air route traffic control centers average forty-three years old.[58]

Because of increasing air traffic and the expense of airport expansion, near misses on the ground at overcrowded airports are becoming one of the most serious safety concerns in civil aviation. The danger arises when airports try to alleviate bottlenecks by adding runways, raising the risk of accidental incursions, where an aircraft or vehicle becomes a collision hazard by venturing onto a runway being used for takeoffs and landings. In the 2007 budget year that ended September 30, 2007, 370 incidents were recorded in the United States.[59] There were twenty-four serious incursions during the period, where a collision was narrowly averted. In Fort Lauderdale, Florida in 2007, two passenger jets missed each other by less than 30 feet. That same year at the same airport a United flight with 133 passengers on board missed a turn on the taxiway and entered an active runway where a Delta jet was about to land with 167 passengers.

The international pilots' union blames poorly designed airports as the primary cause of incursions. An FAA study found that at the well-designed Washington Dulles airport, there were only four incursions between 1997 and 2000, compared to Los Angeles Airport, with a complex layout of multiple intersecting runways and taxiways, which had twenty-nine incursions. The expected increase in the volume of airport traffic greatly increases the potential for near misses and fatal accidents. The dangers caused by poor airport design and high traffic density, complicated operational procedures, overworked controllers, nonstandard markings, and poor comprehension of English, the standard language of aviation, among foreign cockpit crews add to the risks.

The quality of runways at many commercial airports in the United States is unsatisfactory. In 2003, the FAA rated 80 percent of the runways as good, 18 percent as fair, and 2 percent as poor. New runways can increase an airport's capacity by as much as 30 to 60 percent. The median time to open a new runway is ten years.[60]

Maintaining the integrity of the national airport system requires continual updates and a steady and predictable flow of capital. Airport funding comes from several sources:

Airport bonds, 59 percent
Federal grants, 21 percent
Passenger facility charge, 13 percent
State and local funding, 4 percent
Airport revenue, 4 percent

The FAA has estimated that planned capital improvement of $9 billion annually is needed to meet expanding demand. The Airport Council International estimates $15 billion. Congress in 2003 authorized $14.2 billion for the Airport Improvement Program for the years 2004–2007, which amounts to only $3.55 billion per year.

When the Levees Break

The large floods in recent years along the Mississippi River and its tributaries have focused public attention on the nation's artificial levee system—the constructed embankments along rivers designed to hold flooding rivers within their channels. The federal government has advocated and supported their construction for more than a century, although 85 percent of the nation's estimated 100,000 miles of levees are locally owned and maintained. Many are over fifty years old. Over the years, the U.S. Army Corps of Engineers has built levees along the Mississippi River and its tributaries that total 7,000 miles in length. Some of these levees are 30 feet high.

In 2007 the Corps of Engineers reported that 122 levees around the country could fail in a major flood; 37 are in California and 19 in Washington State.[61] The dangerous levees were designed poorly and built of whatever material was close by, such as clay, soft soil, and sand mixed with seashells. These unsuitable and structurally weak materials have been weakened still more over time by tree roots, shifting stones and soil, and rodents. Furthermore, commonly the land the levees are built on has settled, and the river water they restrain constantly probes for weak spots in the levee and under it. Although levee overtopping does occur, it is less of a problem during floods than breaches in the levee resulting from poor construction practices or deterioration. The Corps of Engineers is currently the defendant in a multibillion-dollar class-action lawsuit centering on devastation from Hurricane Katrina in 2005, the costliest natural disaster in U.S. history.

The Corps of Engineers has a good set of engineering guidelines for levees, but it is not always able to follow them because of political influences on engineering decisions. Now largely staffed by civilians, the corps has a backlog of projects it does not have the money to accomplish. Funding for its work has been cut relentlessly for decades. America's flood protection system has long been undermined by bureaucratic turf wars, chronic underfunding by Congress, and a lack of political leadership. In recent years a growing number of experts believe that levee construction is doomed to fail as a means of harnessing large river floods, and the number of these floods is increasing because of climate change (chapter 9).

3

Floods: When the Levees Break

The river belongs to the Nation,
The levee, they say, to the State;
The Government runs navigation,
The commonwealth, though, pays the freight.
Now, here is the problem that's heavy—
Please, which is the right or the wrong—
When the water runs over the levee,
To whom does the river belong?

—Douglas Malloch, *Uncle Sam's River*, 1912

Two of the worst naturally occurring environmental disasters in U.S. history are the 1993 flood in the Mississippi River and the 2005 flooding in New Orleans that resulted from hurricanes Katrina and Rita. Each of these floods was extremely costly, causing tens of billions and even hundreds of billions of dollars in damage, as well as loss of life and property. Worldwide since 1980, hurricane numbers have nearly doubled, and the frequency of floods has more than tripled.[1] These data give rise to the suspicion that climate changes may be the cause of the increases. Both hurricanes and river floods are strongly correlated with climate. In the United States, floods account for about 60 percent of federally declared disasters.

Floods can sometimes be predicted but rarely can be prevented. Preventive measures such as dams and levees are often inadequate and are commonly overridden by surging waters. Analyses reveal that water levels for 100-year floods are profoundly underestimated by flood specialists.[2] At many sites in Iowa and Missouri, the flood of 2008 neared or exceeded the record 200-year or 500-year levels attained in the record Midcontinent flood of 1993. There are numerous examples of repeated inundations over brief periods by floods expected only once in hundreds of years because the data on which predictions are made are inadequate. The records do not match the pattern today in terms of the weather. What have been called 100-year floods may now be 10-year floods due to a combination of changing conditions over time. The floodplain maps issued by the Federal Emergency Management Agency (FEMA) badly need updating.

Nature is usually more powerful than humankind's feeble efforts to control it. As Mark Twain wrote 125 years ago, "You can plan or not plan and it doesn't make a hell of a lot of difference. What makes a difference is how much it decides to rain."[3] On average, floods kill about 140 people each year and cause $6 billion in property damage. Loss of life to floods during the past half-century has declined, mostly because of improved warning systems, but economic losses have continued to rise due to increased urbanization and coastal development.

Floods are expensive calamities that affect all Americans through their taxes. The federal government uses tax money to subsidize reconstruction after floods, so floods affect the wallets of all Americans, whether they live in flood-prone areas or not.

River Floods Are Terrible

Floods can be conveniently divided into two varieties: those that originate on land (excess precipitation) and those that originate at sea (hurricanes). The flooding of the Midwest in the summer of 1993 was the most devastating river flood in U.S. history. More than 50,000 homes were damaged or destroyed in nine states in the Upper Midwest, 74,000 people were displaced, about 50 people died, over 12 million acres of farmland (19,000 square miles) were inundated and rendered useless, and hundreds of thousands of people were without safe drinking water for many weeks. Overall damages totaled $20 billion. The Mississippi River was closed to traffic for two months.

The cause of the deluge of water from the atmosphere was a confluence of atmospheric conditions that occurs occasionally and unpredictably. During the summer of 1993, the jet stream that moves from west to east stalled over the Midwest and formed a barrier to moisture-laden air moving northward from the Gulf of Mexico. The moisture could not continue its northward movement and so dumped its water repeatedly on the hapless people below. Wave after wave of storms rumbled across the river basin. From April through August, the deluge continued almost without letup; rainfall totals in the upper Mississippi River basin were 60 percent higher than average.[4] Added to this was the fact that the previous year was cooler than average, so the ground was still partly saturated and unable to absorb much of the new rainfall. Over 500 river hydrologic stations in the Midwest were above flood stage at the same time. At St. Louis, the river crested at more than 19 feet above flood stage and remained above flood stage at St. Louis for more than two months.

Clearly, there is no way that humans could have stopped the unending rainfall in the spring and summer of 1993. Technology could not have made 1992 warmer so the soil could have absorbed more water, and technology cannot control either the location of the jet stream or the normal northward movement of wet Gulf air.

It is noteworthy that the National Climatic Data Center has documented a 10 percent increase in precipitation and a 20 percent increase in extreme precipitation events (more than 2 inches of rainfall in 24 hours) in the United States since 1900. Since 1958, the number of days with very heavy precipitation have increased 58 percent in the Northeast, 27 percent in the Upper Midwest, 18 percent in the Southeast, 16 percent in the Southwest, 13 percent in the band from the Dakotas to Texas, and 12 percent in the Pacific Northwest.[5] These increases can be attributed to climatic change, an indication that normal floods (those not related to a stagnant jet stream) are likely to become more common.

In populous New York, portions of Lower Manhattan and coastal areas of Brooklyn, Queens, Staten Island, and Nassau County on Long Island will be flooded more often. Much of the city's critical infrastructure, including tunnels, subways, and airports, lies well within the range of projected storm surge and would be flooded during such events.

Attempts at Containment

The chief type of structure along the river that might contain the floodwaters is the levee. The U.S. Army Corps of Engineers over the years has built 3,500 miles of levees (embankments) extending 1,000 miles along the Mississippi and its tributaries in an attempt to prevent the river from spilling onto the floodplain it had built over the centuries.[6] During the 1993 flood, over 1,000 of the levees failed despite continual efforts by people in affected areas to increase levee heights, which average 24 feet, by adding sandbags.

Many dams and reservoirs have been built on tributaries to hold back and store water. Some of the dams burst under the onslaught of the raging waters. Much of the river has been lined with concrete slabs but even these could not restrain the waters.

Engineers, hydrologists, and environmentalists agree that the efforts of the U.S. Army Corps of Engineers to control the Mississippi have only made things worse. They have cut through meanders and shortened the river, causing it to flow more quickly and increase erosion. By restricting the flow of water inside the levees, they have speeded up the flow and increased the pressure on the levees, which increases their chance of failing. And they have altered the natural flow of the river so that the 1993 floods were the worst ever, even though there was less water in the river than during previous major floods. And the levees have been poorly built in many places.[7]

Red River, North Dakota Flood, 2009

Not all river floods are caused or made worse by human foolishness. Some result from the age of the river, the local topography, and the geological history of the area. An example of this is the flood the last week in March 2009 on the Red River in North Dakota. The river, which separates North Dakota and Minnesota and flows north into Canada, is fairly modest in size compared with more famous rivers such as the Missouri and Mississippi. But it has a behavior that is hard to predict. The National Weather Service did, however, rate the area along the border separating North Dakota and Minnesota as an area of extreme flood risk only a week before the flood occurred (figure 3.1). The Weather Service reported that "copious precipitation during the fall, wet soils before freeze-up, and areas of substantial water in the snow pack have produced an imminent risk of major flooding along the Red River. . . . Major overland flooding is expected in eastern North Dakota and northwest Minnesota beginning the week of Sunday, March 22. This will subsequently lead to major, and possibly record, flooding along the Red River of the North and its tributaries." If only all government agencies were that specific and that correct.

The Red River flows across the flat bottom of a dried-up glacial lake that was created from meltwaters at the end of the last glacial episode. The river, only about 9,300 years old, is normally small and shallow, and it flows slowly across a flat landscape. It has not had the time or energy to cut deep channels that can contain water during floods. The river has cut a shallow sinuous valley across one of the flattest expanses

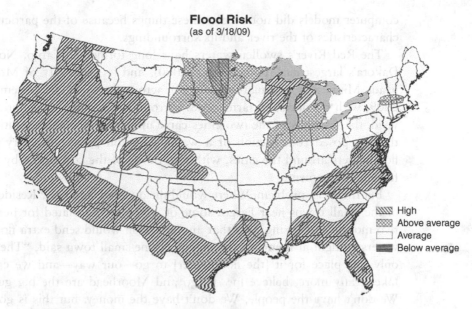

Figure 3.1

Flood risk map of the United States showing the extreme risk of flooding on the Red River in late March 2009. The Red River flows in the middle of the "high risk" area that follows the border between North Dakota and Minnesota. (National Weather Service, March 18, 2009)

of land in the world, similar to the planar bottom surface of ancient Lake Bonneville in Utah where race cars are tested. The average slope of the river channel is about what you would get by slipping a single sheet of paper under an eight-foot sheet of plywood. At Fargo, the site of the 2009 flood, the river bed drops only about 5 inches for each mile of flow, flatter than the slope of the top surface of a pancake on a breakfast plate. At other areas along the river, the slope is only 1.5 inches per mile. This means that when the river is in flood, the water spills in all directions over the surrounding area. Thousands of data points would be needed to make predictions of flood levels, not only a few depth gauges at scattered locations.

Heavy snows or rains on saturated or frozen soil have caused a number of catastrophic floods along the Red River, which often are made worse by the fact that snowmelt starts in the warmer south and waters flowing northward are often dammed or slowed by ice. These periodic major floods have the effect of partially refilling the ancient lake. During the 2009 flood, the river rose quickly but did not crest as high as was feared, and then the water level dropped faster than expected. The best computer models did not predict these things because of the particular characteristics of the river and its surroundings.

The Red River's swollen waters have long tormented Fargo, North Dakota's largest city (population 90,000), and its sister city of Moorhead, Minnesota (population 35,000) across the river. In December 2009, with memories of farmhouses surrounded by miles of water still vivid, the residents of the two cities came up with a $1 billion solution: they proposed construction of a water diversion project to carry the floodwaters around the cities, with two-thirds of the cost borne by the federal government.

But there is a problem: Where would the diverted water go? Residents of the small towns near Fargo, some of which were isolated for nearly two months last spring, fear that any diversion would send extra floodwaters straight at them. As the mayor of one small town said, "There's only one place for it [the floodwater] to go—our way—and we can't take it any more, believe me. Fargo and Moorhead are the big guns. We don't have the people. We don't have the money. But this is going to affect all the little towns." A manager with the local Wild Rice Watershed District, which represents rural communities in six counties, said, "We really want the cities to get their protection, and we're not trying to be confrontational, but we're worried and we need a seat at the table on this."

The mayor of Fargo has a different emphasis. He said he has watched for years as disputes and ambivalence prevented enacting a permanent solution to his city's flood problem. "If this is ever going to get done, ever in my lifetime, this is the opportunity. This is it."

Everyone in the areas affected by the repeated flooding agrees that a solution is needed. But what is fair to the residents of both the big cities and the farming communities?

What should citizens and municipalities along dangerous rivers be doing to prevent another catastrophe like the floods of 1993 and 2009? Although there are no guarantees of safety near riverbanks, some measures can reduce the risk:

• Every state geological survey should publish floodplain maps to inform the public of the locations of flood-prone areas.
• Construction on floodplains should be strongly discouraged. Floodplains are built by rivers that overflow and indicate the areas adjacent to the river where the river is likely to return and construction is ill advised. Owners of existing structures should be encouraged to relocate by new zoning laws and increases in municipal taxes.

• Sellers of real estate should be required to inform prospective buyers of the danger of flooding of the property and how often it has occurred. California has had such a requirement since 1997.

• New structures on floodplains should be required to be as flood-proofed as possible by placing shields around buildings or erecting buildings on stilts that raise the bottom floor to several feet above ground level. The required height would be determined by the height of the greatest recorded flood.

• The government should stop subsidizing the purchase of insurance for those who build on floodplains. The National Flood Insurance Program is the only natural hazard for which the federal government provides insurance. The public should not be required to pay for the foolishness of those who build in dangerous areas. Those who take risks should be prepared to cover their losses if they lose their gamble.

Hurricane Floods Are Worse

The disastrous 1993 flooding in the Midwest was caused by the interaction of two contrasting air masses over the land. But as Mother Nature taught us once again in 2005, even greater disasters can be caused by hurricanes generated by atmospheric and oceanographic conditions far out to sea.

Added to the history of calamitous hurricanes is the well-publicized observation by specialists that climate change over recent decades has caused a global temperature increase. Hurricane specialists have pointed out that ocean surface temperatures have increased dramatically since the 1970s and that increases in ocean surface temperature increase both the severity of hurricanes and possibly also their frequency (table 3.1).[8] The average temperature of waters at ocean surface in July 2009 was the highest ever recorded, at 62.6°F.

Both the number of named storms and the number of hurricanes have increased over the past twenty years. An atmospheric scientist has calculated that the total power released in hurricanes has increased perhaps by 50 percent in recent decades.[9] In other words, things are getting worse for residents of the Gulf and Atlantic coasts, the prime locations for hurricane landfalls (figure 3.2). Coastal scientists have determined, to no one's surprise, that New Orleans is number one on the list of East Coast and Gulf Coast areas most vulnerable to loss of life and property damage in hurricanes.

Table 3.1
Categories of hurricane intensity, the Saffir-Simpson Damage Potential scale

Category	Magnitude	Wind speed (mph)	Storm surge (ft)	Damage
1	Mild	74–95	4–5	Minimal; damage mainly to trees, shrubbery, and unanchored mobile homes. Examples: Cindy 2005; Humberto, 2007
2	Moderate	96–110	6–8	Moderate: some trees blown down; major damages to exposed mobile homes. Example: Dolly, 2008; Ike, 2008
3	Severe	111–130	9–12	Extreme: foliage removed from trees; large trees blown down; mobile homes destroyed; some structural damage to small buildings. Examples: Katrina, 2005; Gustav, 2008
4	Very severe	131–155	13–18	Extreme: all signs blown down; extensive damage to roofs, windows, and doors; complete destruction of mobile homes; flooding inland as far as 6 miles; major damage to lower floors of structures near shore. Examples: Floyd, Isabel, 2003; Charley, 2004
5	Catastrophic	Greater than 155	Greater than 18	Catastrophic: severe damage to windows and doors; extensive damage to roofs of homes and industrial buildings; small buildings overturned and blown away; major damage to lower floors of all structures less than 15 feet above sea level within 1,500 feet of shore. Examples: Gilbert, 1988; Mitch, 1998

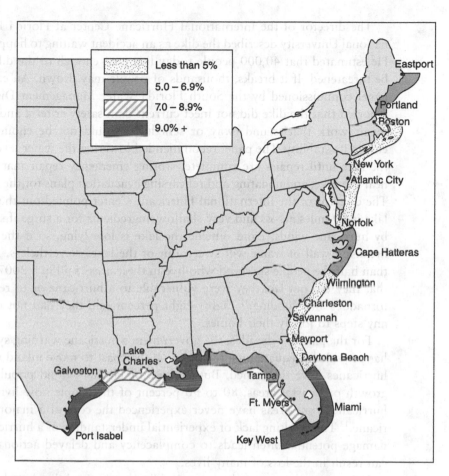

Figure 3.2
Probability of hurricane landfall along the Atlantic and Gulf coasts. (National Oceanographic and Atmospheric Administration)

The second most vulnerable was a surprise: Lake Okeechobee, in south-central Florida.Only those over ninety years old might recall what happened in 1928 when a hurricane sloshed the lake's water into a powerful surge that broke the earthen dam around it. An estimated 2,500 people died in the resulting flood. The Army Corps of Engineers rebuilt the dike, now a massive structure 140 miles long, 45 feet high in places, and up to 150 feet wide at its base. It is part of the extensive system of levees, canals, and other engineering structures that turned the lake's ancient watercourse to the Everglades into farmland. But many are not certain the dike can withstand an assault from a hurricane.

The director of the International Hurricane Center at Florida International University described the dike as an accident waiting to happen.[10] He estimated that 40,000 people today live near enough to the dike to be threatened. If it breaks, thousands of people may drown. An expert panel commissioned by the South Florida Water Management District reported that the dike did not meet current dam safety criteria and that repair work that is underway or scheduled would not be enough to ensure its stability. The panel recommended lowering the water levels in the lake until repairs are complete, storing emergency repair materials near the dike, and updating and rehearsing evacuation plans for the area. The director of the International Hurricane Center pointed out that the lake is 60 miles across and very shallow, ingredients for a surge if swept by hurricane winds. Land outside the lake is low lying, so if the dike breaks, a wall of water will sweep out of the lake. Nevertheless, more than half the people surveyed who live in these areas said in a 2007 poll that they did not feel they were vulnerable to a hurricane or to related tornadoes and flooding.[11] Eighty-eight percent said they had not taken any steps to fortify their homes.

For the past few decades, the government's hurricane warning system has provided adequate time for those on the coast to move inland when hurricanes have threatened. But because of Florida's rapid population growth on coastal areas, 80 to 90 percent of the people now living in hurricane-prone areas have never experienced the core of a major hurricane.[12] The resulting lack of experiential understanding of a hurricane's damage potential often leads to complacency and delayed actions that can result in the loss of many lives.

U.S. coasts are among the most rapidly growing and developed areas in the nation. Fifty-six percent of the coastal residents reside along the Atlantic and Gulf coasts—the low-lying, and therefore most dangerous, areas. Coastal counties constitute only 17 percent of the total land area of the United States (excluding Alaska) but accounted for 53 percent of the total population in 2003, and twenty-three of the twenty-five most densely populated counties were coastal.[13] In addition to the permanent residents, the holiday, weekend, and vacation populations swell in some coastal cities 10- to 100-fold.

Three types of adaptation to hurricane storms (and sea-level rise) are available. One is to move buildings and infrastructure farther inland. Another is to accommodate rising water through changes in building design and construction, such as elevating buildings on stilts. The third is to try to protect existing development by building levees and river flood control structures. This last method is the one that was supposed to protect New Orleans from Hurricane Katrina.

Hurricane Katrina
In the late summer of 2005, Hurricane Katrina roared ashore in New Orleans, generating the city's twenty-eighth major flood, causing $134 billion in damage, the most costly natural disaster in U.S. history. Barge shipping was halted, as was grain export from the Port of New Orleans, the nation's largest site of grain exports. The extensive oil and gas pipeline network was shut down by the loss of electrical power, producing shortages of natural gas and petroleum products. Total recovery costs for the roads, bridges, and utilities, as well as debris removal, have been estimated at $15 billion to $18 billion.

The storm surge was 20 feet high, creating a disaster zone covering 90,000 square miles in three states, an area larger than Kansas. Eighty percent of New Orleans disappeared under water that was up to 15 feet deep, 1,300 people died (1,800 in the three-state affected area), and the United States faced a humanitarian disaster on a scale not seen since the Great Depression. With a second inundation from Hurricane Rita in September, it took fifty-three days from Katrina's landfall to pump the city dry; because half the city lies below mean sea level, the water would not drain out by itself and had to be pumped uphill. Approximately 1.5 million people aged sixteen and older left their homes. As of 2009, only three-quarters of Louisiana residents had returned.[14] The hurricane destroyed or damaged 300,000 homes on the Gulf Coast and led to billions of dollars of waste in the emigration that followed.

New Orleans was a catastrophe waiting to happen, with extensive and repeated warnings from scientists, engineers, and the media. The population of nearly half a million lived in a bowl between the natural levees of the Mississippi River and the built levees along Lake Pontchartrain. Katrina brought severe winds, record rainfalls, and storm water damage, followed by the collapse of major canal floodwalls that allowed water to fill the bowl in about 80 percent of the city. Rescue operations were remarkably inadequate, evacuation was incomplete, levees collapsed, and those remaining in the city were in desperate straits. As Jeffrey Mount, director of the Center for Watershed Sciences at the University of California, Davis, has said, "The dark secret that no one wants to share is that there are two kinds of levees: those that have failed and those that will fail."

As of 2010, much of Katrina's devastation in southern Louisiana still remains, and it is not clear how much of it will ever be repaired.[15] Large parts of the city remain empty tracts, and mainstays of the economy in medicine and education have not recovered. Reduced staffs and below-normal enrollment have hindered the full recovery of universities. Hospital closures have left a major gap in health care. Capacity is down by 25 percent, and the city continues to experience a major loss of skilled medical jobs. Consequently, physicians continue to leave the area five years after Katrina struck.

Because of delays in recovery funds, slow insurance payments, and not enough labor to work on repairs, rehabilitation of existing housing has been slow. Planned reconstruction is just beginning, and some communities may be lost forever. Reconstruction is expected to take at least ten years. In August 2009, four years after Katrina, 33 percent of the public schools were still closed, as were 48 percent of the child care centers. There are still 63,000 vacant and abandoned properties, 20,000 families still live in temporary mobile homes and apartments, and a housing shortage has driven up rent by nearly 50 percent. In May 2009, FEMA canceled its plan to phase out its trailers as temporary shelters and will donate them to the 3,450 families still living in them. The donation has helped alleviate a major shortage of affordable housing, a situation that has become increasingly desperate as the Federal Housing Administration proceeds with its plan to demolish 4,500 public housing units and replace them with mixed-income, mixed-use development.

In 2009 the inspector general in the Department of Homeland Security reported that the government remains unable to provide emergency housing after large-scale catastrophes such as Katrina and must do more to prepare survivors of such disasters for permanent relocation. He told the Congress that "FEMA does not have sufficient tools, operational procedures, and legislative authorities to aggressively promote the cost-effective repair of housing stocks."[16]

An additional headache for residents of the area affected by Hurricane Katrina is the spread of pollutants over the area.[17] Southern Louisiana is heavily industrialized, and the storm surge spread noxious materials over the populated areas of New Orleans, including arsenic, lead, petroleum hydrocarbons, and industrial chemicals.

Climatologists anticipate that sea-level rise, combined with sediment compaction and consequent high rates of subsidence in the three-state area affected by Katrina, will make much of the existing infrastructure more prone to frequent or permanent inundation.[18] They anticipate that

27 percent of the major roads, 9 percent of the rail lines, and 72 percent of the ports in the area are built on land at or below 4 feet in elevation, a level within the range of projections for sea-level rise in this century. And, as noted in chapter 9, this level is likely to be reached before the end of the century. Before then, increased storm intensity may lead to more service disruption and infrastructure damage. More than half of the area's major highways (64 percent of the interstates, 57 percent of arterials), almost half of the rail miles, twenty-nine airports, and virtually all of the ports are below 23 feet in elevation and subject to flooding and damage due to hurricane storm surge.

Engineers have been trying for 200 years to restrain the Mississippi River's natural urge to move sideways, change its meander pattern, and otherwise display its power. Artificial levees were built to protect cities and farmland from flooding. In effect, engineers have been trying to fit a straitjacket to a moving target, and they have not had great success. In 2007 officials in Louisiana proposed to spend what will amount to hundreds of billions of federal dollars in a far-reaching plan to cause a major rerouting of the southern end of one of the world's major waterways.[19] If the rerouting plan is approved, it would be one of the great engineering challenges of this century. Whether such an ambitious plan could be completed in time to save annually threatened communities in southern Louisiana is questionable. Given the inevitable political and financial delays that always accompany ambitious federal projects, the area might well be underwater before rerouting occurs.

The plan would allow the Mississippi to flow out of its levees in more than a dozen places in Louisiana and would create new waterways at seven or more sites that would carry a volume of water similar to or greater than that of the Potomac River (2 percent of Mississippi River discharge at New Orleans). Sediment from these new rivers would carry the water and land-enhancing sediment into eroding coastal areas. Other plans call for mechanically pumping sediment to rebuild marshes and barrier islands. Hundreds of miles of new or reconstructed levees would add flood protection.

It is noteworthy that abatement plans in the United States are designed to withstand floods or storms that would occur statistically once every 30 to 100 years. (As noted earlier, these estimates seem to underestimate flood frequencies.) The Netherlands, a country that has half its land at or below sea level, builds its levee and dike systems to withstand a 1,250-year flood, though rising sea levels and increasing storm intensities have made even this degree of protection suspect. The past may be an inadequate guide to the future when coastal flooding is concerned.

Most of southeastern Louisiana was built over many thousands of years by the deposition of sediment from the Mississippi River, which naturally changed course and flooded innumerable times over the millennia. The sediments that were deposited as the floods subsided were much greater than the sediment lost to hurricane waves and created the land that New Orleans sits on, as well as every patch of ground to the south.

Then people moved into the area, and, as people are wont to do, deliberately upset the natural balance because of the human inclination that always seems to be to conquer nature rather than learn to live with it. Over time, thousands of acres of trees, bushes, and other vegetation were paved over, decreasing the land's ability to absorb rain. Levees and concrete barriers were built along the river to prevent the flooding that had built the land, replenishing the swamps and marshes that had provided partial protection from Gulf hurricanes. The artificial restraints at the river's sides increased water volume in the channel, causing rapid erosion of Louisiana's land at its contact with the Gulf of Mexico. Since the 1930s, an estimated 1,900 square miles of land have washed into the Gulf, an area the size of a football field lost every forty-five minutes. The area of land lost in the past seventy years is twice the size of Rhode Island. Louisiana accounts for 80 percent of the nation's coastal land loss, with rates ranging between 25 and 35 square miles per year.

Not all of southern Louisiana's land-loss problems have resulted from human activities. Below the delta surface are many faults—breaks in the earth's crust parallel to the coast—and the land on the south side of the faults is dropping relative to the landward side at a rate of 0.2 inch each year.

Increased hurricane intensities in recent years, due at least in part to the burning of fossil fuels and climate change, have made the problem worse. In August and September 2005, the one-two punch of Hurricanes Katrina and Rita removed 200 square miles of Louisiana wetlands.[20]

Environmentalists and ecologists have pointed out that the vast earthen levee walls damage any wetlands they cross. Healthy tidal wetlands are not compatible with levee construction, and without healthy wetlands, the land loss will continue. Since 1956, the New Orleans metro area has lost 23 percent of its wetlands. Scientists estimate that 2.7 miles of wetlands can reduce storm surge by 1 foot.[21] The importance of wetlands has been increasingly recognized during the past decade, and the country as a whole gained wetlands at a rate of about 32,000 acres per year (50 square miles) between 1998 and 2004.

Should New Orleans Be Abandoned?

Despite more than 290 years of effort to protect New Orleans, the overall vulnerability of the city and surroundings to hurricanes has increased. The bowl-shaped location that is half below sea level along a hurricane-prone flat coastline cannot be changed. The vulnerability to disaster is increased by accelerating subsidence, rising sea level, storm surges, and the apparent increased frequency of stronger hurricanes. These natural phenomena have been enhanced by human location decisions; extraction of groundwater, oil, and natural gas; canal development; loss of barrier wetlands; global warming; and the design, construction, and failure of protective structures.

Given the enormous, increasing, and probably insurmountable obstacles facing the residents of southern Louisiana, it may be time to consider a solution that no politician or government has publicly considered: abandoning New Orleans and the surrounding smaller communities. The American public, not as emotionally attached to the city as its residents are, is willing to consider this step as indicated by a poll in 2005 after Katrina hit: more than half the respondents (54 percent) said that flooded areas of New Orleans below sea level should be abandoned and rebuilt on higher ground, a sensible approach if the city is to be saved.[22]

Extreme rainfalls in the Mississippi River basin are becoming more common, the city is sinking up to 1 inch per year and is already 17 feet below sea level in places, the ground is subsiding as the mud it is built on compacts, and breaks in the earth's crust parallel to the shoreline are lowering the ground surface. In its 291-year history, major pre-Katrina hurricanes or floods have swamped the city twenty-seven times—about once every eleven years. Home insurance is prohibitively expensive.

The existing levee system is not only inadequate and sinking, but also poorly designed and built. In 2006, two independent investigations and the Corps of Engineers' own $25 million study gave a detailed picture of flaws in the planning, design, and construction of the levee system. The Corps of Engineers blamed the deficiencies on political influences on engineering decisions over the years and the lack of adequate financing. An engineering professor at Louisiana State University examined the levee system and concluded that it was not capable of withstanding even a category 1 hurricane, much less a category 3 or higher. He noted that "New Orleans' future is very hard to predict. The big unknown is global warming. If sea level rises by another meter [3.25 feet] in the next 50 to 60 years, if we see far more of these major hurricanes, we could well reach a point where we see the need to abandon these cities like New Orleans."[23]

A report released by the National Academy of Engineering and the National Research Council concluded that no levee or floodwall is big enough or strong enough to fully protect New Orleans from hurricanes or other extreme weather events.[24] The report recommended relocating neighborhoods to less flood-prone areas of the city or elevating homes above the 100-year flood mark as the best strategies to prevent the type of water damage caused by Hurricane Katrina.

Modern New Orleans is a city whose early inhabitants ignored the natural setting that was not favorable for human settlement and built in a mosquito-infested bowl-shaped swamp squashed between two vast bodies of water. Much of the city is below sea level, and continual pumping has caused the ground to subside. Since 1878, the city has sunk 15 feet, one of the highest rates of subsidence in the United States.

Added to this is the loss of the coastal wetlands that serve as a buffer against hurricane waters. In one day of Katrina, Louisiana lost 15 square miles of wetlands, three-quarters of its annual loss. Coastal degradation was a problem long before Katrina appeared. After the Army Corps of Engineers "tamed" the Mississippi in the 1940s, the wetlands, deprived of the river's sediment, began to sink noticeably below sea level. Their health further deteriorated as extensive canals were dug, first to explore for oil and gas and then to pump them out. Adding insult to injury, a beaver-sized rodent, the nutria, introduced in the 1930s for its fur, turned out to have a voracious appetite for marsh plants. More than 1,500 square miles of wetlands have been lost since 1950.

The long odds against saving the city led geologist Robert Giegengack to tell policymakers a few months after Katrina hit, "We simply lack the capacity to protect New Orleans."[25] He agreed with Wilson Shaffer, a storm-surge modeler at the National Weather Service, who wrote in 1984 that "there are no high areas near the city that wouldn't flood in extreme cases. High ground is several tens of miles away."[26] Giegengack recommended selling the French Quarter to Disney, moving the port 150 miles upstream, and abandoning the city to the Gulf waters that are certain to cover it in the not-too-distant future. Others have suggested rebuilding it as a smaller, safer enclave on higher ground.

The brutal hurricane that hit Galveston, Texas, in 1900 may be a guide to the future. Most survivors of that deadly storm moved to higher ground. After the hurricane, Galveston, which had been a large and thriving port, was essentially abandoned for Houston, transforming that sleepy backwater into the financial center for the entire Gulf South.

Galveston devolved into a smallish port and tourist center that is easy to evacuate when hurricanes are on the horizon.

Economic arguments to save the New Orleans area include the large number of oil pipelines that discharge there, the many petroleum refineries and chemical plants that line the river, the presence of a large ship-building and repair industry, the presence of two large universities, and the fact that the Port of New Orleans is the busiest in the United States by gross tonnage. Added to the economic considerations are the equally important but largely emotional responses by people who would be displaced and those who feel the city's distinctive culture and cuisine are worth saving. And of course there are the all-important national, state, and local political considerations that are nearly always a barrier to major changes in anything, however justifiable on rational scientific grounds.

In 2010, it seems most likely that the city will not be abandoned in the foreseeable future and that ever greater amounts of money will be spent in coming decades in a futile effort to thwart natural processes. The money that will be spent will no doubt dwarf the amount it would cost the federal government to oversee a phased withdrawal from the region around New Orleans. But one thing is certain: Mother Nature will have her way in the end. She always does, and her prowess has already affected Florida, for many years the mecca for snowbirds from the frigid Northeast. According to the Census Bureau, the number of people moving from other states to Florida declined by 87 percent between 2005 and 2007. In 2007, as many people moved out of Florida as moved in.[27]

Will Your Insurance Cover It?

America's coastal home owners and property insurers are in trouble.[28] Nearly all of the most costly natural catastrophes in the United States were caused by hurricanes (table 3.2) as a result of flooding or by wind damage, and because of the increasing frequency of hurricanes, the price of private property insurance has skyrocketed for those located near the Atlantic and Gulf coasts. Between 2001 and 2006, premiums for some home owner policies in Alabama rose more than tenfold, while the average cost for a policy in Florida rose 77 percent; Virginia 67 percent; Louisiana 65 percent; Mississippi 63 percent; South Carolina 56 percent; and Texas 50 percent. One retiree in Florida saw the premiums on his 1968 house increase from $394 in 2000 to $5,479 in 2007. Another paid

Table 3.2
The ten most costly catastrophes in the United States

Event	Cost (billions)
Hurricane Katrina, August 2005	$100
Terrorism: World Trade Center/Pentagon, September 2001	30
Hurricane Wilma, October 2005	29.1
Hurricane Andrew, August 1992	26.5
Northridge earthquake, California, January 1994	20
Hurricane Charlie, August 2004	13
Hurricane Ivan, September,2004	13
Hurricane Frances, September 2004	12
Hurricane Rita, September 2005	11.3
Hurricane Hugo, September 1989	7

Source: Wikipedia (2009).

$2,066 in 2006 and $4,700 in 2007.[29] Florida's population fell by 58,000 from April 2008 to April 2009, the first decline in more than fifty years. Some major insurers are refusing to issue new policies in hurricane-prone areas, and some existing policies are not being renewed, a change that has affected hundreds of thousands of policy holders.

More than half of all Americans live within 50 miles of the coast, and the percentage has been increasing. The lure of open water is apparently too strong for many people to resist. Many additional risk takers who wish to locate along the coast cannot do so: they cannot get mortgages because lenders require that mortgaged homes have insurance.

Insurance companies, the world's most serious futurists, are in a panic over the increasing costs of hurricane damage. Although rate hikes and discontinued policies are nothing new in hurricane-battered Florida and the Gulf Coast, the problems are now affecting property owners in the Northeast and mid-Atlantic states. Allstate, which for some years has refused to write home owners' policies in Florida, Louisiana, and coastal Texas, has stopped issuing new policies in New Jersey, Connecticut, Delaware, coastal Maryland, coastal Virginia, and New York City. A host of other insurers are dropping coverage, refusing to insure properties, or limiting coverage along the Atlantic coast from Maine to the Carolinas.

Whether global warming is at work or not, damage costs will increase because of rising property values and development. For example, a direct hit on Miami by Hurricane Andrew would cost twice as much today as it did in 1992. Private insurers say they can no longer handle losses on the scale of Hurricane Katrina, and such calamities are likely to increase in frequency as global warming becomes more severe, as it certainly will. We are in a new era of catastrophic losses from natural disasters.

About half of the $100 billion of damage from Katrina was privately insured. States and the federal government are being expected to be insurers of last resort and to subsidize the cost of insurance to property owners. Katrina helped put the federal flood insurance program $23 billion in the red and prompted federal relief spending of more than $100 billion. A spokesman for Allstate said, "Our view is that there are some events that have the potential to be so large as to exceed the capabilities of the insurance industry, as well as the funding and financing capability of individual states. Those are events that have the potential to be $100 billion. These events are so enormous, no entity has the ability to manage it."[30] A spokesperson for Citizens Property Insurance, the state-chartered insurance corporation in Florida that is the state's largest insurer, said, "What everybody knows is that no one could charge enough in the high-risk areas of Florida to cover the potential damage in a one-in-a-hundred year storm. We just can't do it."[31]

But using governments as backstop insurers encourages migration to the coast and makes potential hurricane catastrophes more expensive. To what extent should individuals be held responsible for making sound, informed decisions about where to live? Those who live inland in vulnerable states are not happy about having to pay some of the cost of insurance for coastal dwellers. To what extent are Americans their brother's keeper where property insurance is concerned?

Many people who live in flood-prone areas are eligible for low-cost federal flood insurance. FEMA is responsible for the maps that mandate which Americans require flood insurance, and these maps require frequent updating as the climate changes and existing flood protection is no longer adequate. An example is our nation's capital. Washington, D.C., was built on swampland more than 200 years ago. When the Army Corps of Engineers recently reexamined America's outdated flood control components, it found that levees built into the landscape to protect historic landmarks, world-renowned museums, and federal buildings in Washington were likely to fail in the next big flood. If a major storm hit, parts of Washington would be under 10 feet of water, causing $200 million in damage.[32] Flood maps were redrawn, causing some homeowners to purchase flood insurance for the first time.

Tornadoes

Tornadoes are a natural disaster related to the increase in severe hurricanes.[33] Meteorologists know that hurricanes tend to generate tornadoes when they make landfall. Not surprisingly, the Gulf Coast is particularly susceptible to these tornadoes, and their occurrence has increased in frequency over the past four decades. Compared to the previous peak tornado period from 1948 to 1964 when there was an average of six tornadoes per year, the current Gulf Coast hurricane pattern is generating an average of fifteen tornadoes per year. Researchers have devised a statistical model that predicted almost perfectly the number of tornadoes that would be spawned by Hurricanes Katrina in 2005 and Ike in 2008. The connection between different disastrous meteorological events is reminiscent of the often repeated maxim of ecologists: Everything is connected to everything else.

4

Garbage: Trash Talk

Nothing ever goes away.
—Barry Commoner, biologist

If there were an entry in the annual Guinness World Records book for milestones in garbage, the United States would certainly be represented. Our production of municipal solid waste, known colloquially as trash or garbage, increased from 2.7 pounds per person per day in 1960 to 3.7 in 1980 to 4.6 pounds per person per day in 2007 (table 4.1).[1] This is twice the amount produced by other industrialized countries such as Canada, United Kingdom, Germany, France, or Japan and 20 percent more than Norway, Switzerland, or Denmark. We are the world's most wasteful society. The figure of more than four-and-a-half pounds of discards per day for each of us has remained fairly constant since 1990, but total tonnage has continued to increase as the population has increased.

The Stuff We Throw Out

The bulk of garbage consists of manufactured items that are essential to the functioning of a modern society—paper, plastics, metals, and glass (figure 4.1). Together they form 58 percent of our trash. Paper forms nearly one-third of the stuff we throw out—computer paper, junk mail, books, newspapers, magazines, telephone directories, and many other types of printed matter. Decades ago we disposed of paper by burning it in metal trash cans, but city ordinances today prohibit such fires.

The percentage of plastics (containers and packaging) in our garbage pails has increased steadily, from 7 percent in 1988 to 12 percent in 2007, and is now tied for second in abundance with yard trimmings and food scraps. The excess plastic packaging of the things we buy,

Table 4.1
Trends in the generation of municipal solid waste

Year	Pounds per person per day	Total MSW generation (millions of tons)
2008	4.5	250.0
2007	4.6	254.1
2006	4.6	251.3
2002	4.6	239.4
2000	4.6	238.3
1990	4.5	205.2
1980	3.7	151.6
1970	3.3	121.1
1960	2.7	88.1

Source: Environmental Protection Agency, Quantity of Municipal Solid Waste Generated and Managed, 2009.

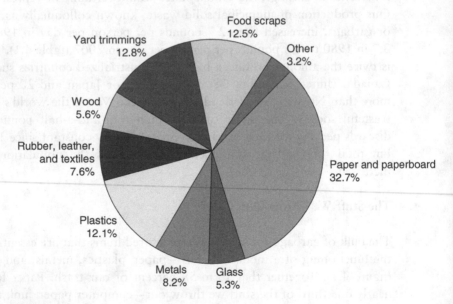

Figure 4.1
Percentages of materials in municipal solid waste in 2007. (Environmental Protection Agency, 2008)

particularly supermarket food products, is a national disgrace and could easily be reduced without harming the quality of the food.

Only 12 percent of the daily garbage is food scraps because half of American homes have garbage disposal units under the kitchen sink.[2]

Electronic Equipment

An increasing amount of discards from American homes is computers. The technology of computers advances rapidly, leading to rapid obsolescence of existing models. There are hundreds of millions of obsolete computers in the United States. Seventy percent of them will end up in landfills. The Environmental Protection Agency (EPA) says computers are the nation's fastest-growing category of solid waste.

It has been estimated that about 75 percent of obsolete electronic equipment is being stored, but about 13,000 computers are disposed of every day in the United States.[3] Circuit boards, batteries, and color cathode ray tubes contain lead, mercury, chromium, and other hazardous material. If the computers are simply dumped, these toxins can be released into the soil and water through landfills or in the form of toxic incinerated ash. Trashing old computer equipment is not an environmentally friendly disposal method. Fortunately, many voluntary and nonprofit organizations are dedicated to computer recycling. Reusable materials include steel, glass, plastic, and precious metals. Those expensive ink cartridges, floppy disks, CDs, speakers, keyboards, and cords also contain materials that can easily be reused.

Babies' Bottoms

Disposable diapers first appeared in 1949 and, like computer hardware, are a growing source of household discards. They now are the third largest consumer item in landfills, behind newspapers and bottles. In families with several young children, disposable diapers can fill the garbage pail for years because babies may need to have their diapers changed ten or more times a day. The average baby uses about 7,000 diapers from birth to potty training, producing about 4,000 pounds of waste to a landfill if he or she wears disposable diapers. The Bay Area of San Francisco, with 7.2 million residents, generates 600 million disposable diapers a year—83 diapers per resident.[4] They weigh 180 million pounds; for the entire United States, the number is 8 billion pounds.

Twenty billion diapers are sold each year in the United States. If a parent uses only reusable cotton diapers, about sixty of them will suffice from birth to potty training. However, few parents today use only cloth diapers. Most diaper-using households use disposable diapers, but some parents use both types.[5] Fifty-five percent use both cloth and disposables, 33 percent use only disposables, and 12 percent use only cloth diapers. An estimated 27.4 billion disposable diapers, weighing 3.4 million tons, are used each year in the United States. Disposable diapers cost parents around two dollars a day, roughly the same amount that a commercial diaper service charges for handling cloth diapers.

Disposable diapers such as Pampers or Huggies, the best-selling brands, contain 0.3 to 0.5 ounces of sodium polyacrylate, a superabsorbent solid polymer that can absorb up to 100 times its weight in liquid— 2 to 3 pounds (or 0.96 to 1.44 quarts) of urine. Of those who use cloth diapers, 90 percent say they are concerned about the impact of disposables on the environment. And well they should be. The 20 billion disposables used each year add 3.5 million tons of poop and plastic to landfills. A diaper can take more than 500 years to decompose in a landfill.

Cloth diapers are environmentally superior to disposables.[6] Disposables use twenty times more raw materials, three times more energy, and twice as much water as reusables. They generate sixty times more waste. In addition, it requires 4.6 barrels of oil to produce one disposable diaper, according to the American Petroleum Institute.

Toilet Paper

Related to the issue of cleaning human bottoms is toilet paper. The soft, fluffy toilet paper widely preferred in the United States for home use can be made only from tree farms or by logging wild forests. Toilet paper can easily be made from recycled paper, but only at the cost of a coarser final product. Manufacturers admit that the primary factor that keeps them making toilet paper out of freshly cut trees is that standing trees yield longer fibers than recycled material does. Longer fibers make for softer, fluffier toilet paper. Plush toilet paper also requires more water for its manufacture.

The popularity of plump, soft toilet paper is a distinctly American phenomenon. Only five percent of the toilet paper purchased in the United States for home use is composed entirely of recycled fibers. In contrast, 20 percent of the at-home toilet paper manufactured in Europe and Latin America includes recycled content. Customers in other countries are also more willing to accept a rougher toilet tissue. American consumers demand a soft and comfortable conclusion to their toilet experience. Georgia-Pacific made more than $144 million in 2008 selling more than 24 million packs of its "Quilted Northern Ultra Plush" three-ply toilet paper. Furthermore the United States uses more toilet paper than any other country in the world: an average of 23.6 rolls per person per year, which is thirteen times as much as Asians and fifty-seven times as much as Africans.

Nevertheless, increased roughness of toilet paper for American consumers appears to be on the horizon. In August 2009, Kimberly-Clark, the maker of popular brands like Kleenex, announced that within two years, 40 percent of the fiber used by its North American division would be from either recycled sources or stock certified by the Forest Stewardship Council, an industry group promoting responsible forest management.

Grocery Bags

In the past few years, a furious controversy has emerged about the relative greenness of the 100 billion plastic and 10 billion paper bags used in the United States each year, with the debate centering on the bags dispensed at the checkout counters in supermarkets, although many other types of stores use them as well. Many communities in the United States have tried to pass laws banning or taxing plastic bags, but only a few cities and small towns have totally banned plastic bags.

Plastic bags first appeared in the late 1970s and were touted as saviors of vast stands of forests. Worldwide an estimated 90 billion plastic bags are discarded, fifty to eighty billion of them in the United States. The UN says that plastic bags are the second most common type of litter worldwide, just behind cigarette butts. Recent legislation in New York, Rhode Island, Delaware, and California may significantly decrease this percentage.[7] These new laws require all large grocers and retailers that offer plastic bags to their customers to provide collection bins for them, which will dramatically increase recycling opportunities in the two most populous states in the nation. Recycling of plastic bags and film reached 12 percent in 2007, a record high, driven by greater consumer access to recycling programs, primarily at large grocery and retail stores, as well as by new markets for these recycled materials.

What are the pros and cons of plastic versus paper bags? Favoring paper are:

• Plastic bags are made from a derivative of natural gas, a fossil fuel whose use creates carbon dioxide gas that increases global warming. And fossil fuels are increasingly expensive.
• Plastic bags may take a thousand years to decompose in a landfill. Few of them are recycled.
• Paper bags decompose much more rapidly than plastic bags and do not form unsightly litter hanging from tree branches or cause the death of sea creatures that get caught in them or eat them.

Favoring plastic are:
• Contrary to the impression one gets from publications, paper is no more "natural" than plastic. Nature produces trees, not paper, and many millions of trees are cut down every year to produce paper bags. Huge gas-guzzling and pollution-generating machines log, haul, and pulp trees. The entire paper-making process is heavily dependent on chemicals, electricity, and fossil fuels. Although plastic bags are produced from natural gas, which, like paper, was formed from the tissues of living organisms (one-celled marine plants and animals), these organisms died naturally millions of years ago. To produce paper, living organisms, the trees, are killed.
• Paper bags require more energy and fossil fuels to manufacture than plastic bags and generate more solid waste.
• Plastic bag production uses less than 4 percent of the water needed to make paper bags.
• The EPA reports that making paper bags generates 70 percent more air pollution and 50 times more water pollution than making plastic bags.
• Forty-one million carbon dioxide–absorbing trees are saved each year by using plastic rather than paper bags.
• Despite their numerical abundance, plastic bags are thin, flexible, and compressible, and therefore occupy only 2 percent of landfill space. The EPA says that nearly 100 landfills are closed every year because they are full, and plastic bags take up much less space than paper bags do.
• Manufacturers of plastic bags will be using 40 percent recycled content by 2015, cutting greenhouse gas emissions and waste annually.

So there is a cost to the environment regardless of which kind of bag is used. Despite the lack of a clear conclusion about the relative greenness of paper and plastic bags, most people seem to believe paper is better.

Cloth bags are sold in many supermarkets in the United States for between one and five dollars. These reusable bags are most commonly made from cotton, but the cotton-farming process is extremely fossil fuel intensive because of the machinery used. In addition, conventionally grown cotton uses more insecticides than any other single crop: world-wide, cotton growers use more than 10 percent of the world's pesticides and nearly 25 percent of the world's insecticides. Cotton is also responsible for 25 percent of all chemical pesticides used on American crops. Most of the cotton grocery bags are woven outside the United States where labor is less costly, but that increases the use of fossil fuels to get them from the foreign factories to our shores.

In summary, there is no free lunch. Everything that is manufactured has environmental costs, and when we attempt a life cycle analysis to determine relative greenness, the answer is often not clear. The opinions of the public are commonly influenced more by the relative strengths of advertising campaigns than by an objective evaluation of the facts.

Litter: Do We Care?

Littering has decreased by almost two-thirds in America since 1950 but is still widespread, as everyone knows. Tobacco products are the biggest problem, accounting for 38 percent of litter, with paper products at 22 percent, plastic at 19 percent, metal at 6 percent, and glass at 4 percent. The biggest growth category has been plastic because of the increase in plastic packaging at the expense of glass, paper, and metal.

Who Wants My Garbage?

Good waste management attempts first to reduce the amount of waste produced, then to recycle whatever is recyclable, to compost what is compostable, and finally to dispose of the remaining waste in an environmentally responsible manner. The final waste goes to either a waste processing facility, typically one that turns it into energy, or, more often, to a landfill. The environmental impacts are considerably different.

Landfills

There are three ways to dispose of trash: it can be buried (landfills), burned (incinerators), or partially recovered and recycled.

Fifty-four percent of household garbage goes to landfills, a percentage that has not changed since 2000, and the number of landfills has remained constant at 1,754 since 2002, although the size of some has increased.[8] At the national level, landfill capacity appears to be adequate, although it is limited in some areas.

A secure landfill is a carefully engineered depression in the ground into which wastes are put. The aim is to avoid any water-related connection between the wastes and the environment. Ideally, a landfill should be located on impermeable (waterproof) rock and have strict construction requirements. The base of the landfill should be sloped, and it should have a plastic liner at its base. Today's landfill liners are made of a tough plastic film one-tenth of an inch thick. But the liners do not last forever; most are guaranteed for only fifty years or less. Landfill operators are liable for landfill problems for thirty years after closure of the facility. After that, the public must pay to rectify problems.

Overlying the liner is a system of pipes and sealed pumps to collect leachate, the liquid that drains through the wastes. Garbage commonly contains hazardous liquids such as drain cleaners, bug sprays, turpentine, weed killer, and many other materials that must be prevented from mixing with our water supply. Each year every American household sends an average of 1.5 quarts of hazardous liquid to landfills, and there are more than 100 million households in the United States.

Many of these household chemicals can degrade the plastic liner and cause it to leak. Eighty-two percent of landfills leak.[9] Studies show that a ten-acre landfill will leak up to 10 gallons per day.[10] However, if the rock beneath it is impermeable, the leachate (contaminated water that drips through the waste) will not be able to move downward and contaminate groundwater. Collecting pipes funnel leachate to a wastewater treatment plant. But experience has revealed that leachate collection systems can become clogged in less than ten years by mud, microorganisms growing in the pipes, or chemical reactions that cause the precipitation of minerals in the pipes. The pipes may also become weakened by chemical attack by acids, solvents, oxidizing agents, or corrosion and may then be crushed by the tons of garbage piled on them.

Each day's waste should have an impermeable cover of clay or other material to serve as an umbrella over the landfill to keep water out and keep leachate to a minimum.

Landfill problems result not only from liner and pipe failures. More important are problems with older landfills that were not built to today's standards. Many are built on permeable soil and rock, others have no liner at their base or no pipes to collect leachate, and others were located too close to streams or lakes. The average landfill does not meet current mandated standards, but even the best-designed and best-built landfill is likely to leak eventually. Sharp objects in the garbage may puncture the plastic liner. Vermin burrowing through the waste pile in search of food may tear it. Acids in the leachate may eventually breach the liner. Eventually all pipes clog and all buried pumps fail. Landfills are not the best method of garbage disposal.

Three companies own more than two-thirds of all landfills in the United States. Landfills are valuable because it gets harder every year to open a new one because of public opposition; another reason is the expense of finding a location and building it to federal standards. It takes an average of ten years to open a new landfill.[11] It is much cheaper to expand an existing one.

Landfill Gas

When the organic portion of garbage decomposes in the absence of oxygen, as in a sealed and impermeable landfill, gas is produced that is typically composed of 35 to 55 percent methane, 25 to 45 percent carbon dioxide, and up to 10 percent water vapor.[12] But some landfills produce gas that is 85 percent methane, a greenhouse gas twenty times more potent than carbon dioxide, so keeping it from dispersing in the atmosphere is environmentally significant. Gas generation from a landfill can continue at stable rates for many decades and, when purified, the methane (natural gas) can be a valuable energy source. Municipal solid waste landfills are the largest source of human-related methane emissions in the United States, accounting for about 34 percent of these emissions.[13]

Landfill gas is extracted from landfills using a vacuum system and a series of wells drilled into the waste pile. This system directs the collected gas to a central point where it can be cleaned of hydrogen sulfide, volatile organic compounds, and carbon dioxide that landfill gas is likely to contain. It can then generate electricity, replace fossil fuels, fuel greenhouse operations, or be upgraded to pipeline quality gas. In mid-2008, 469 projects were functioning in the United States, and according to the EPA, 520 landfills exist that could economically support a project (figure 4.2).[14] Those operating in 2008 were supplying 11 billion kilowatt-hours of electricity per year (enough for about 870,000 households) and 77 billion cubic feet of gas (enough to heat 534,000 homes), and they prevented emissions from 182 million barrels of oil (U.S. consumption is about 8 billion barrels per year) or the greenhouse gas emissions from 14.3 million cars (there are 251 million cars in the United States).

Landfill gas has safety and economic benefits as well. Whether or not the gas is captured, landfills must be vented to prevent accumulations of methane because the gas is explosive at concentrations of greater than 5

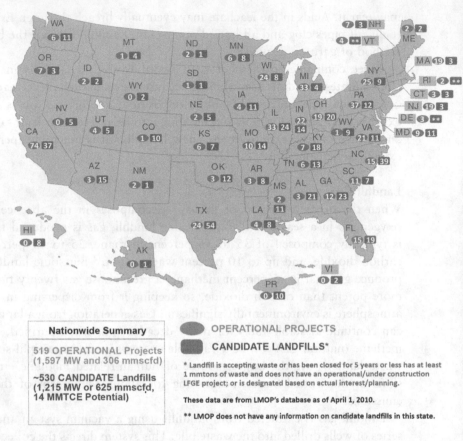

Nationwide Summary

519 OPERATIONAL Projects
(1,597 MW and 306 mmscfd)

~530 CANDIDATE Landfills
(1,215 MW or 625 mmscfd,
14 MMTCE Potential)

OPERATIONAL PROJECTS

CANDIDATE LANDFILLS*

* Landfill is accepting waste or has been closed for 5 years or less has at least
1 mmtons of waste and does not have an operational/under construction
LFGE project; or is designated based on actual interest/planning.

These data are from LMOP's database as of April 1, 2010.

** LMOP does not have any information on candidate landfills in this state.

Figure 4.2
Operational (469) and candidate (520) landfill gas energy projects in the United States as
of December 2008. (Landfill Methane Outreach Program)

percent in landfill air. The economic benefit is clear. For example, at the
world's largest landfill in New York City, enough methane is recovered
to satisfy the energy needs of 12,000 households.

Closing a Landfill

The size of a landfill cannot be increased indefinitely. At some point, the
final covering of dirt will occur, and a new means of getting rid of the
city's garbage must be found. For many municipalities the solution is to
give it to someone else for disposal. The amount of garbage moved across
state lines for immoral purposes is almost unbelievable (figure 4.3).
Many of the eighteen-wheelers on interstate highways are carrying
garbage to far-away places.

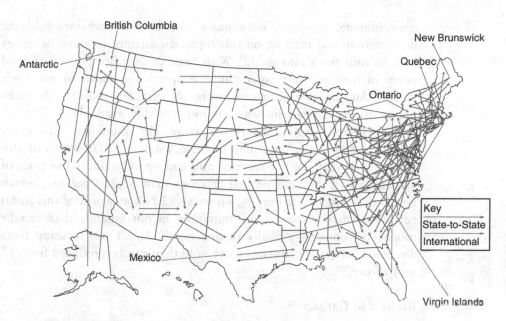

Figure 4.3
Simplified map of interstate traffic in garbage. Trash-shipping routes from New York and New Jersey, the biggest garbage exporters, are not shown to avoid clutter. (Edward W. Repa, "Interstate Movement of Municipal Solid Waste," *NSWMA Research Bulletin* May, 3, 2003)

New York City closed its major landfill on Staten Island in 2001 and now must ship its refuse elsewhere. The city has 2,230 garbage collection trucks that carry the trash to transfer points, from where it typically leaves the state. Some trash goes as far as New Mexico. It takes 450 tractor trailers, traveling 135,000 miles in combined round trips and burning 33,000 gallons of diesel fuel, to transport one day's worth of the city's garbage to out-of-state locations.[15] Large cities pay as much as $100 per ton to one of the thousands of haulage contracting companies. In 2009 New York City paid $309 million to export more than four million tons of waste, mostly to landfills in distant states. Federal, state, and local governments spend billions each year on waste disposal. It accounts for the majority of environmental expenditures.

Landfill Mining

"Raiding the dump" for food and salable materials is an urgent necessity for many people in impoverished countries, and it may have economic benefits in industrialized nations as well. In October 2008, environmental scientists, economists, and landfill operators held the first international meeting on this topic, an attempt to show delegates how to turn trash into gold.[16] With both commodity prices and land prices increasing, every square mile is worth too much to use for a landfill. And because of innovation by the recycling industry, the technology to process landfill waste is more readily available.

For example, Americans throw away 317 aluminum cans every second of every day. About half of these, totaling 570,000 tons of aluminum each year, are not recycled and end up in landfills. The price of aluminum in July 2008 peaked at more than $2,250 per ton, which means that America is burying up to $1.83 billion worth of this metal per year. There is now more aluminum in our landfills than can be produced from ores globally in one year.[17] And 1 ton of scrap from discarded computers contains more gold than can be produced from 13 tons of ore.

Really Bad Garbage

Not all garbage is created equal. The worst stuff is hazardous or toxic waste, defined as substances harmful to the environment. They may be poisonous, radioactive, flammable, explosive, corrosive, carcinogenic, mutagenic (damaging chromosomes), teratogenic (causing defects in the unborn), or bioaccumulative (accumulating in the bodies of plants and animals and thus in food chains). Such toxic or poisonous wastes are produced during industrial, chemical, and biological processes. Thousands of landfills contain hazardous wastes that were stored in metal barrels by industrial firms and buried by simply digging a hole and throwing them in. Many of the barrels are now rusted and leaking harmful liquids into the nation's water supply.

Industrial concerns are not the only producers or disposers of hazardous wastes. Common household items such as paints, cleaners, oils, batteries, and pesticides contain hazardous components. Clues to their toxic character are words on their labels such as *danger, warning, caution, toxic, corrosive, flammable,* or *poison.* Such products should be disposed of in a safer manner than chicken bones and computer paper, although there is no completely safe way to dispose of toxic wastes. Among the safer methods are these:

• *Land disposal* Waste is buried in landfills that should be permanently sealed to contain the waste. The landfills may be lined with clay or plastic, and the waste may be encapsulated in concrete. However, there is no guarantee that a leak will not occur. As noted above, plastic tears and concrete dissolves over time.

• *Incineration*   Burning of many hazardous industrial wastes such as tar, paint, pesticides, and solvents at a temperature of 1,500°F prevents the formation of dioxins, which affect the human endocrine system, nervous system, and reproductive functions.

• *Chemical or biological*   Chemicals are added to the wastes to make them less toxic or bacteria eat the waste, resulting in a less toxic residue.

Proper disposal of toxic waste is a major problem in industrialized countries. Approximately 4 billion tons of hazardous waste are shipped within the United States each year, with more than 250,000 shipments entering the transportation system daily. Disposal can cost $2,500 per ton and takes up valuable landfill space. Moreover, because it is never totally safe, nobody wants a toxic waste dump in their back yard or even within an hour's driving distance. This is the reason that 90 percent of the hazardous waste deposited on land is injected deep into the ground in permitted wells; the remaining 10 percent is treated and disposed of in a manner to minimize risk to human health and the environment.

Another way to solve the toxic waste disposal problem is to pay another country to take over the problem. Many Third World nations, desperate to earn money, have become the unfortunate targets of waste brokers—the middlemen between producers of waste and possible dump sites. Transporting (often by sea) waste to another country runs the risk of spillages, such as happens to oil tankers. Also, the receiving country often lacks the expertise and technology to deal with the toxic waste safely. Every country should take responsibility for its own hazardous waste. If a country is smart enough to produce such stuff, it should be smart enough to dispose of it safely.

Superfund: The Worst of the Worst

Thousands of toxic waste dumps and landfills are found across the United States, the accumulation of many decades of largely irresponsible disposal practices by today's standards. Studies in both the United States and Europe have shown that mothers living within 1 or 2 miles of a hazardous waste landfill site are one-third more likely to have babies with birth defects.[18] The farther away from such a landfill that the mother lives, the less the likelihood that her child will have a birth defect that affects the heart, nervous system, or blood vessels.

The American public was deeply concerned by such data, and in 1980 Congress passed what has become known as the Superfund law. According to the legislation, polluting industries were to pay the bulk of the cost of cleaning up the nation's toxic waste dumps. But in 1995, their lobbying groups succeeded in having their payment responsibility repealed, removing hundreds of millions of dollars annually from cleanup use. Since 1995, the Superfund program has been chronically underfinanced. Today it cleans up hazardous waste sites with money that comes directly from polluters and money appropriated by Congress. Appropriations are necessary because it is not possible to identify the polluter at many sites. And at other sites, the polluter cannot be located or is no longer in business. The stimulus bill that was passed in 2009 provides $600 million for Superfund work, nearly doubling the amount available in 2008.

Since the Superfund Law was enacted, Congress has appropriated an average of $1.2 billion per year for its implementation, and states have also contributed billions. Nevertheless, as of mid-2007, twenty-seven years after the Superfund law was passed, 80 percent of the hazardous waste sites in the EPA's Superfund program remain contaminated, and nearly half of the population lives within 10 miles of one of the 1,600 sites on the National Priorities List (NPR).[19] New Jersey leads the NPR list with 116 sites—one for each 65 square miles. California and Pennsylvania are tied for second place with 96 each.

It is nearly impossible to determine the number of hazardous sites that have been cleaned up because the EPA in 1990 changed its definition of success to one that does not indicate remediation. According to the revised definition, construction completeness, not remediation, is the standard. The EPA is apparently at peace when any necessary physical construction at a site is complete (for example, pumps are in place to remediate groundwater), even if final cleanup levels or other requirements for the site have not been met. It can be years before cleanup goals are achieved at some sites. Between 1992 and 2000, EPA's "construction-complete" sites ranged between sixty-one and eighty-eight, but the number of construction-complete sites has dropped into the forties since then.[20] One suspects that hazardous waste sites will persist throughout this century.

According to the EPA, as many as 350,000 contaminated sites will require cleanup over the next thirty years, assuming that current regulations and practices remain the same. The national bill for this cleanup may amount to $250 billion.[21] Humans are the only creatures in nature that deliberately poison their nest. It is truly amazing what a highly evolved brain can accomplish in a short time.

Brownfields

A *brownfield* is defined in the United States as land that has been previously used for industrial or commercial purposes and is contaminated by low concentrations of hazardous waste, but has the potential to be reused once it has been cleaned up. The land is not compromised severely enough to be a Superfund site. For example, a site classed as a brownfield may have been occupied by a dry cleaning establishment, an agricultural supply business, a paint store, or a gas station. Contaminants in the soil or subsurface at such locations may be solvents, pesticides, heavy metals, hydrocarbons, or asbestos. If the levels of subsurface contaminants are too high, brownfield land might sit unused for decades before someone believes it is possible to clean the property economically for a proposed use. The dividing line between a brownfield site and a Superfund site may not always be clear.

Remediation of a brownfield site may use bioremediation, using naturally occurring microbes in soils and groundwater to expedite the cleanup. Alternatively, phytoremediation may be useful. Plants that naturally accumulate heavy metals are artificially rooted in contaminated soil. The plants are later removed and disposed of as hazardous waste. Another method is in situ oxidation, that is, using oxidant chemicals to enhance the cleanup. Such methods typically require years to be effective. The technique used depends, of course, on the nature of the contaminant.

In 2009, 425,000 brownfield sites awaited redevelopment.[22] More than 150 cities had successfully redeveloped 922 brownfield sites, returning more than 10,000 acres to economic productivity. Impediments to brownfield redevelopment include shortages of cleanup funds, liability issues, and the need for environmental assessments. As cities expand and pristine land for development becomes scarce, remediation of brownfield land is becoming more popular.

Incineration

Incinerators cost several hundred million dollars to build and last for twenty-five to thirty years. They are safe and effective, but because of public opposition and their high cost, none have been built since 1997. There are 113 trash-burning power plants in the United States, and almost all were built at least fifteen years ago. Eighty-seven of them are used to generate electricity. Twenty-two are permitted to incinerate hazardous waste. The older ones in operation are known to create health hazards, releasing mercury and several noxious heavy metals into the air, as well as dioxins, and this has given incineration a bad name.

About 13 percent of America's waste is incinerated, a disposal method that reduces the volume of trash by 90 to 95 percent, leaving 5 to 10 percent ash residue that is taken to a landfill. The ash is usually nontoxic if the burning temperature is high enough to destroy harmful chemicals. Gases emanating from the furnace pass through a scrubber, where they are sprayed with lime to remove sulfur dioxide and hydrogen chloride that form acids if released into the air (acid rain). The flue gas is then channeled to a baghouse, where the fly ash, which contains the pollutants, is trapped before being released through the smokestack into the atmosphere. This procedure improves the quality of the emissions with 99.9 percent efficiency.

There has been much concern about suspected hormone-disrupting and cancer-causing compounds called dioxins being released from incinerators. Dioxins form when the chlorine molecules in paper, wood, and plastics interact with organic materials. The compounds are formed in the smokestack at temperatures between 300°F and 650°F when the flue gases are being cooled. If temperatures higher than 1,500°F are achieved throughout the furnace for at least half a second, the organic compounds in the gases will be destroyed, preventing the formation of dioxins. However, not all incinerators reach 1,500°F, so dioxins can be formed and released during the incineration process.

Plasma Gasification

Within the past few years, a method of garbage disposal called plasma gasification has been developed that has the promise of solving both landfill problems and those associated with some incinerators. Plasma gasification obliterates the trash, producing only hydrogen, carbon monoxide, and glass as by-products. The two gases can be converted into fuel (syngas), and the glass can be sold for use as household tiles, roof shingles, concrete, or road asphalt.

Construction costs are about the same as for incinerators. One plasma facility can handle at least 2,000 tons of trash per day, the output of a city of 1 million people. Only nine cities in the United States have populations larger than this. Dropping off a ton of garbage at a gasification plant costs several times more than sending it to a local landfill, but if the trash has to be trucked to out-of-state locations to a landfill, the economic considerations can be very different.

In plasma gasification, an electric current is used to ionize an inert gas—sometimes nitrogen, sometimes plain air. The ionized gas is called a plasma, and it is raised to temperatures that can exceed 27,000°F, hotter than the surface of the sun (10,000°F). Once the garbage has been zapped, the gas produced is cleansed of harmful traces of hydrogen chloride, which forms hydrochloric acid in the air. The decontaminated syngas is burned like natural gas, producing enough electricity to power the plant itself and offered for resale to the electrical grid. Mercury and heavy metals must be removed from the glassy slag before it is sold, as they must also from an incinerator.

The largest plasma gasification plant in the world was scheduled for completion in Saint Lucie, Florida, in 2009 but was not yet in service in June 2010. It is expected to generate 160 megawatts of electricity, enough to power 36,000 homes from a daily diet of trash.[23] The facility is being built next to a landfill that contains 4.3 million tons of trash, which the facility will excavate at a rate of 1,000 tons per day. Added to this volume will be 2,000 tons of garbage trucked in from the surrounding area. The plant cost $450 million to build, and the operating company believes it will recoup its investment within twenty years through the sale of electricity and slag.

Recycling

Approximately 33 percent of municipal solid waste (MSW) is recycled, a percentage that has increased steadily since 1965 (figure 4.4). The percentage did not exceed 15 percent until the early 1980s, and the growth since then reflects a rapid increase in infrastructure and increasing market demand for products made using recycled materials. It is a selling point for an increasing number of Americans. Recycling and reuse businesses now employ about as many people as the auto industry does.[24]

According to the editor of *Resource Recycling* magazine, "Without recycling, given current virgin raw material supplies, we could not print the daily newspaper, build a car, or ship a product in a cardboard box. They are key ingredients to industrial growth and stability."[25]

Paper forms nearly one-third of our trash, and we recycle more than half of it (table 4.2). The product most recycled is automobile batteries; nearly everyone turns in their dead automobile batteries when they buy a new one, and Americans recycle more than half their steel cans and yard trimmings. The steel, aluminum, and mixed metals that were recycled in 2007 eliminated the equivalent of 4.5 million cars from the road

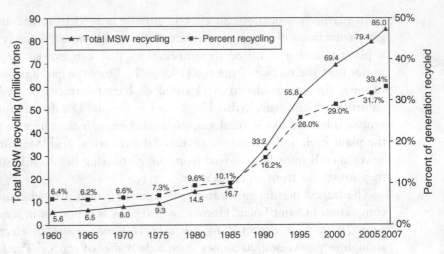

Figure 4.4
Recycling rates of municipal solid waste since 1960. (Environmental Protection Agency, 2008)

Table 4.2
Recycling rates of selected products, 2008

Item	Percentage recycled
Auto batteries	99.2
Office-type papers	70.9
Yard trimmings	64.7
Steel cans	62.8
Aluminum beer and soda cans	48.2
Plastic soda bottles	36.6
Tires	35.4
Plastic HDPE bottles	29.3
Glass containers	28.0
PET bottles and jars	27.2

Source: Municipal Solid Waste Generation, Recycling, and Disposal in the United States: Facts and Figures for 2008, Environmental Protection Agency (2009).

for one year.[26] Overall, the 85 million tons of municipal solid waste recycled in the United States in 2007 reduced an amount of greenhouse gas emissions equivalent to emissions from 35 million passenger cars.

Plastic Bottles

Americans buy an estimated 28 billion plastic water bottles each year, and the number is growing. Only 24 percent of them are recycled. More than one-third of plastic soda bottles were recycled in 2007, in part because of the increasing number of bottle bills being passed in states, commonly over strong objections by grocery and beverage industry lobbies. The purchaser of a soda bottle pays a small fee, usually five cents (ten cents in Michigan), when the product is bought, and gets the money back when the bottle is returned. As of mid-2010, eleven states have bottle deposits on soft drink bottles, and six of those states extend the deposits to water bottles as well. Many other states are considering bottle bill proposals. According to the Container Recycling Institute, the recycling rate for plastic bottles of carbonated beverages in 2008 was 80 percent in the eleven states with bottle deposit bills compared to 27 percent in the other thirty-nine states. For noncarbonated beverages such as water, the recycling rate was 35.2 percent in states with bottle deposit laws and 13.6 percent in the states without such laws. As usual, hitting people in their wallet has a pronounced effect.

The first bottle bill was passed in Oregon in 1971, but the nickel deposit it mandated has never been increased and has become the standard charge nationwide in states that passed similar laws. The 1971 nickel is now worth about a penny, and redemption rates have become depressed over the years. This has not occurred in Michigan, where the redemption rate remains at 97 percent. The difference between a nickel and a dime seems to have great significance to Americans. The amount of deposit on these bottles should be increased nationwide.

In most states, if the bottle is not redeemed, the mountain of nickel deposits paid by purchasers remains with the beverage distributing company to which the deposit was paid. The company is rewarded for the purchaser's lack of environmental concern with a large percentage increase in the price that was paid for the soft drink. In New York, these unredeemed deposits total about $100 million each year. The money should be recovered by the state and used for environmental purposes, a proposal that the grocery and drink industries oppose.

Bottle laws do stimulate recycling and reduce litter. But they have a huge flaw: they place a deposit on beer and carbonated beverages only, not on water. Bottled water bottles are excluded in forty-four states. In 2007, Americans drank more than 30 billion single-serving bottles of water (7 billion gallons), for which they paid no deposit. Most end up in landfills. Less than one-fourth are sent to be recycled.

In October 2009, North Carolina banned all rigid plastic containers from landfills. This includes any bottles with a neck smaller than the container itself. Such a standard will prohibit not only soda and water bottles, but ketchup, milk, salad dressing, vinegar, pancake syrup, and other containers as well.

Most soda and water bottles are made of PET plastic (polyethylene terephthalate). North Carolina, which is building the nation's largest facility to recycle these bottles, already has the second largest recycling facility in the United States for HDPE plastic, the other commonly used plastic bottle resin, which can be found in detergent bottles and milk jugs. PET and HDPE form 96 percent of all plastic containers. But companies specialize in recycling either PET or HDPE; only 5 percent of companies recycle both types of plastic.

After bottles are collected, they are taken to a materials recovery facility where they are condensed into large bales for shipping. Each bale weighs from 800 to 1,200 pounds and can contain 6,400 to 9,600 beverage, food, or nonfood bottles. The bales are shipped to a plastic reclaimer, where a machine rips apart the bales, and the pieces are then machine-shredded into tiny flakes. The flakes are washed, dried, and melted, and the melted plastic is extruded into pellets and sold to be developed into various plastic products.

In many PET applications the pellets are spun into a very fine thread-like material that can then be used to make carpets, clothing, or filling for jackets and quilts. The thin plastic is also a good insulator. HDPE pellets are melted and extruded into plastic lumber or pipe and can be blow-molded into plastic bottles, or injection-molded or thermoformed into plastic containers, garden products, sheet, and packaging.

Recycling a single plastic bottle conserves enough energy to light a 60 watt light bulb for 6 hours. Recycling 1 ton of plastic bottles saves 200 cubic feet of landfill space.

The Biota company in Colorado believed it had a solution to plastic bottle refuse: a biodegradable bottle. It made water bottles out of biodegradable plastic made from cornstarch, and used 20 to 50 percent less energy to manufacture them than to make petroleum-based bottles. Approximately 17 million barrels of oil are used to make plastic water bottles, enough to fuel about 100,000 cars. The company said that while traditional plastic bottles take many hundreds of years to degrade in a landfill, its bottles can biodegrade within eighty days in a commercial composting operation. The bottles are safe on store shelves because they degrade only when they have been emptied and placed in composting conditions, where high heat and humidity, as well as microorganisms, eat them. Unfortunately, Biota declared bankruptcy in 2007 for reasons unrelated to its eco-friendly bottle.

Tires

Another seemingly indestructible product used by almost every adult in the United States is the automobile tire. Worldwide, about 1 billion tires are sold annually, and eventually all are discarded. However, forty-four states will not accept them in landfills because they have large volumes but 75 percent void space. In addition, their concavity traps methane gas generated by other garbage, and when they catch fire they burn uncontrollably, spewing black pollutants. In August 1998, a grass fire ignited 7 million tires in the San Joaquin Valley of California, sending a plume of soot and noxious gas thousands of feet into the air and burning for two and a half years. Cleanup took years and cost $19 million.

Every year Americans discard 300 million tires whose 0.2 inch of tread has worn off—about one tire for every man, woman, and child in the country. About 45 million tires are used to make 25 million retreads. Thirty percent of scrap tires are recycled for commercial purposes, such as construction materials in roads and sidewalks, insulation, and fuel for power plants and cement kilns. About half of the reclaimed tires in the United States are burned for fuel. But burning a tire yields only one-sixth of the energy used to make it, and 85 percent of a tire is carbon, making tires a significant source of carbon dioxide emissions. They can be burned at temperatures high enough to destroy most organic pollutants.

Environmental Benefits

Recycling has environmental benefits at every stage in the life cycle of the products we use. Any time you make new stuff out of old stuff, less energy, fewer chemicals, and less water are used. Recycling also does more than reduce greenhouse gas emissions; it reduces air and water pollution associated with making new products from raw materials. Recycling trash leads to better health for all Americans, as well as generating a more sustainable economy.

Garbage in the Ocean

Most of the trash in the ocean stems from human activities at the shoreline, particularly recreational activities and smoking. Each year the Washington-based Ocean Conservancy reports the results of a worldwide systematic collection of trash found on beaches, the most recent report in 2010. Along the North American coastline, 55 percent of the beach trash stemmed from shoreline and recreational activities, 37 percent from smoking-related activities, 5 percent from fishing and boating activities, 1 percent was medical and personal hygiene materials, and 2 percent was from dumping.[27] Cigarette butts were the most common item found (2.2 million); second were 1.1 million plastic bags.

Many millions of tons of trash are present in the oceans, a major contributor to the death of sea life. Experts estimate that hundreds of thousands, and perhaps millions, of animals are killed each year by the artificial debris. The Ocean Conservancy reported finding 336 animals entangled or trapped in marine debris, with fishing-related items accounting for 65 percent of the cases where animals were trapped. Birds were the most commonly affected (41 percent), with fish (26 percent) and invertebrates (16 percent) second and third, respectively. Animals choke or become poisoned when they eat trash and drown when they become entangled in bags, ropes, and old fishing lines.

International shipping regulations prohibit the dumping of plastics and metal containers at sea, but it is difficult to enforce such rules. Inspectors cannot be on every ship, or even on very many of them. No one checks when a boat comes into a harbor whether the amount of garbage onboard is consistent with the size and occupancy of the boat or the length of time it has been at sea or, indeed, whether it has any garbage on board at all. But it is suspicious that there are five garbage debris fields, composed 90 percent of plastic, floating in the Pacific Ocean, one of which is twice the size of Texas.[28] The garbage patch doubles in size every decade.

5
Soil, Crops, and Food: Dirt and Nutrition

Food is not rational. Food is culture, habit and identity. For some, that irrationality leads to a kind of resignation.

—J. F. Foer, *Eating Animals*, 2009

In these days of indigestion
It is oftentimes a question
As to what to eat and what to leave alone;
For each microbe and bacillus
Has a different way to kill us,
And in time they always claim us for their own.

—Roy Atwell, *Some Little Bug Is Going to Find You Some Day*, 1915

Most Americans take food for granted. We are so used to warehouse-size supermarkets, shelves groaning under the weight of mostly unnecessary and unhealthy products, and the continual availability of out-of-season fruits and vegetables that we rarely think about the farms that produced them, the countries they came from, or the soil in which they grew. There is an almost total disconnect between food producers and food consumers. Few of us are aware that most farmers today must go to the supermarket to buy food, just like the 98 percent of Americans who are not farmers. Farmers journey to the supermarket as often as the rest of us because their farms do not produce a variety of products.

The Department of Agriculture defines a farm as any operation with the potential to produce at least $1,000 worth of agricultural goods in a given year. Based on 2006 prices, an operation could be considered a farm for growing 4 acres of corn or a tenth of an acre of berries (a square area only 66 feet by 66 feet) or for owning one milk cow. Consequently, most U.S. establishments classified as farms produce very little, and most agricultural production occurs on a small number of much larger operations called *factory farms*.[1]

Nearly all farmers specialize in only one crop or animal, such as corn, soybeans, wheat, chicken, cattle, or swine. The idyllic vista of a farm that grows a variety of grains, vegetables, fruits, and a few chickens exists only in our imaginations or in films that romanticize farming as we believe it existed one or two hundred years ago.

As little as we think about farmers, crops, and agriculture, we think even less about the soil in which the crops grow or the food that the farm animals eat. Perhaps this is a blessing. Who wants to consider the facts that the topsoil is now half as thick as it was for the early settlers and that the crops that grow in it have lost many of their nutrients, that there are poisonous pesticide residues in the food on our table, or that the farm animals we eat have been fed hormones and antibiotics for most of their lives?

As unhealthy and unpleasant as these facts are, they should cause us to ask questions such as these:

• How did agricultural soil lose its nutrients? Does this affect the crops that grow in them? Can the soil be restored to health?
• Why do most farms grow only one type of crop?
• Can we avoid eating pesticides three times a day?
• Do the hormones and antibiotics the farm animals eat affect the health of the humans who eat the animals?

In this chapter I look at these and other questions that affect our environment and our health.

How Are Crops Nourished?

Agricultural soil is in a group with air and water as the most essential commodity for human survival. And we are running out of it according to a study by an earth scientist at the University of Washington.[2] Six inches of soil are needed for crop production, a thickness that takes many hundreds or perhaps thousands of years to form, and human farming activities are causing it to erode an average of 10 to 100 times faster than this. Soil loss is not a problem in floodplain agriculture because annual floods deposit new and nourishing soil each year, but most agriculture in the United States is in upland areas where lost soil is not replaced.

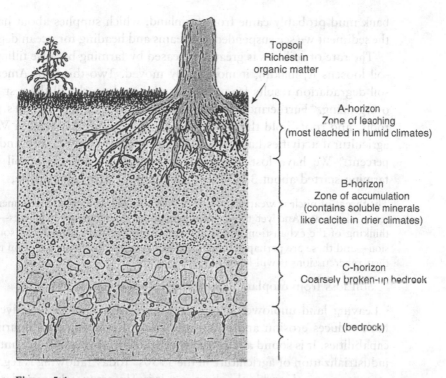

Topsoil
Richest in
organic matter

A-horizon
Zone of leaching
(most leached in humid climates)

B-horizon
Zone of accumulation
(contains soluble minerals
like calcite in drier climates)

C-horizon
Coarsely broken-up bedrock

(bedrock)

Figure 5.1
A generalized soil profile. Individual horizons vary in thickness, and some may be locally absent. The A-horizon is dominated by mineral grains in various stages of decay, releasing plant nutrients in the process. Water and air are present between the mineral grains. The B-horizon is enriched in either red hematite (iron oxide) in humid climates or calcite (calcium carbonate in dry climates).

Soil performs many important roles that are essential for human survival, especially the topsoil (A-horizon) (figure 5.1). It gives plants a foothold so they can root, nourishes them, provides homes for the myriad of small creatures whose presence is necessary for healthy crops, and filters rainwater.

But excessive cultivation, monoculture, erosion, and exposure have led to the degradation of much of America's cropland. The Department of Agriculture reports that 54 percent of the cropland is eroding above soil loss tolerance rates and that about 1.8 billion tons of soil are lost from cropland each year because of improper management, deforestation, overgrazing, and industrial activity.[3] Approximately 40 percent of the mud in U.S. streams comes directly from cultivated land, and 26 percent comes from the erosion of stream banks. Much of the stream bank mud probably came from farmland, which supplies about half of the sediment we see suspended in streams and heading for ocean depths.[4]

The rate of soil loss is greatly increased by farming because tilling the soil loosens it, making it more easily moved. Two-thirds of America's soil degradation results from crop farming and nearly all the rest from overgrazing.[5] Furthermore, tilling rips from the soil last season's plant roots, and roots hold the soil in place. In the Upper Mississippi Valley, agricultural activities have boosted erosion rates between 200 and 700 percent.[6] We have lost about one-third of the nation's topsoil since farming started about 350 years ago.[7] As the Dalai Lama wrote,

The threat of nuclear weapons and man's ability to destroy the environment are really alarming. And yet there are other almost imperceptible changes—I am thinking of the exhaustion of our natural resources, and especially of soil erosion—and these are perhaps more dangerous still, because once we begin to feel their repercussions it will be too late.[8]

Soil loss from cropland can be reduced in several ways.

• Leaving land unplowed every seven years is a biblical directive that both reduces erosion and allows the land to rejuvenate its nourishing capabilities. It is sound advice and was practiced from antiquity until the industrialization of agriculture in the 1950s. Today fallowing is regarded as uneconomical, and in the short run, it is. Unfortunately, consideration of only the short run is the guiding light of most of agriculture in the United States today.
• Crop rotations are effective in controlling erosion and were used by American farmers from the founding of the country until the advent of industrial monocultures following World War II. Changing the crop from year to year can reduce erosion rates by as much as 90 percent, depending on the crops chosen.[9]
• Another effective way to reduce erosion is no-till cultivation.[10] Whereas traditional plowing overturns up to 8 inches of soil, no-till reduces the depth of overturning by 75 percent, keeps crop and plant residue on the surface longer, and can reduce soil erosion by 90 to 95 percent. No-till works best in sandy soils. If the soil contains too much clay, traditional plowing is required to stir the soil sufficiently for adequate crop yields. The downside to no-till is an increase in weeds. No-till cultivation in planted cropland grew from 6 percent in 1990 to 16 percent in 2000 to 22 percent in 2008.[11]
• If crop residue such as corn stalks is left on the field all winter after harvesting, erosion is reduced and soil fertility is increased. The

drawback to this approach is an increase in crop pests that feed on the residue.

• Wind erosion can be reduced by windbreaks, but they reduce the acreage available for crops. For every 1,300 feet of windbreak, approximately 1 acre is taken out of crop production.[12] Windbreaks also put crops on the side of them in shade.

The federal government encourages soil conservation through its Conservation Compliance Program. Under this program, farmers who grow crops on highly erodible land must apply an approved soil conservation system or risk losing eligibility for federal income support, conservation, and other payments. Eighty-six percent of all cropland and about 83 percent of highly erodible cropland is located on farms that receive farm program payments.

Soil Organisms

Good agricultural soil contains not only minerals formed from the decay of underlying rocks, but an amazing array of living creatures as well. Fertile soil is alive: an acre of healthy topsoil contains about 54 tons of organic matter, of which about 20 tons is composed of living organisms such as bacteria, fungi, insects, and earthworms, while the rest is in various stages of decay. The mobile organisms turn over and continually aerate the soil and maintain soil porosity so nutrients in soil moisture can move through the soil. Worm excretions contain a balanced selection of minerals and plant nutrients in a form accessible to plant roots. Earthworm casts are five times richer in available nitrogen, seven times richer in available phosphates, and eleven times richer in available potash than the surrounding upper 6 inches of soil. The weight of worm excretions may exceed 10 pounds per year.

Soil microorganisms play an important role in the cycling of major plant nutrients as well. Some species of bacteria attach to the roots of some plants (legumes) and convert gaseous nitrogen that plants cannot use into dissolved nitrogen in a form accessible to plants. Soil-dwelling fungi have a symbiotic relationship with the roots of vascular plants in which the fungi increase the availability of minerals, water, and organic nutrients to the plant. Healthy crops rely on soil microorganisms for their survival.

The best way for a crop's uptake of soil nutrients to be returned to the soil can be determined by seeing what happens in an undisturbed natural environment. Mother Nature's method has been honed over hundreds of millions of years and is a good guide to maintaining a healthy soil environment. A plant environment such as a forest can maintain itself for thousands or, in the case of tropical rain forests, millions of years without decline. Several rules are clear:

• Multicropping is the rule. Nature never attempts monoculture, which depletes the soil of selected nutrients.
• Nature mixes plants and animals. Crops are always raised in conjunction with livestock.
• The balance between growth and decay is strictly maintained. What grows dies, and in dying is returned to the soil.

Soil Pollution
Given that the soil nourishes the plants that grow in it, and given that a great variety of living organisms in the soil contribute to a plant's health, it is little short of astonishing that farmers in the United States are so willing to spray poisons on their crops and into the soil. The United States consumes 35 percent of the world's pesticides.[13] Cereal crops are sprayed an average of five or six times a season; potatoes thirteen times; apple trees eighteen times; and peaches are sprayed with forty-nine assorted pesticides and fungicides on a weekly basis from March until harvesting in July or August. Apples and peaches are the most contaminated produce in the supermarket.[14]

And these artificial poisonous chemicals are not the only harmful concoctions routinely added to the soil. Even the "pure" NPK (nitrogen, phosphorous, potassium) fertilizers that nearly every commercial farmer and gardener uses contain poisons. An analysis of twenty-nine commercial fertilizers by the U.S. Public Interest Research Group revealed that each of them contained twenty-two toxic heavy metals. In twenty of the products, levels exceeded the limits set for waste sent to public landfills.[15]

Most pesticides destroy a broad range of living organisms, many of them either harmless or beneficial—like ladybugs, praying mantises, and earthworms—along with the undesirable pests. The appalling devastation of soil organisms caused by agricultural pesticides was well described by a farmer who switched from conventional to organic farming. He examined his alfalfa field after applying a potent insecticide and found "there was nothing but dead bugs, dead birds, dead worms—and little alfalfa shoots. Hardly any of the dead things were crop pests. We stood in amazement. It was shocking. It was the last time I ever did anything like that."[16]

Pesticides and the Human Body

Every day we are exposed to perhaps 10,000 chemicals, mainly in food but also by air that we inhale. And every year in the United States, about 700 new chemicals are introduced into the environment and are not checked for toxicity in a very thorough way. The key point to keep in mind is that the cumulative effect of the various pesticide poisons in the human body is unknown. Tens of thousands of different pesticides are used in American agriculture. At least fifty-three of them are known to be carcinogenic and have devastating effects on the human body. The Environmental Protection Agency (EPA) estimates that there are 10,000 to 20,000 physician-diagnosed cases of pesticide poisoning among agricultural workers each year in the United States. Because many cases are undiagnosed, the true number of cases may well be much higher. Farmers who work with pesticides get Parkinson's disease, have damaged immune systems, have lower quality and quantity of sperm in men, and develop several types of cancer more often than the general public does. A British study concluded that more than 10 percent of those who are regularly exposed to organophosphate pesticides (most commercial pesticides) will suffer irreversible physical and mental damage.[17]

The Centers for Disease Control has found 116 pesticides and other artificial chemicals in human blood and urine, chemicals that are passed on from mother to child through placental fluids and breast milk. Exposure to toxic chemicals during pregnancy affects not only a woman's children but her grandchildren and great-grandchildren as well. With rare exceptions, assurances from the federal government that the amounts people swallow are not harmful, is not something you would want to bet your life on. But we do. There is no way to verify such assurances. The only guaranteed safe exposure is no exposure.

There are no health data for many pesticides for toxicity, cancer-causing potential, or endocrine disruption. More than 2,000 chemicals enter the market every year, and most of them do not go through even the simplest tests to determine toxicity. And when testing is done, the chemicals are tested individually, despite the fact that chemicals ingested in combination can be much more harmful than the chemicals are individually. Because of financial and time constraints, thorough testing of chemicals in combination is impossible. According to a survey by an environmental organization, about a quarter of substances labeled as inert on container labels are not, because they are classified as hazardous under federal statutes.[18]

It is normal for some of these poisons to be on or in the food we eat. The Food and Drug Administration (FDA) determined in 2005 that three-quarters of the fresh fruits and vegetables in nonorganic supermarkets contained pesticide residues, as did almost all milk and cream samples.[19] Rinsing solid food in water will remove some pesticide residues that lie on the surface, but other pesticides are formulated to bind chemically to the surface of the crop and require washing in vinegar or dishwasher detergent to be removed. Produce such as celery, lettuce, grapes, and strawberries is likely to have absorbed the poisons, so it is impossible to wash them off. The amount of poison in a single strawberry is normally very small, but strawberries are like potato chips: nobody eats just one.

Why Eat Organic Food?

In response to the clear dangers of eating industrially produced food, the organic food movement began in the 1970s. Initially sales were confined to health food stores, but they soon entered the mainstream market. Major supermarket chains now sell 46 percent of the nation's organic foods.[20] The several supermarket chains that carry only organic food items sell 44 percent of all organic foods in the United States. Organic food sales grew 18 percent between 2006 and 2007, 15 percent between 2007 and 2008, and another 5 percent in 2009, continuing the rapid annual growth of past years. A 2008 survey found that 69 percent of U.S. food shoppers bought something organic and 28 percent buy organic products weekly. From 2007 to 2008, organic fruit and vegetable sales were up 6 percent, dairy up 13 percent, nondairy beverages up 32 percent, breads and grains up 35 percent, and meat and fish up 12 percent. Other categories include packaged and prepared foods up 21 percent, snack foods up 11 percent, and sauces and condiments up 23 percent. Organic food's share of retail food and beverages was 3.5 percent in 2008. The price premium for organic products decreases rapidly as volume increases. Fruits and vegetables are the dominant organic products sold (38 percent of total), and more than two-thirds of fruits and vegetables are less than 30 percent more expensive than nonorganic varieties. Beef, fish, and poultry form only 2 percent of the organic food sales.

More than 80 percent of domestic organic food is produced on relatively small family farms that cover only about 0.6 percent of America's

Table 5.1
Percentages of certified organic crops planted on U.S. cropland, 2008

Crop	Percentage
Vegetables	8.7
Fruits	3.1
Millet	2.1
Rice	1.8
Oats	1.5
Barley	1.2
Rye	0.9
Wheat	0.7
Corn	0.2
Soybeans	0.2
Sorghum	0.2

Source: U.S. Department of Agriculture, "Organic Production".

crop acreage, with California, Wyoming, and Texas leading the way. Vegetables are the favored crop (table 5.1). Every state has some organic farmland. However, imports of certified organic products have increased as demand in the United States for organic foods has exceeded domestic supply.

Organic crops are grown using natural fertilizers such as animal manures or natural plant materials, and without the use of artificial chemical pesticides. Pests are controlled using the insects that prey on them and by crop rotation. Surveys have shown that yields from organic fields are comparable to those of conventional systems over the long run.[21] In addition, organic foods are more nutritious. As a bonus, organic farming uses less fossil fuel, keeps more water in the soil to resist drought, causes less erosion, and maintains better soil quality.

Organic fruits and vegetables are not totally free of artificial pesticides but they harbor considerably fewer of them (table 5.2). The president's cancer panel in 2010 recommended that consumers turn to organic food if possible, a stunning condemnation of the nation's food system. Conventionally grown produce has pesticide residues far more often than organically grown produce, and the amounts of pesticide they bear are higher two-thirds of the time. The conventional crops also contain multiple pesticides eight times more often than organically grown produce does.

There are several possible explanations for why produce grown organically contains artificial pesticides. Some of the produce may have been

Table 5.2
Percentages of produce found with pesticide residue

	Conventional	Organic
Pears	95	25
Peaches	93	50
Strawberries	91	25
Spinach	84	47
All fruit	82	23
Grapes	78	25
Bell pepper	69	9
All vegetables	65	23
Lettuce	50	33

Source: consumer reports.org, 2002. consumerreports.org/main/content/aboutus
.jsp?FOLDER%3C%3Efolder.

grown in soil previously used for conventional agriculture that had not yet been completely cleaned of pesticides, which takes a few years. Perhaps the pesticides drifted in from a neighboring field where crops using pesticides were being grown. Perhaps the contamination was obtained in trucks or warehouses during transportation or storage. Although organic produce is not 100 percent free of pesticides it is nevertheless much better in this regard than crops grown in the conventional manner. Lesser amounts of poison no doubt translate into safer eating.

Opponents of organic farming methods cite the ever-present hunger and starvation that plagued humanity throughout history—maladies they say are being ended by today's industrial farming methods. They believe that without the multitude of artificial chemical fertilizers and pesticides produced by industrial chemists, people would once again be undernourished. They believe that organic farming is synonymous with starvation. This is clearly untrue. The hunger that most of the world suffered from until the twentieth century, and that 840 million people still do, resulted from the same causes as hunger in Third World countries today: inequitable distribution of land, people too poor to purchase food, lack of support by their government, and armed conflict.

In any event, organic farming is not a return to farming the way our grandfathers did it. It is a sophisticated combination of old wisdom and modern ecological understanding and innovations that help harness the yield-boosting effects of nutrient cycles, beneficial insects, and crop synergies. It is heavily dependent on technology, but not necessarily the technology that comes from a chemical factory. The federal government seems finally to have formally recognized the obvious benefits of organic farming and has overcome the objections of the agrochemical industry. The farm bill enacted in May 2008 contained:

• A guarantee of $78 million for organic agriculture research, five times more than the former funding level
• The possibility of $100 million more through the year 2012
• $22 million to help qualified farmers and handlers achieve organic certification
• $5 million to collect specialized organic marketing data

Nevertheless, all is not well in the organic farming community, and the main problem has no easy solution. The problem is nonorganic seeds and pollen carried by the wind and insects. Organic crops are becoming "polluted" with genes from neighbors' nonorganic and genetically engineered crops. In 2002 a survey by the Organic Farming Research Foundation found that 8 percent of organic farmers had lost certification because of contamination by genetically engineered crops.[22] One hundred forty-two cases of contamination were reported worldwide between 1997 and 2006, twenty-two of them in the United States. In one-third of the cases, the contaminated product was corn. Dozens of lawsuits by farmers against agrobiotech companies have been filed. The few that have been settled have been won by the farmers.[23]

Is Your Food Contaminated?

The food safety regulatory system in the United States is a shared responsibility of more than 3,000 local, state, and federal agencies that for the most part are poorly financed, poorly trained, and disconnected. In the federal government, responsibility is split among fifteen agencies operating under at least thirty statutes. The U.S. Department of Agriculture (USDA) has 7,600 inspectors and is responsible for the safety of meat, poultry, and processed egg products; most of the rest, 80 percent of the food supply, is the responsibility of the FDA. Hence, cheese pizzas are inspected by the FDA, while pepperoni pies go to the USDA. The FDA's budget has decreased regularly over the years, and the number of inspectors and inspections has decreased correspondingly. There were nine times more food safety inspections in the 1970s than there are today, even as the volume of imported food from Third World countries has grown dramatically.

Almost every year there is a national food scare because of products contaminated with harmful bacteria. Sometimes the culprit is spinach, sometimes tomatoes, sometimes lettuce, sometimes peppers, and most recently peanut butter, pistachio nuts, cookie dough, and beef. Sometimes the cause is unsanitary conditions in an American processing plant; other times the bacteria came from one of the many other nations whose food we import. The number of produce-related outbreaks of food-borne illness more than doubled between 1999 and 2004. One-quarter of the population is sickened, 325,000 are hospitalized, and 5,000 die each year from something they ate.[24] There are an estimated 76 million cases of food-borne illness every year, most of which are unreported or are not traced to the source. Fruits and vegetables are the most likely products to be the source of the contamination because they are commonly eaten uncooked; live and harmful organisms can enter the body unimpeded (see table 5.3).

Actually every type of food we eat contains items that are not part of the product. The FDA accepts these items as impossible to economically remove. A certain percentage of "natural contaminants" is considered acceptable because there is no other practical alternative. These include rodent hair and excretions, maggots, parasites, and fly eggs, among other less-than-desirable animal products. The FDA considers these defects to be only "aesthetic" or "offensive to the senses" rather than harmful. It is explained in the FDA booklet *The Food Defect Action Levels: Levels of Natural or Unavoidable Defects in Foods that Present No Health Hazards for Humans* (2009). Tomato juice, for example, may average

Table 5.3
Illnesses caused by contaminated foods, 2009

Food product	Number of outbreaks	Reported cases of illness
Leafy greens	363	13,568
Eggs	352	11,163
Tuna	268	2,341
Oysters	132	3,409
Potatoes	108	3,659
Cheese	83	2,761
Ice cream	74	2,594
Tomatoes	31	3,292
Sprouts	31	2,022
Berries	25	3,397

Source: "The Ten Riskiest Foods Regulated by the U.S. Food and Drug Administration," Center for Science in the Public Interest, 2009.

10 or more fly eggs in a small juice glass; pizza sauce is allowed 15 or more fly eggs and one or more maggots in the same size glass. Ground pepper is allowed up to 270 insect parts per ounce. An 18-ounce jar of peanut butter may contain approximately 145 bug parts or 5 or more rodent hairs. (Who does the counting?) The most recent edition of the booklet states that "the defect levels do not represent an average of the defects that occur in any of the products; the averages are actually much lower." That's comforting to know. Perhaps the peanut butter sandwich I had for lunch today had only 9 bug parts and 2 rodent hairs because I ate only a small part of the contents of the peanut butter jar. It seems that processed foods contain extra proteins not listed on the label.

Processed foods have recently come under scrutiny for salmonella and other pathogens. The salmonella bacterium causes more than 1 million illnesses each year in the United States. Some of the 100 million frozen pot pies made and sold by ConAgra Foods sickened an estimated 15,000 people in 2007.[25] The company was unable to determine which of the twenty-five pie ingredients was contaminated and decided to change previous policy and make the consumer responsible for killing the pathogens during the cooking process by advising on the package "food safety" instructions that the internal temperature needs to reach 165°F as measured by a food thermometer in several spots. The USDA reports that less that half the population owns a food thermometer and only 3 percent use it when cooking high-risk foods.

An increasing percentage of Americans are justifiably concerned about the apparent lack of safety of the food supply. A Gallup poll in 2007 revealed that the percentage of Americans expressing confidence in the federal government's ability to ensure the safety of our food supply dropped to 71 percent from a high of 85 percent in 2005.[26]

State oversight varies greatly among states. In the scare over salmonella-laced peanuts in 2008, forty-two Minnesotans were reported sick compared with three Kentuckians. It seems doubtful that fourteen times as many Minnesotans eat peanuts than residents of Kentucky. In 2008 forty-two Minnesotans became ill from jalapeno peppers compared to two in Kentucky. It is unlikely that Kentuckians have better immune systems. From 1990 to 2006, Minnesota health officials uncovered 548 food-related illness outbreaks, while those in Kentucky found 18. The differences between the two states arise because health officials in Kentucky and many other states fail to investigate many complaints of food-related sickness, while those in Minnesota are more diligent. The public health commissioner in Kentucky blamed tight budgets and admitted that his state has had a historically poor record of reporting food-borne illnesses.[27]

In an attempt to partially fill the gap in inspections, some industries have instituted their own inspection programs to reassure consumers. They are paying other government agencies to do what the FDA rarely does: muck through fields and pore over records to make sure food is handled properly. In California, the leafy greens industry, which grows spinach and lettuce, pays the state so that inspectors can inspect farm fields for safety. Arizona created a similar leafy greens agreement, and there are plans to expand it nationwide. Other food industries have expressed interest in the leafy greens model. They realize they cannot sit back and wait for the bacterial outbreaks that have hit the leafy greens people in recent years. Since the program started in 2007, there have been no new outbreaks tied to California leafy greens.

The FDA inspects less than 1 percent of the food it is responsible for, much of it imported fruits and vegetables, commonly from countries whose standards of cleanliness are much lower than those in the United States, particularly for seafood and fresh produce.[28] More than 130 countries ship food to the United States, many of them in the Third World and with substandard inspection systems (table 5.4). The share of import violations attributed to low-income countries increased from 12 percent in 1998, to 15 percent in 2002, to 18 percent in 2004, while the share from lower middle-income countries has also risen. During the same period, violations by wealthier countries have declined.

In 2006 we imported 15 percent of our food from developing countries. The value of imported food increased by more than 60 percent between 2002 and 2007. More than 80 percent of the seafood consumed in the United States is imported, mainly from Asia, where aquaculture practices often involve using raw domestic sewage or livestock manure as feed, as well as the use of pesticides and antibiotics not approved in the United States.

Imports from China are of particular concern because of their rapid increase in recent years. China's imports formed 6 percent of our food imports in 2008 and were valued at $5.2 billion, up from only $1.2 billion in 2000.[29] The country is now the third largest source of food imports to the United States, behind Canada and Mexico. In 2008, fish and shellfish, mostly products of aquaculture, formed 41 percent of their imports, with juices and other products of fruits and vegetables adding 34 percent. China accounted for 60 percent of the U.S. supply of apple juice, more than 50 percent of the garlic supply, and 10 percent of the shrimp and catfish supply.

Table 5.4
Food imports denied by the FDA, selected countries, July 2006–June 2007

	Number of refused food shipments	Most frequent food violation and counts	Total value of food imports, 2006
Mexico	1,480	Filth (mostly on candy, chiles, juice, seafood)—385	$9.8 billion
India	1,763	Salmonella (spices, seeds, shrimp)—256	1.2
China (excluding Hong Kong)	1,368	Filth (produce, seafood, bean curd, noodles)—287	3.8
Dominican Republic	828	Pesticide (produce)—789	0.3
Vietnam	533	Salmonella (seafood, black pepper)—118	1.1
Indonesia	460	Filth (seafood, crackers, candy)—122	1.5
Japan	508	Missing documentation (drinks, soups, beans)—143	0.5
Italy	482	Missing documentation (beans, jarred foods)—138	2.9
Denmark	543	Problems with nutrition label (candy)—85	0.4

Source: A. Martin and G. Palmer, "China Not Sole Source of Dubious Food," *New York Times*, July 12, 2007.

Filth and unsafe additives were the chief reasons for FDA refusal to accept China's products, with labeling problems and veterinary drug residues third and fourth, respectively. The main problems from other nations are labeling and lack of needed registrations rather than health reasons. Addressing the safety risks from Chinese food imports is difficult because of the vast array of products from China and its weak enforcement of food safety standards, heavy use of agricultural chemicals, and considerable environmental pollution.

Former FDA commissioner William Hubbard noted, "The public thinks the food supply is much more protected than it is. If people really knew how weak the FDA program is, they would be shocked." In reference to imported food, he said, "The word is out. If you send a problem shipment to the United States it is going to get in and you won't get caught, and you won't have your food returned to you, let alone get arrested or imprisoned."[30] For many years, FDA inspectors have simply returned the small percentage of contaminated food imports they have caught to their country of origin, and many of them turn up again at U.S. ports or borders, making a second or third attempt at entry. The odds strongly favor their admittance.

A poll by Consumer Reports in 2007 revealed that 92 percent of Americans wanted imported foods to be labeled by their country of origin. Such labeling was vigorously opposed by food industry groups, and their lobbying of congressional representatives was successful in delaying the implementation of such a law for many years. But the public's wishes were realized late in 2008 when such labeling became mandatory. There are exemptions, however. Meat and poultry sold in butcher shops and fish sold in fish markets, which total 11 percent of all meat and fish, are exempt because the law is worded to cover only large establishments that sell a certain minimum amount of fresh produce. There also are exemptions for processed foods. Imported ham and roasted peanuts, for example, do not need to be labeled unless they were canned or packaged in another country. Mixtures such as mixed frozen vegetables are also exempt.

Can We Improve on Nature? Genetically Modified Food

Plant breeders have been modifying the genetic makeup of crops for centuries. Today, almost nothing we eat is a natural product that could survive on its own in the wild in competition with naturally occurring vegetation. By selectively breeding individual plants within a single species, cross-breeding plants from different but related species, or inducing genetic mutations using chemicals or radiation, conventional plant breeders have promoted desirable traits to increase agricultural productivity.

Living organisms contain genes, complex chemicals that for unknown reasons have physical meaning. One arrangement of genes translates into tomato, another into fruit fly, a third into tiger. Certain snippets of the gene pool (genome) in broccoli are expressed as florets, another snippet as stalks, a third snippet means green. It is nothing short of magical. Geneticists and biochemists are now able to take a gene that they have determined to have a certain meaning in an organism and splice it into a completely different type of organism. Grapevines have been injected with genes from silkworms to make the vines resistant to a disease spread by insects. A strawberry can be given a flounder gene that makes it frost resistant. The body of a fish might be turned into rows of corn kernels. The visible divisions between species can be overcome if the researcher wishes it.

Use of Genetic Modification in Agriculture

The chief use of genetic engineering in agriculture has been to make crops resistant to herbicides and insecticides. This enables a farmer to blanket-spray the crops with herbicides and insecticides produced by the agrobiotech companies without harming his crops, which have been engineered to be resistant to these chemicals. Weeds and insects will die but the crops are unaffected. The appeal of this to the farmer is obvious in the farm labor it saves.

Adoption of genetically engineered soybeans and corn by farmers in the United States has been rapid, although they still form only 9 percent of global crop production.[31] Genetically modified crops were introduced in 1996, and by 2009, 63 percent of America's corn was GM, as were 93 percent of soybeans. Both of these crops are used mostly for animal feed. Wheat is consumed largely by humans, and resistance to GM foods around the world halted Monsanto's plans to develop GM wheat. As of 2008, Europe's supermarket shelves contain almost no biotech produce, and top retailers shun GM foods. Most food sold in American stores contains GM ingredients. There is no evidence that the public's health has been harmed by eating GM food, although several animal studies have resulted in harm or death to the animals.[32]

The Downside of Genetically Modified Crops

Hybridization of GM crops with nearby weeds can transfer herbicide resistance or other beneficial crop traits to weeds. This has already occurred in fifteen weed species.[33] In addition, weeds and insects develop resistance to certain pesticides through evolution. Although this phenomenon can and has affected both conventional and GM varieties, the evolutionary pressure to develop resistance is stronger in GM crops because of greater reliance on a single gene or herbicide.

Studies over the past decade have revealed that early results with GM crops were as the agrobiotech companies predicted: a decreased need for pesticides and better crop yields as weed and insect problems declined. However, the pattern reverses within a few years as superweeds appear and insect resistance evolves. Reports of weeds resistant to the herbicide sprayed on GM crops have increased from 50 in 1980 to 310 in 2008.[34] Additional pesticides are then needed as Mother Nature changes in response to the attack on her plant and animal species and finds ways to combat them. Pesticide use has increased by 383 million pounds since GE crops were planted by farmers.

The use of GM crops actually increases the need for newer and more pesticides. As Rachel Carson noted fifty years ago, we are engaged in a biowar we cannot win.

Perhaps even more frightening is the harm that pesticide poisons do to the human body, as I noted earlier in this chapter. The American Academy of Environmental Medicine has warned that the public should avoid GM foods. Based on many animal studies, the academy stated that "there is more than a casual connection between GM foods and adverse health effects. There is causation." It is noteworthy that there have been no clinical trials of any GM crop. As Canadian geneticist David Suzuki said, "The experiments simply haven't been done and now we have become guinea pigs. Anyone that says 'Oh, we know that this is perfectly safe,' I say is either unbelievably stupid or deliberately lying."[35]

How Much Food Can We Grow?

Agricultural productivity in the United States has increased at an average annual rate of 1.6 percent since 1948, and in 2006 it was 152 percent above the 1948 level.[36] Farmers have doubled or tripled the yield of most major grains in two ways: growing more plants per acre and harvesting more human food (wheat, corn) or animal feed (corn, soybeans) per plant. Many farmers now plant 30,000 or more corn seeds per acre, about three times the planting density common in the 1940s. The farming tactics of closer spacing, widespread use of artificial chemical fertilizers, irrigation, and pesticides tend to create big plants that grow fast, but they do not absorb a comparable amount of many soil nutrients. Agriculture is basically an effort to move beyond the limits set by nature.

The bounty of increased production has come at the cost of decreased nutritive value. Food scientists have compared the nutritional levels of modern crops with historic, and generally lower-yielding, ones.[37] Today's food produces 10 to 25 percent less iron, zinc, protein, calcium, vitamin C, and other nutrients such as beta-carotene, phosphorous, copper, and selenium. Today's vegetables are 5 to 40 percent lower in nutritional value than in our grandparents' day. Plants cultivated to produce higher yields tend to have less energy for other activities, such as growing deep, robust roots that more aggressively absorb nutrients from the soil. And high-yielding crops generate fewer phytochemicals—health-promoting compounds like antioxidants and vitamins. Nutrient deficiencies in Americans are common, even among the well fed. The yield/nutrient dilemma is simply another example of human interference in natural processes causing harm. In fact, natural processes tend to lead to the most healthful outcomes.

Many Americans who worry about the lack of nutrients in the processed foods they eat take vitamin supplements, believing that they will improve their health. They consume cheap processed foods that make them sick and then spend billions on expensive products to keep them healthy. Today about half of all adults in the United States use some form of dietary supplement.[38] However, it is not clear that ingesting nutrients in pill form has the same benefits as eating nutrient-rich fruits and vegetables. In the past few years, several high-quality studies have failed to show that extra vitamins in pill form help prevent chronic disease or prolong life. As Eric Klein, chairman of the Cleveland Clinic's Urologic and Kidney Institute, noted, "The public's belief in the benefits of vitamins and nutrients is not supported by the available scientific data."[39]

Scientists suspect that the benefits of a healthful diet come from eating the whole fruit or vegetable, not just the individual vitamins it contains. This seems reasonable from an evolutionary viewpoint. There may not be a single component of broccoli or green leafy vegetables that is responsible for most of its health benefits. The reductionist approach of considering the effect of one or two chemicals in isolation may be an ineffective way to recognize the collaborative benefits of eating the entire plant that contains the vitamins.

Will Climate Change Affect Crop Production?

Several factors connect agricultural productivity and climate change:

• Average temperature increase
• Change in rainfall amount and distribution
• Rising atmospheric concentrations of carbon dioxide
• Pollution levels such as tropospheric ozone (smog)
• Change in climatic variability and extreme events

Few studies of climate change on agriculture have considered all of these factors:

Temperature   An increase in average temperature can lengthen the growing season, adversely affect crops in regions where summer heat already limits production, increase soil evaporation rates, and increase the chances of severe droughts.

Rainfall   Changes in rainfall can affect soil erosion rates and soil moisture, both of which are important for crop yields. Predicting future changes in rainfall, whether in total amount or intensity, is impossible, especially at regional scales.

Carbon dioxide fertilization   Increases in carbon dioxide levels can boost the growth of most crops. But this beneficial effect can be either strengthened or weakened depending on temperature effects and nutrient availability. Unfortunately, increased carbon dioxide also benefits weed growth.

Smog   Higher smog levels limit the growth of crops. Ozone levels near the ground are determined by both auto emissions and temperature. Such changes may offset any beneficial yield effects that result from elevated carbon dioxide levels.

Climatic variability and extreme events   Changes in the frequency and severity of heat waves, drought, floods, and hurricanes are an important uncertainty about future climate change. Such changes are anticipated but their effects are uncertain.

There will no doubt be significant regional transitions associated with shifts in agriculture as a result of climate change. Some areas of the agricultural belt will benefit, and others will suffer declining yields, but agriculture in developed countries is less vulnerable to climate change than agriculture in developing nations is, especially in the tropics, where farmers may have limited ability to adapt.

Can Terrorists Harm Our Food Supply?

We live in a world in which terrorism is a threat. The weapons used have been mostly those common throughout human history, upgraded from clubs and crossbows to hand-held firearms to missiles, bombs, and other explosives.

One weapon that has not yet been used but may well be on the horizon is bioterrorism, the sabotage of a nation's health or food supply, a topic explored in some depth in a recent book.[40] Plant killers of various kinds abound, they act quickly, and many of them are unaffected by our existing array of poisons. A new strain of the potato blight that caused the famine in Ireland in the 1840s hit North America in 1994 and is resistant to all fungicides yet developed. In 1996, the fungal disease known as karnal bunt invaded the U.S. wheat belt, ruining more than half of that year's crop and forcing the quarantine of more than 290,000 acres. A virulent new strain of the black stem wheat fungus that destroyed 20 percent of the American wheat crop fifty years ago has spread from North Africa to India and is expected to invade the United States soon. We do not yet know how to kill it. These naturally occurring enemies of our crops evolve every now and then and cause havoc in agricultural areas. Terrorists could easily "seed" America's crops with such enemies.

Fungi are not the only fast-acting creatures that successfully attack crops. Insects can also destroy crops with devastating speed, and spread disease as well, facts that have frequently been noted by military strategists. The French, German, Japanese, and American military establishments have either contemplated or used food-eating beetles, plague-infected fleas, cholera-coated flies, and yellow fever–infected mosquitoes as weapons. The dispersal and biting capacity of (uninfected) mosquitoes was tested by the U.S. government during World War II by secretly dropping them over U.S. cities. According to biodefense experts, a terrorist with a few inexpensive supplies could introduce human disease or crop enemies to the United States with almost no chance of being caught.

There are recent examples of insect attacks on plants in the United States not caused by terrorists. The Asian longhorned beetle, which arrived in 1996, and the emerald ash borer, found in 2002, together have the potential to destroy more than $700 billion worth of forests, according to the USDA. This amount is twenty-five times the direct damage caused by the attacks of September 11, 2001. An insect-borne disease that destroyed enough orchards to cut the sale of orange juice by 50 percent for five years would cost the U.S. economy $9.5 billion, approximately the cost of building the World Trade Center from scratch.

The invasion of the incredibly fast-spreading Japanese kudzu plant in the southeastern United States and the invasion of predatory organisms loosed from the ballast of ships in the Great Lakes are other examples of the damage that can be done to plants and animals by invasive foreign species. About 40 percent of insect pests in the United States are nonnative, as are 40 percent of our weeds and 70 percent of our plant pathogens.[41] It takes little imagination to envision the devastation that could be loosed in the United States by crop terrorists.

Fossil Fuels and Crop Production

Few Americans are aware of how dependent farming in the United States is on fossil fuels. Seventeen percent of the oil we use each year is used in some aspect of agriculture.[42] Most of the fuel, 83 percent, is used in the production phase, to manufacture fertilizers and pesticides; to power tractors, cultivation, and harvesting equipment; to dry grain; and to keep irrigation pumps working. Contrary to popular belief, transportation and delivering food from producer or processor to the point of retail sale account for only 4 percent of the food system's greenhouse gas emissions, even though the produce we eat has traveled an average of 1,500 miles in eighteen-wheel trucks.[43] The fact that New York City gets its produce from California rather than from the "Garden State" of New Jersey next door does not add greatly to greenhouse gas emissions. Eating food produced locally means you are eating fresher and probably more nutritious food, but it has little effect on the production of greenhouse gases.

The environmental impact of food depends in part on how it was transported, not only on the distance traveled. For example, trains are ten times more efficient than trucks are. So you could eat potatoes trucked in from 100 miles away, or potatoes shipped by rail from 1,000 miles away, and the greenhouse gas emissions associated with their transport would be roughly the same.

The largest reduction in emissions from agriculture would come from a reduction in the consumption of beef and dairy products (figure 5.2). Raising livestock accounts for 18 percent of greenhouse gas emissions—more than from cars, buses, and airplanes. Producing meat is highly inefficient because it involves eating higher on the food chain. It takes more energy and generates more emissions to grow grain, feed it to cows, and produce meat and dairy products for human consumption than to feed grain to humans directly. Producing a pound of beef creates 11 times as much greenhouse gas emission as a pound of chicken and 100 times more than a pound of carrots. And a large percentage of the emissions associated with meat and dairy production are methane and nitrous oxide. Compared with carbon dioxide, methane has considerably greater warming potential but a shorter atmospheric lifetime. Overall, methane is 20 times as potent as carbon dioxide over a 100-year horizon. Nitrous oxide, in contrast, is characterized by a relatively low warming effect but a very long atmospheric lifetime, which causes it to have a global warming potential 310 times that of carbon dioxide. In 2006, carbon dioxide,

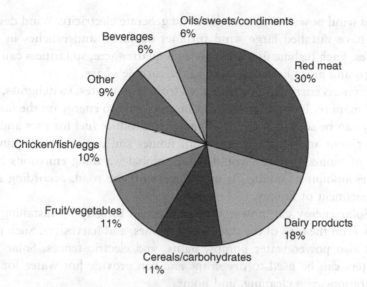

Red meat and dairy are responsible for nearly half of all
greenhouse gas emissions from food for an average U.O. household.
(Percentages total 101% because of rounding of numbers.)

Figure 5.2
Greenhouse gas emissions from various types of food eaten in an average American household, (*Environmental Science and Technology*. May 15, 2008, 3482)

methane, and nitrous oxide constituted about 84.8, 7.9, and 5.2 percent of all U.S. greenhouse gas emissions, respectively.[44]

Methane emanates from both ends of cattle as a by-product of ruminant digestion and is also released by the breakdown of the mountains of manure the animals produce. Nitrous oxide is generated by the breakdown of manure. Only 5 to 25 percent of the nutrients farmers feed their livestock are converted into meat. Our diets account for up to twice as many greenhouse emissions as driving our automobiles. Becoming a vegetarian will help the planet (and your health) more than buying a hybrid car. Vegetarians in Hummers do more for the planet than do meat eaters who drive hybrids or collect recyclable soda cans.

Food Production and Renewable Energy

Agriculture accounts for 18 percent of greenhouse gas emissions, but more than four-fifths of the fossil fuel used in agriculture is used during production on the farm. Hence, alternative sources of energy have a bright future in our sunny, flat, and windy Midwest. Farms have long used wind power to pump water and generate electricity. Wind developers have installed large wind turbines on farms and ranches in many states. Each turbine used only about a quarter-acre, so farmers can plant crops and graze livestock right to the turbine's base.

Biomass energy can be generated from crop wastes, switchgrass, trees, and manure. Biomass wastes can be converted to energy on the farm, or they can be sold to energy companies to produce fuel for cars and farm equipment and heat and power for homes and businesses. Tripling our use of biomass energy would reduce global warming emissions by the same amount as taking 70 million cars off the road, according to the Department of Energy.

Solar energy can power farm equipment directly by installing solar panels on the tops of the tractors, combines, and harvesters. Such panels can also power water pumps, lights, and electric fences. Solar water heaters can be used to dry crops and can provide hot water for dairy operations, pen cleaning, and homes.

Eating Animals

Americans eat massive amounts of meat and milk products—1 million animals per hour. The environmental and human cost of raising cattle the way it is done on nearly all American livestock farms is staggering:[45]

• Approximately half the energy used in agriculture is devoted to raising livestock. To produce the yearly average beef consumption of an average American family of four requires over 260 gallons of fossil fuel. There are 112 million households in the United States.

• It requires 8.5 times more fossil fuel to produce 1 calorie of meat protein than to produce 1 calorie of protein from grain for human consumption.

• It requires about 100 times more water to raise 1 pound of beef than 1 pound of wheat or corn.

• Seven pounds of grain are required to produce 1 pound of beef. That amount of grain would provide a daily ration of grain for ten people.

• Fifty-six million acres of land in the United States are growing hay for livestock. Only 4 million acres are producing vegetables for human consumption.

• More than 90 percent of the pesticides Americans consume are found in the fat and tissue of meat and dairy products.

• The saturated fat content of beef and milk products is a major contributor to high blood pressure, heart disease, and some types of cancer.
• The meat in our diets causes more emissions of greenhouse gases carbon dioxide, methane, and nitrous oxide into the atmosphere than either transportation or industry.

Beef consumption is a serious drain on the consumption of fuel and water, as well as a waste of grain and agricultural land. But meat consumption is considered a sign of affluence around the world, and affluence in the Third World is growing rapidly. In developing countries like China, India, and Brazil, consumption of red meat has risen 33 percent in the past decade. It is expected to double globally between 2000 and 2050. In 2007, 56 percent of the world's beef was produced in developing nations.

Food that makes people sick should not be considered food at all. However, some executives in the meat industry do not believe that eating meat harms a person's health. James Hodges, executive president of the American Meat Institute, believes that "meat products are part of a healthy, balanced diet and studies show they actually provide a sense of satisfaction and fullness that can help with weight control. Proper body weight contributes to good health overall."[46] I suspect that a nationwide poll would reveal that few adults would be willing to risk their life for a little sense of satisfaction and fullness, which could easily be experienced by eating a healthful diet consisting of fruits, vegetables, grains, legumes, nuts, and seeds.

Chickens are fed ground-up parts of cows, sheep, and other animals. Some of that chicken feed spills out and is swept up as chicken litter, which is then fed to cows. USDA regulations allow chicken feces to be used as feed for cows. Dead farm animals are fed to chickens, and then chicken feed spills onto the floor where, combined with chicken poop, it gets swept up and fed to vegetarian cows. It's enough to make a person sick. From a health standpoint, eating meat and its derivative dairy products is analogous to smoking cigarettes.

A properly balanced vegetarian diet is likely to be the healthiest diet and is followed by about 2.5 percent of Americans. Two-thirds of them are women. Vegetarians live longer and have lower rates of heart disease, lower blood pressure, and lower rates of hypertension, adult-onset diabetes, dementia, and prostate and colon cancer.[47] Albert Einstein demonstrated his insight in a field far removed from physics when he said, "Nothing will benefit human health and increase the chances for survival on earth as much as the evolution to a vegetarian diet."[48]

Fish, the Bounty of the Sea

Seafood consumption by Americans has exploded during the past thirty years. Consumption in 2008 was 16.3 pounds per person per year, near the all-time high of 16.6 pounds set in 2004. Prior to World War II, fish in the oceans were thought of as inexhaustible; their ability to evade traps and their rate of reproduction were great enough to maintain their numbers against the technology then available to catch them.

This changed in the 1950s as military hardware was adapted to serve the commercial fishing industry, and today ocean fish are being hunted at a rate that exceeds their ability to maintain their numbers. The ocean is becoming a marine "dust bowl." The large predatory fish that Americans enjoy—Atlantic salmon, swordfish, halibut, tuna, marlin, cod, flounder—are nearing extinction. A few years ago, a single bluefin tuna sold for a record $173,600 in Tokyo. According to fishery experts, only 10 percent of the original numbers of these large fish are left in the sea.[49] Two-thirds of the world's fish stocks are either fully exploited or overexploited.[50]

The world fish catch has stagnated, and these large fish are being replaced on American dinner plates by fish lower on the food chain. Today's seafood is often yesterday's trash fish and monsters. Many of the fish listed on restaurant menus today were not present a few decades ago. Part of this has resulted from renaming. The slimehead was renamed in the 1970s as "orange roughy" to remove the gag reflex associated with the original name. With this tasty-sounding name, the slimehead was widely overfished. Joining the slimehead on the renamed list are goosefish (renamed monkfish), Patagonian toothfish, and Antarctic toothfish (both renamed Chilean sea bass, although they are not sea bass), snakehead (renamed channa), and whore's eggs (sea urchin). It is reminiscent of the renaming of bull testicles as 'Rocky Mountain oysters.'

Decreases in numbers are also occurring to tiny fish such as anchovies, herring, sardines, mackerel, and other small to medium-size fishes. Thirty-seven percent of the world's marine fish catches are ground up and fed to farm-raised fish, pigs, and poultry. Pigs and poultry in the United States consume six times the amount of these forage fish than human fish eaters do.[51] This, of course, reduces the food supply for the remaining large fish, many of which are visibly emaciated and starving. Scrawny predators have turned up on coastlines all over the world. Not only whales and fish species have been affected. Fish-eating birds, as well as other large mammal predators such as seals and sea lions, have also been found emaciated from lack of food, vulnerable to disease, and without enough energy to reproduce. Human predation has exceeded the sustainable yield—the amount of fish that can be removed from the sea without harming the species.

As we have done with meat from land animals, so also have we done with flesh from fish. We have added endocrine-disrupting chemicals to the diet of both freshwater and marine fish that live near coastal sewage outlets. The amount of estrogen coming out of swine and cattle is estimated to be more than ten times higher than the amount the human population puts out.[52] Scientists at the National Institutes of Health have found that male fish in contaminated rivers have been feminized to the extent that only 20 to 30 percent of them are able to release fish sperm. Those that do produce sperm produce up to 50 percent less than do normal male fish. Also, the ability of the sperm to produce viable offspring is reduced.[53]

Marine fish have suffered the same effects. In the Pacific Ocean near Los Angeles, nearly 1 billion tons of treated and filtered sewage are released into the ocean every day. Thirteen percent of fish caught near the sewage pipelines had ovary tissue in their testes. Two related studies found that two-thirds of male fish near one of the pipelines had egg-producing qualities.[54] No one knows whether eating the amounts of pollutants found in the fish will cause reproductive problems in humans.

Other chemicals found in marine fish include antidepressants, flame-retardant chemicals, and PCBs. Many artificial chemical products such as prescription medications flushed down the toilet pass through sewage plants and are ingested by marine life when the effluent is piped into the sea.

The catch of wild fish has stagnated at about 100 million tons since 1990, but the production of farmed fish has skyrocketed, from about 15 million tons in 1990 to 60 million tons in 2007. China produces more than two-thirds of the world's farmed fish, with other East Asian countries producing most of the rest. The growth in aquaculture far outpaces that in all other animal food sectors. In many fish aquaculture farms the number of fish raised in limited areas can lead to parasitic and other infections. Antibiotics are added to the fish food to combat this, as is done with farmed cattle. The potential problem this creates is that the drug will be ineffective when given to humans who become ill.

In addition, just as factory-raised cattle are different from pastured, grass-fed cattle, factory-farmed fish are an entirely different source of protein than wild fish. For example, farmed salmon contain much lower amounts of healthy omega-3 fatty acids.

From Farms to Factories to Families

Nearly everything humans eat has been commercially processed. The Food and Drug Administration lists approximately 2,800 international food additives and about 3,000 chemicals that are added to our food supply. Each of us consumes between 10,000 and 15,000 added artificial chemicals in a day of eating.[55] The average American eats his or her body weight in food additives each year.[56]

Processors have several objectives during their operations. First among these is the prevention of microbial illness—the elimination of bacteria that may already be present and preventing the growth of new organisms. However, only 1 percent of the additives used in food processing is for inhibiting the growth of harmful microorganisms.[57] The other 99 percent are added to entice consumers to eat more or increase profitability for processors and food purveyors, or both. They include acidity regulators to stabilize taste, antioxidants to retard or prevent rancid flavors from developing in fats and oils exposed to oxygen, flavor enhancers to encourage overeating, thickeners to make spaghetti sauce and ketchup more enticing, carbon monoxide to maintain a red color in meat packages for weeks, and numerous things to enhance eye appeal. And water is injected into meat and fish to increase profits for producers and sellers.

Grains from the farm are refined, which means removing the nutritious bran coating from the kernel, which turns brown seed to white, so that most of the flour and rice are white instead of brown. Most consumers prefer their processed grains to be white, perhaps because people associate white with cleanliness. Following stripping of most of the nutrients from the original grain, the product may be fortified, meaning that some of the nutrients that were lost during earlier processing have been artificially replaced. The fortification is not going to make the grain nutritionally whole again. If it did, the bran coating would not have been removed in the first place.

Before a food is canned or frozen, it is usually blanched by heating it quickly with steam or water. This process destroys water-soluble vitamins, as does subsequent cooking in water. The loss of nutrients during cooking in water depends on the temperature, duration of cooking, and the nutrient. Vitamin C and those in the B group are less stable than A, D, E, and K during food processing and storage. The least stable vitamins are folate, thiamin, and vitamin C. Most stable are niacin (vitamin B_3), vitamins D and K, biotin (vitamin B_7), and pantothenic acid (vitamin B_5).

Cooking in water also causes the loss of the more water-soluble minerals (nutritious chemical elements) such as sodium, potassium, and calcium. Food should be microwaved, steamed, roasted, or grilled rather than boiled. Better yet, vegetables and fruits should be eaten raw after cleaning rather than cooked. Cooking destroys enzymes in the food that help the body digest it.

Most vegetables are peeled or trimmed before cooking to remove the tough skin or outer leaves in spite of the fact that the bulk of nutrients tend to lie close to the skin of most vegetables. The skin of a baked potato or cucumber, for example, should be eaten, not discarded.

The processing of dairy products is secondary to the question of whether milk and milk products should be eaten at all. It is significant and informative that with the exception of humans, only infants of mammal species drink milk. After completing their youngest stage of growth, mammals abandon milk for the adult types of food suited for their species. Of course, only humans can produce other milk products such as cheese, butter, and ice cream. Dairy farmers produce about 163 billion pounds of dairy products each year, and the average person consumes an amazing 600 pounds of them, 70 percent of which is milk and cream.

Nutritionists have long associated dairy consumption with harmful effects to the human body.[58] Milk has been linked to anemia, allergies, obesity, and ovarian cancer. Fifty million American adults are lactose intolerant, including 15 percent of Caucasians, 70 percent of African Americans, and 80 to 97 percent of Asians, Native Americans, and Jews of European descent. Our fellow mammals have set a good example for us by abandoning milk after infanthood. We should pay attention to their message. Unfortunately, the USDA has the dual responsibility of assisting dairy farmers while promoting healthy dietary choices for Americans. A clearer conflict of interest is hard to imagine.

What Is Food?

Assuming the plants have not been poisoned by artificial pesticides or chemical fertilizers, the animals have not been stuffed with hormones and antibiotics, and the plant and animal products have not been heavily processed, potatoes, broccoli, red peppers, and squash are food. So are grapes, apples, dates, blueberries, honey, and salt, as well as whole grain bread, beans, brown rice, nuts, oatmeal, olive oil, eggs, ocean salmon, and forage-fed land animals.

Potato chips, refined sugar, margarine, soft drinks, corned beef hash, pop tarts, and spam are not food, naturally occurring unprocessed edible products. White bread, farmed fish, and cattle from feedlots are not real food. Most processed "foods" are not food—Cheerios, salami, nondairy coffee whitener, commercial pancake flour, bacon, and the myriad of other edible supermarket products that your grandmother would not recognize. Americans spend about 90 percent of their food budget on processed foods.

The food humans evolved with—fruits, vegetables, and animals that forage their entire lives—are today only a minor part of the human diet. Advertising is designed to sell processed products, not real food. Only those regarded by many people as food faddists push organic tomatoes, spinach, or raspberries, and their scattered voices are drowned out by mass market merchants and nutritionists.[59] America is addicted to fast foods and junk foods. People eat to fill their bellies and entertain their tongues, not to nourish their minds and bodies. Food has deteriorated from something that nourishes the body to a chemically altered sensory addition fabricated in a factory.

Americans read about nutrients, not food. The common words Americans are besieged with that are related to eating are not *carrots, pears,* and *rice,* but chemical terms such as *good* and *bad cholesterol, low fat, antioxidants, triglycerides, carbohydrates, fatty acids, free radicals,* and other terms that only a chemist could define. To these chemical terms are added pseudochemical words devised by advertisers such as *probiotic* (contains live micro-organisms) and *bioflavinoids* (plant metabolites). As Canadian humorist Stephen Leacock has noted, "Advertising may be described as the science of arresting the human intelligence long enough to get money from it."

Ethics is apparently not a word in the vocabulary of companies' advertising arms. They are deliberately deceptive, and their only aim is to entice consumers into purchasing their products. Terms such as "Heart Healthy," "A Better Choice," and "Good for You" appear on products that are often none of these things.[60] In late August 2009, a consortium of major processed food makers initiated its Smart Choices Program that they claimed was "designed to help shoppers easily identify smarter food and beverage choices." A few months later, after a public outcry, the FDA announced the labeling might be "misleading" and said it intended to investigate. Three of the four companies subsequently pulled out of the Smart Choices campaign.

Among the supposed Smart Choice products were Cocoa Krispies, Froot Loops, and other products that are almost 50 percent refined sugar. The chairman of the Smart Choices program said that the labeling system was based on federal dietary guidelines and sound nutrition, and that "Our nutrition criteria are based on sound, consensus science."[61] Despite being chased out of the Smart Choices campaign, Kellogg's in October 2009, began shipping boxes of Cocoa Krispies emblazoned with the claim, "Now helps support your child's immunity."

Consider the content of one so-called health bar. In order of abundance, the label cites sugar (sucrose), rolled oats, dextrose (another sugar), wheat flakes, rice, dried lemon (sulfited), soybeans, fructose (another sugar), corn syrup (yet another), partially hydrogenated peanut and soybean oil (industrially processed fat), nonfat milk, almonds, malt (more sugar), sorbitol (and yet more sugar), and flavoring. Percentages of each constituent are not given. Anyone who will buy that product as a health bar will buy anything.

Health claims by the manufacturers of processed foods are bogus and center on only one or two of the properties of their product rather than its overall effect on health. The FDA commissioner points out that "products with symbols stating they provide a high percentage of daily vegetable requirements and other nutrients neglect to mention they represent 80 percent of your daily fat allowance. There are those with zero percent trans fats on the front [label] but don't indicate that they contain very high percentages of saturated fats."[62]

No one knows the long-term effect of processed foods on the human body, but it certainly cannot be good. There must be a reason the United States leads the world in arthritis, cancer, constipation, diabetes, gout, heart disease, high blood pressure, high cholesterol, and obesity. If processed products were healthy, they would occur naturally. The food industry concentrates on making what you eat exceptionally tasty, which typically means adding lots of sugar, salt, and fat to their product. Humans should eat natural food, not the artificial concoctions hawked by the companies that fill supermarkets with unhealthy and unnecessary edible products. In his book *In Defense of Food: An Eater's Manifesto*, Michael Pollan writes that the first rule is to avoid any food products with more than five ingredients and those that contain unfamiliar ingredients. Good advice.[63]

6

Fossil Fuels: Energy from the Past

Energy is eternal delight.
—William Blake, *The Marriage of Heaven and Hell: The Voice of the Devil*,
1793

We Americans want it all: endless and secure energy supplies; low prices; no
pollution; less global warming; no new power plants (or oil and gas drilling,
either) near people or pristine places. This is a wonderful wish list, whose only
shortcoming is the minor inconvenience of massive inconsistency.
—Robert J. Samuelson, columnist, 2007

Presidents going back to Richard Nixon have been talking about energy
independence. It has an appeal to the American public equal to world
peace, mom, and apple pie. It gives people a feeling of control, or possible control, of our energy future. And energy supplies are critical to
the nation's continued development. But the concept of self-sufficiency
for the United States is an illusion. It is unlikely to happen during the
lifetime of anyone now living.

Petroleum: Can We Live without It?

Oil is the dominant fuel in the U.S. energy market, meeting 37 percent
of our total energy needs. American oil production peaked in 1970 at
3.5 billion barrels per year and has since declined to 1.8 billion barrels
per year. Imports of oil, our major energy source, have been increasing
for many decades as our domestic sources have declined and our energy
needs have increased. We use about one-quarter of the world's oil but
have only 2.4 percent of global reserves (figure 6.1).[1] Imports have been
increasing for decades and are now more than 60 percent of the oil we
use, and the percentage is projected to increase into the future. The
United States is an old oil province whose fields are in decline. Nothing

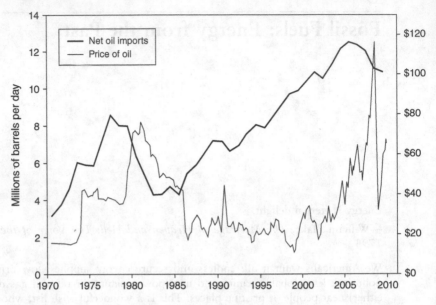

Figure 6.1
Changes in the amount of oil imports and the price of a barrel of crude oil, 1970–2010.

lasts forever. No amount of domestic exploration, in either the lower forty-eight or in Alaska, can change this. Additional domestic oil fields, both large and small, will no doubt be found, as has recently occurred deep beneath the Gulf of Mexico, but the increase in America's dependence on external sources of oil cannot be stopped.

Sources of America's Imported Oil
Because of the emphasis of American foreign policy on the Middle East, many Americans believe the Persian Gulf region is the dominant source of our imported oil. And the area does contain 75 percent of the world's proven reserves.[2] However, although the dependence of the United States on Middle Eastern oil was great in decades past, today it is less than 20 percent of our imports—perhaps 8 percent of all the oil we use (table 6.1). The European Union imports about the same percentage of its oil as does the United States, but 45 percent of it comes from the Middle East, and we are politically obligated to support their oil needs.

Most of the world's oil is controlled by the twelve members of the Organization of Petroleum Exporting Countries (OPEC), many of whom believe that the United States and Western civilization in general are a corrupting influence on human behavior and want to destroy them.

Table 6.1
Crude oil imports, 2009 (top 15 countries)

Country	Percentage
Canada	22.0
Mexico	13.9
Saudi Arabia*	13.1
Venezuela*	11.9
Angola*	7.1
Nigeria*	7.0
Iraq*	6.6
Brazil	4.2
Algeria*	2.8
Ecuador*	2.8
Colombia	2.7
Kuwait*	2.5
Russia	2.0
United Kingdom	0.6
Congo (Brazzaville)	0.5
Total	62.2

Source: "Crude Oil and Total Petroleum Imports Top 15 Countries," Energy Information Administration, 2010.
* = OPEC countries.

(table 6.2). This is the reason the United States has decreased its petroleum dependence on OPEC since the Arab oil embargo in 1973.

The Cost of Imported Oil

Most Americans consider the cost of oil to be the published cost of a barrel (42 gallons) of oil, which reached a record high in 2008 (equaling the cost of a barrel of milk but below the cost of a barrel of bottled water; figure 6.1) before dropping to "more acceptable" levels later in the year because of poor economic conditions in the developed world. But the price of an imported barrel of oil is only a small part of the true cost of oil to the American taxpayer. The principal reason we are not fully aware of the true economic cost of our dependence on imported oil is that it is largely in the form of what economists call externalities—costs caused by the production or consumption of an item that are not reflected in its pricing.[3]

The main externalities related to oil are the federal tax breaks to domestic oil corporations,[4] which average about $4 billion a year, and the most expensive item, the military costs of securing supply lines and

Table 6.2
Proven oil reserves in countries with at least 10 billion barrels. Summary of Reserve Data as of 2008

Country	Reserves		Production		Reserve life[1]
	109 bbl	109 m3	106 bbl/d	103 m3/d	years
Saudi Arabia*	267	42.4	10.2	1,620	72
Canada	179	28.5	3.3	520	149
Iran*	138	21.9	4.0	640	95
Iraq*	115	18.3	2.1	330	150
Kuwait*	104	16.5	2.6	410	110
United Arab Emirates*	98	15.6	2.9	460	93
Venezuela[2]*	95	15.1	2.7	430	88
Russia	60	9.5	9.9	1,570	17
Libya*	41	6.5	1.7	270	66
Nigeria*	36	5.7	2.4	380	41
Kazakhstan	30	4.8	1.4	220	59
United States	21	3.3	7.5	1,190	8
China	16	2.5	3.9	620	11
Qatar*	15	2.4	0.9	140	46
Algeria*	12	1.9	2.2	350	15
Brazil	12	1.9	2.3	370	14
Mexico	12	1.9	3.5	560	9
Total of top seventeen reserves	1,243	197.6	63.5	10,100	54

Notes: 1. Reserve-to-production ratio (in years), calculated as reserves / annual production. (from above). * = OPEC countries.
2. Source: BP Statistical Review, 2009.

sea routes worldwide ($100 billion per year), to which can be added the cost of the current war in Iraq that currently totals $750 billion. If there were no oil in the Middle East or if the United States were self-sufficient in energy, the military costs would disappear. The United States would be as concerned about Middle East dictatorships and the people they oppress as it currently is about the people of Myanmar and Darfur—verbal hand-wringing but nothing that costs much money. This is the way international politics typically operates.

It is noteworthy that in the thirty-five years since the Arab oil embargo of 1973, which had terrible effects on the American economy, President Jimmy Carter has been the only president who pushed for a change in our major energy source, but he was unsuccessful. No other presidents or Congresses of either political party have campaigned for it, despite the clear and persistent danger of another embargo or disruption in America's oil supply. The reason is obvious: lobbying pressure against the change by oil companies and automobile companies. The wishes of the public in poll after poll for decades have indicated the public's desire for such a change, but to no avail.

This may now be changing with the Obama administration. The president campaigned for the development of alternative energy before his election and has pushed for development money since then. But it will be many years before sources such as wind, solar, or geothermal are a major part of America's energy picture. As the International Energy Agency noted in 1999, "The world is in the early stages of an inevitable transition to a sustainable energy system that will be largely dependent on renewable resources."[5]

The Cost of Gasoline

Americans use 150 billion gallons of gasoline each year and were appalled when the price of a gallon of gasoline rose to $4 (figure 6.2).

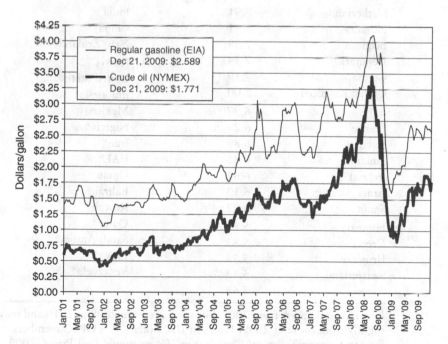

Figure 6.2
Price of regular gasoline in December 2009 dollars. (Department of Energy, 2009)

But motorists in most Western countries were jealous of our low price. Almost all of them pay more, the Netherlands leading the pack at $7.91 a gallon in 2009 (table 6.3). The explanation for the relatively low price in the United States is not that we pay less for a barrel of oil than others, or pay less for refining it, or because we do not import as much oil as nations in the European Union. The difference in price results from the high taxes that these other countries place on gasoline (table 6.4). Decades ago the governments in most developed countries decided to derive a significant part of their national income from the sale of gasoline and diesel fuel. In some countries, as much as 75 percent of the price at the pump is tax. In the United States in 2008, the price at the pump varied wildly from $1.60 to $4.10 and averaged about $3.15; the federal gasoline tax per gallon was 18.4 cents (unchanged since October 1993), and the average state tax was 28.6 cents, for a total tax in 2008 of 47

Table 6.3
Prices of a gallon of regular gasoline in countries around the world

Eritrea	9.58	Poland	5.34
Turkey	9.24	Japan	5.34
Netherlands	7.91	India	3.79
Norway	7.91	China	3.58
Belgium	7.38	New Zealand	3.03
Denmark	7.34	Russia	3.02
Germany	7.19	United States	2.68
United Kingdom	7.00	Indonesia*	2.23
Italy	6.97	Mexico	2.12
France	6.89	Nigeria*	1.67
Sweden	6.89	Iran*	1.51
Finland	6.81	UAE*	1.40
Ireland	6.62	Oman	1.17
Israel	6.13	Bahrain	1.02
Spain	6.02	Kuwait*	0.87
Hungary	5.87	Qatar*	0.83
Greece	5.83	Saudi Arabia*	0.61
Uruguay	5.75	Libya*	0.57
Switzerland	5.72	Venezuela*	0.19
Austria	5.68		

Note: Most data are from 2009, but some are from 2008 or 2010 and may differ slightly from those in 2009. Prices in U.S. dollars. * = OPEC members.
Source: German Technical Corporation, International Fuel Prices, 2009, 2010.

Table 6.4
Tax on gasoline and diesel fuel in some industrialized countries

Country	Gasoline tax (€ per gallon)	Diesel tax (€ per gallon)
Netherlands	4.11	2.29
Germany	3.73	2.33
Belgium	3.56	2.54
United Kingdom	3.73	3.77
France	2.25	2.59
Italy	3.35	2.59
Japan	$2.34	$4.00
United States (federal plus state)	$0.39	$0.45

Note: On November 23, 2010, 1 euro = $1.35
Source: "EU Fuel Prices, 2010." Available at http://energy.eu

cents, about 15 percent of the cost at the pump. This percentage varies between 10 and 20 percent, depending mostly on the price of a barrel of oil.

The cost of gasoline is composed of three parts, each of which is quite variable. According to the Energy Information Administration, the relative costs in the United States in 2009 were price of crude oil, 61 percent; refining, 11 percent; marketing and distribution, 10 percent; and federal and state taxes, 18 percent. These percentages can vary greatly from month to month as the price of a barrel of crude oil changes.

No new oil refineries have been built in the United States since the 1970s, but the equivalent of one new refinery has been added to capacity each year for many years as a result of additions to existing refineries and other improvements that allow greater output from the same facilities. As a result, refining capacity for crude oil has not been stagnant.

World Oil Supplies

Some investigators believe that the situation in the United States is a harbinger of world oil production. They believe that the peak of world oil production has been reached, or is about to be. However, many other researchers do not believe this.[6] They point out that while it is true that the big firms are struggling to replace reserves, the reason is that they no longer have access to the vast deposits of cheap and easy oil that are left in Russia and the OPEC nations. As late as the

1970s, Western corporations controlled 85 percent of the world's oil production. These companies now produce only 6 percent.[7] Today 78 percent of petroleum reserves are held by state-owned companies that are increasingly assertive about protecting their natural resources.

Large parts of Siberia, Iraq, Saudi Arabia, West Africa, the nations that were part of the former Soviet Union, and the soon-to-be-accessible Arctic Ocean likely contain significant amounts of petroleum. The U.S. Geological Survey has estimated that the several petroleum basins in the Arctic may contain 90 billion barrels of oil, one-third of it in Alaska. Added to this possible bounty are new technologies developed by the petroleum industry that are revitalizing existing fields formerly regarded as economically depleted. Between 1970, when America's oil production began to decline, and 2007, world oil production increased by 70 percent; it jumped 9 percent between 2000 and 2007. Global proven oil reserves have increased almost every year since 1980; they increased 10 percent from 1990 to 2000 and another 12 percent between 2000 and 2007.[8] Refining capacity increased by 10 percent between 1990 and 2000 and by another 7 percent between 2000 and 2007. There is reason for optimism about the amount of oil still present and accessible beneath the earth's surface.

However, world oil consumption increases almost every year, and it jumped 12 percent between 2000 and 2007, mostly as a result of increasing development in the Third World, particularly China. China now accounts for about 40 percent of the world's recent increase in demand for oil and uses twice as much now as it did in 1998. Its oil consumption increased by 15 percent from 1998 to 2007 and is expected to increase by 3 percent annually over the next two decades. China plans to add 120 million vehicles to its automobile fleet by 2015, ultimately requiring 11.7 million barrels per day of new oil supplies.[9] Auto sales in China are booming, with the strongest growth in the sale of gas-guzzling large luxury cars and SUVs, where sales are expected to surge. The situation is much like that in the United States before the current surge in gas prices. As one Chinese entrepreneur said, "In China, size matters. People want to have a car that shows off their status in society. No one wants to buy small."[10] He owns a Hummer, an urban assault vehicle for which demand in China has been so strong that since 2008, Chinese consumers can buy a similar military-style vehicle called the Predator at more than twenty-five new dealerships. In 2009, the Hummer brand was sold by General Motors (now largely owned by the U.S. government) to a Chinese company.

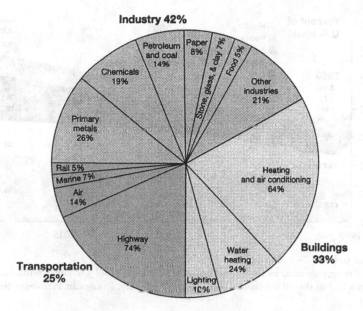

Figure 6.3
Uses of energy in the United States. (Energy Information Administration)

The world now consumes 86 million barrels of oil every day; 21 million of them are used in the United States, mostly for transportation, heating, and air conditioning (figure 6.3). Clearly civilization is in a race between increasing oil production and the world's growing demand for this liquid gold. So long as the world, including the United States, relies on petroleum as a major energy source, exploration needs to continue in every area where new sources might be found.

The ratio of proved oil reserves to annual production has held steady at roughly 40:1 for more than twenty years, but the remaining reserves are increasingly concentrated in more politically and technically challenging terrain, such as Saudi Arabia, Russia, Venezuela, and Nigeria. Increases in world oil consumption will have to be met primarily from the oil-rich Middle East, a region with a history of wars, illegal occupations, coups, revolutions, sabotage, terrorism, and oil embargoes. To this list of instabilities can be added powerful fundamentalist Islamist movements that are at war with Western Judeo-Christian civilization. They have their sights set on its oil supplies. In December 2004, al-Qaeda issued a fatwa that said in part, "We call on the mujahideen in the Arabian Peninsula to unify their ranks and target the oil supplies that do not serve the Islamic nation but the enemies of this nation. Be active

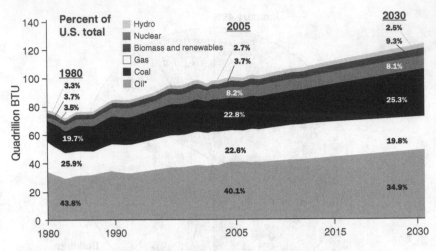

Figure 6.4
Future energy demand in the United States. Excludes ethanol and other biofuels, which are included in the "Biomass and Renewables" category. (American Petroleum Institute, 2008).

and prevent them from getting hold of our oil and concentrate on it particularly in Iraq and the Gulf."[11]

Since this fatwa was issued, there have been successful attacks on Iraqi oil pipelines, and Saudi Arabia has foiled planned attacks on its oil facilities. Political disruptions in Saudi Arabia or other major oil exporters that restrict oil exports could cause an oil crisis at any time.

There is no way to know in advance the amount of undiscovered and producible oil that remains in the ground around the world, but demand is predicted to increase in the United States and in the rest of the world (figure 6.4). A glass that is half empty is also half full. And there may well be more liquid in the unopened refrigerator, as well as solids and semisolids that can be liquefied (tar sands, oil shale, coal). Petroleum is certain to be a major source of energy for many decades to come. It may not be desirable (chapters 8 and 9), but it is inevitable.

Tar Sands
When the world thinks of oil reserves, the first nation that comes to mind is Saudi Arabia, which has about 25 percent of the world's liquid reserves. But there are rocks at and near the surface in Venezuela and western Canada that hold oil reserves potentially as large as those in Saudi Arabia, but not in liquid form. The rocks, called tar sands, consist of

grains of quartz sand held together by 10 percent of a tarry hydrocarbon material called bitumen that can be extracted and liquefied using steam. The tar was once liquid and able to flow, but the volatile material has evaporated over the eons, leaving the tarry residue that requires an expensive refining process to be converted into syncrude (synthetic crude oil). About 10 percent of the Canadian tar sands crop out at the surface and can be strip-mined, much like some Appalachian coal deposits; 90 percent are below ground.

Converting the tar into syncrude requires more energy and water than producing conventional oil, and the refining process produces lots of environmental damage: sulfur, acid rain, and 10 to 15 percent more greenhouse gases than the refining of conventional oil. Greenhouse emissions from tar sand mining are predicted to triple by 2020.[12]

Mining tar sands generates enormous volumes of liquid waste that are stored in toxic lakes in the sparsely inhabited area of northern Alberta where the tar sands occur.[13] The sludge seeps into the rivers and groundwater systems. One of the mining companies admitted that its Tar Island Pond leaks approximately 423,000 gallons of toxic fluid into the Athabasca River every day. The tailings ponds are growing constantly and in February 2008 covered more than 31 square miles. Canada's National Energy Board has warned that current production trends indicate that the volume of fine tailings ponds produced by the two major mining companies alone will exceed 260 billion gallons by 2020. Tailings dikes have major failures every six months, on average, worldwide. The largest dam in the world is a toxic sludge reservoir behind one of the Syncrude company's earthen dikes. It is visible from space.

Toxic materials from tar sand activities are not restricted to the lakes. A recent study commissioned by a health organization in Alberta identified high levels of carcinogens and toxic substances in fish, water, and sediment downstream from tar sands projects. Mercury levels in sediments are as much as double historical means, and arsenic levels are seventeen to thirty-three times the acceptable levels in moose meat from the region. In a Native American community near the tar sands project, the resident doctor has been concerned about the high numbers of normally rare types of cancers in the town. Environmental concerns related to tar sand mining have not been the top priority in Alberta or in the nation's capital. Tens of thousands of new jobs in resource extraction have been generated by exploitation of the tar sands, and in 2006 every Albertan received a $340 check from the government because of the fiscal surplus generated by the oil from tar sands.

The amount of recoverable oil from Canadian tar sands is estimated by the Alberta government to be 174 billion barrels, equivalent to about 10 percent of the bitumen in place. However, oil companies report they can easily recover 60 percent of the bitumen using proven technology, meaning that the estimate of 174 billion barrels is extremely conservative. The United States imported 780,000 barrels a day of tar sands oil in 2008, 60 percent of production. Oil output from the tar sands was 1.2 million barrels in 2008 and is expected to reach 3.3 million barrels a day by 2015. At this rate of production, the reserves would last over 400 years. Environmentally conscious President Obama, not wanting to irritate our major oil supplier, has pointed out that tar sands oil production can be made clean by capture of the carbon dioxide produced during refining. There is no suitable technology for this yet.

Production costs for Albertan tar sand are $25 per barrel, compared to $15 in the deep waters of the Gulf of Mexico and $5 in the Middle East. The minimum of 174 billion barrels of oil recoverable from the tar sands in Canada is somewhat less than Saudi Arabia's claims of 260 million barrels of traditional crude oil, although only the Saudis have access to actual field data, and many Western companies question the Saudi claim.

Venezuela, currently America's third largest supplier of imported oil, may have even more recoverable tar sand than Canada, although it lies deeper than the Canadian deposits and must all be recovered by underground mining. The published estimate of 300 million barrels of bitumen is largely guesswork. President Chavez has more than once threatened to stop sending Venezuela's oil to the United States, but oil is fungible, making such threats almost meaningless.

Alternatives to Petroleum in Motor Vehicles

For most uses, oil, natural gas, and coal are interchangeable, a prominent exception being fuel for motor vehicles, where only refined petroleum (gasoline or diesel fuel) or natural gas can be used. Nearly all the world's fleet of more than 800 million cars and trucks (96 percent) are fueled by gasoline or diesel fuel, products that most scientists believe are major causes of global warming and air pollution. About a quarter of the motor vehicles are owned by Americans.

Some cities have adapted their municipal vehicles to use compressed natural gas; they can then be refueled at central municipal locations. Natural gas is less suitable than gasoline or diesel fuel for most personal and commercial cars and trucks. There are very few refueling stations for the general public, and the fuel tank in the car must be heavy and bulky to contain the compressed natural gas, a major disadvantage for long-distance driving. Additional weight causes a great reduction in the number of miles per gallon of gasoline (or miles per cubic foot of natural gas).

Because of the need to reduce the nation's dependence on petroleum, automobile companies are making major efforts to produce hybrid gas-electric and all-electric cars. In hybrid vehicles, the small gasoline engine recharges the batteries. Hybrids have been available for about ten years, but because they typically cost $3,000 to $4,000 more than a conventional gasoline-powered car, they are still only one-tenth of 1 percent of the cars on the road—360,000 of 250 million cars.

Within the next two to three years, hybrid cars will be available that build on the efficiency advantages of today's hybrid cars. By adding a large and powerful battery, today's hybrid will be converted into a plug-in hybrid in which the battery can be recharged by plugging its cable into a standard electric outlet. Cars may have a retractable tail like vacuum cleaners. The large battery increases the distance the car can travel without recharging and further reduces energy use.

Within the past few years there has been a resurgence of interest in all-electric vehicles powered by an array of large batteries in the vehicle. The batteries can last for ten years, but the range of such vehicles is only 100 miles before the batteries need recharging, a serious limitation to their use in a large country such as the United States. Unlike standard car batteries that are based on lead, the batteries for electric vehicles are fueled by lithium, an element whose abundance and availability are uncertain. The majority of extractable lithium reserves are believed to be in Chile and Bolivia, but smaller deposits occur in the United States, Russia, Argentina, Brazil, Australia, and Tibet. And it has been predicted that once lithium-ion batteries are in wide use, lithium will be recycled from old lithium batteries to provide a stable source of the metal.

Because powerful lithium-ion batteries are very expensive, rechargeable electric cars will likely cost $6,000 to $8,000 more than hybrids. Automotive engineers are working feverishly to produce a low-cost powerful battery. If battery costs are not significantly reduced, electric cars will not be a significant part of America's automotive fleet. Nevertheless, research is ongoing and has substantial federal financial support.

Electric motors are more efficient than gasoline motors: they need less energy to do the same amount of work. At current electricity prices, running an electric vehicle would cost the equivalent of 75 cents a gallon.

So even accounting for the fuel burned at a power plant to generate the electricity for the battery (about half of the time it will be a coal-burning plant), plug-in hybrids will significantly reduce fuel use and carbon dioxide pollution.

The 254 million privately owned vehicles in the United States have an average life span of more than a dozen years, and the average age of America's passenger cars was 9.2 years in 2007 (trucks 7.2 years). Therefore, we will require conventional fuels for at least a decade, even if every new vehicle produced from today onward runs on some alternative.

In mid-2009 Congress passed what became known as the cash-for-clunkers bill, which lasted for 53 days. Its aim was to get highly polluting cars off the road (and boost car sales in a down economy) more quickly than would otherwise happen. Under the program, car buyers could trade in their clunkers that were no older than 1994 models and receive vouchers worth up to $4,500. To be eligible for the voucher, the clunker had to have an average fuel economy of less than 19 miles per gallon, about half of the cars on America's roads. The new car or truck then had to get better gas mileage than the clunker that was being traded in. The amount of the voucher depended on the difference in the fuel economies of the old and new cars. A new car getting at least 4 miles more per gallon than the old car was eligible for a $3,500 voucher. A new car getting at least 10 miles more per gallon could get a $4,500 voucher.

A similar plan in Germany boosted auto sales by 21 percent the first month of the program compared to the previous year. During the same period, sales in the United States slumped 41 percent. But the German plan had additional incentives not present in the American plan, so the U.S. plan did not have the dramatic effect of the German one, which has an additional tax of 50 cents per gallon on gasoline and a new tax based on carbon dioxide emissions. In the United States, talk of a new tax, even a small one, is anathema to most politicians. Taxes in the United States are the lowest among Western countries, and most Americans want to keep it that way.

Unconventional Oil

In 2008 the U.S. Geological Survey released a study of the Bakken Formation, a layer of oil-bearing rock located in North Dakota, Montana, Wyoming, and two Canadian provinces.[14] The oil is located in shale, a rock with low porosity and very low permeability. The survey estimated the amount of oil recoverable with today's technology at several billion barrels, much more oil than occurs in Saudi Arabia. If even a small part of this oil is recovered, it could satisfy America's energy needs for many years. Producing this oil requires extensive fracturing of the impermeable rock to release the oil, but this technology is well developed. The Bakken is one of the largest on-land oil formations in the United States, now made economical by the skyrocketing price of oil in recent years. It is an illustration of the truism that we never run out of anything in the earth. As technology improves and the value of a product increases, additional supplies of the product become economically obtainable.

Oil Shale

According to the Department of Energy, up to 2 trillion barrels of oil, eight times the reserves of Saudi Arabia, are locked within finely grained brown rock known as oil shale in Colorado, Utah, and Wyoming. The oil in America's oil shale is enough to meet current demand for the next 250 years. It is probably the largest unconventional oil reserve on the planet. Technology makes mining and refining the immature oil (kerogen) in this rock economically profitable when the price of a barrel of petroleum exceeds $30—and no one believes the price of a barrel of oil will ever be below $30 again.

The Department of Energy believes the oil shale region can produce 2 million barrels a day by 2020 and 3 million by 2040. Other government estimates have suggested an upper range of 5 million.[15] At that level, oil shale in the western United States would rival the largest oil fields in the world. Production levels could be maintained without the steep depletion rates that have affected other oil fields, perhaps for centuries. It is not clear that environmentally conscious President Obama will permit oil shale development on the federal lands where most of the oil shale is located. If and when he does, it would still be at least ten years before oil production could be expected.

Environmental Considerations

Mining of oil shale has a negative impact on the environment, as does all other mining. They include mountains of waste material, acid drainage, the introduction of metals into surface water and groundwater, increased erosion, sulfur gas emissions, carbon dioxide gas, and air pollution caused by the production of particulates during processing, transportation, and support activities.

In addition, ecological damage occurs wherever humans interfere in the natural setting. Rapid development of alternative energy resources is much more environmentally friendly than oil shale development.

Other Petroleum Products

About 5 percent of petroleum in the United States is used to manufacture things we use every day, such as plastics, house paint, toothpaste, disposable diapers, pills, lipstick, clothing, deodorants, DVDs, credit cards, shoes, rugs, computer and television screens, mattresses, and tires. It is not clear how these would be manufactured without petroleum as a feedstock. But because petroleum production at some level will no doubt continue far into the future, a solution to this problem is not a pressing issue.

Natural Gas: Better Than Petroleum

Natural gas is 96 percent methane (CH_4) and supplies 24 percent of the energy used in the United States and 23 percent of its electric power; industry uses the largest amount (figure 6.5). We produce 84 percent of our needs and import 16 percent but as was the case with petroleum, the percentage imported has increased steadily; it was only 4.7 percent in 1987.[16] Prices have quadrupled between 1995 and 2008. We use 27 percent of the world's production. Russia is the Saudi Arabia of natural gas, with more than a quarter of the world's reserves (table 6.5).

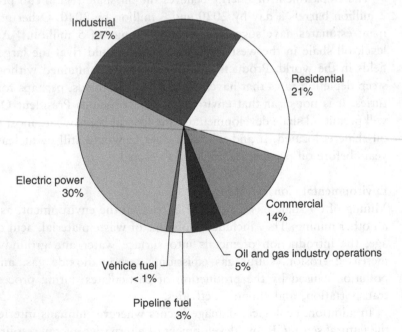

Figure 6.5
Use of natural gas in the United States, 2008. (Energy Information Administration, 2009)

Table 6.5
Distribution of world reserves of natural gas

Country	Percentage of world reserves
Russia	25.2
Iran	15.7
Qatar	14.4
Saudi Arabia	4.0
United Arab Emirates	3.4
United States	3.4
Nigeria	3.0
Venezuela	2.9
Algeria	2.5
Iraq	1.8
Indonesia	1.7
Norway	1.7
Turkmenistan	1.5
Malaysia	1.4
Australia	1.4
Egypt	1.2
Kazakhstan	1.1
China	1.1
Uzbekistan	1.0
Kuwait	1.0
Canada	0.9
Others	11.7

Source: British Petroleum (BP Statistical Review of Energy, 2008).

Natural gas is a relatively clean fossil fuel because methane combustion yields fewer pollutants than either oil or coal, with markedly lower emissions of carbon dioxide, nitrogen oxides, sulfur, and particulates. Compared with coal, natural gas allows a 50 to 80 percent reduction in greenhouse gas emissions, depending on application. It could serve as a transition fuel until nonpolluting sources of energy come online. In addition to its use as a fuel in power plants for generating electricity, natural gas is used to make cleaner transportation fuels such as ultra-low-sulfur diesel, and is used in the manufacture of artificial fertilizer to grow crops, some of which are used in ethanol production. It is currently the source of hydrogen in fuel cells for cars.

There is not yet an organization for natural gas like OPEC for petroleum, but an informal group of natural gas producers, the Gas Exporting

Table 6.6
Sources of natural gas imports for the United States

Country	Percentage
Canada	82.1
Trinidad	9.7
Egypt	2.5
Nigeria	2.1
Algeria	1.7
Others	1.9
TOTAL	100.0

Source: Energy Information Administration (2008).

Countries Forum, meets each year and will probably formally declare its solidarity in the near future. Many of the nations in the forum are the same ones that belong to OPEC—Iran, Qatar, United Arab Emirates, Nigeria, Algeria, and Venezuela. It is not a group that the United States would like to depend on for an essential resource.

America's natural gas imports are mostly from Canada and are delivered by pipeline (table 6.6). In part, the dominance of Canadian imports reflects a desire to avoid sources in the volatile Middle East. But mostly it reflects the fact that shipping gas by tanker ship across the ocean is a complicated and expensive process. To permit large amounts of natural gas to be transported across the ocean requires liquefying it. Liquefaction converts gas at room temperature to 1/600 of its original volume. Natural gas liquefies at −258°F. It must be cooled at the shipping terminal, pumped into heavily insulated cargo holds, and then heated as it is discharged at the receiving terminal. New receiving terminals for liquefied natural gas (LNG) are being built along America's coastline as fast as possible because of our soaring rate of imports. As of February 2009, nine terminals were in operation, 7 more were under construction, sixteen more were federally approved but not yet under construction, and nine more were proposed.[17] In early 2007, less than 10 percent of global natural gas production was transported as LNG.

Unconventional Sources of Natural Gas
Production of natural gas in the United States from conventional sources (subsurface wells) has been declining for a number of years, but our domestic supply has been fairly level because production of gas from unconventional sources has been rising. It now supplies more than 40

percent of our natural gas needs.[18] The major forms of unconventional gas are tight gas in sandstones and shales and coal bed methane, all of whose production has been rising.

Tight gas is gas produced from sandstones that have very low porosities and permeabilities. Even gas will not flow through them. The rock must be fractured repeatedly to release the gas. Until recently tight gas was uneconomical to produce, but technological advances and increasing prices for natural gas have made it the major unconventional source (66 percent).

Natural gas from shales forms 13 percent of America's supply of unconventional gas. In June 2009 the Potential Gas Committee, the authority on gas supplies, reported that the nation's estimated gas reserves are 35 percent higher than estimated previously, mostly because of an increase in the estimated amount of natural gas from shale rocks. Gas prices need to be at least $4.00 per thousand cubic feet to justify the cost of developing shale bed gas. In June 2009 the price was in the range of $4.00 to $4.50 range; in December 2009, it was up to $5.70 because of increased worldwide demand.

Like tight sandstones, extensive fracturing is required to make the shale productive. The fluid that is forced into the gas-bearing rock to fracture it is not plain water, but is loaded with more than two hundred chemicals to enhance the "fracking" ability of the fluid. The Endocrine Disruption Exchange has determined that many of them are dangerous to human health. Thirty percent of fifty-four tested chemicals used in the fluid are carcinogenic; 74 percent can cause respiratory damage, and 54 percent pose a danger to the blood and cardiovascular systems.[19]

Recent evidence indicates that people living in areas where unconventional gas is being exploited are suffering health problems as a result.[20] The black rock layer in the Catskill Mountains of New York, western Pennsylvania, and eastern Ohio known as the Marcellus Shale extends over 44 million acres and contains at least 350 trillion cubic feet of natural gas, enough to supply the entire U.S. demand for ten to fifteen years. But residents in the drilling area say their well water has become discolored or foul smelling, their pets and farm animals have died from drinking it, and their children have suffered from diarrhea and vomiting. Bathing in well water can cause rashes and inflammation, and ponds bubble with methane that has escaped during drilling, they say. Escaped methane has caused some well and tap water to become flammable.

According to a home owner living 500 feet from a drilling operation, "The smell and the rotten taste [of his water], you couldn't take a shower in it because the smell stayed on your skin, you couldn't wash clothes in it." Another resident said her five children were sick for months, until the family stopped using tap water for drinking or cooking. "They're fine all day at school, they come home, they get a drink of water, and that's when they got sick. They would have very, very severe stomach cramps, and double over, and throw up or have diarrhea." Another neighbor worried about her water when a relative showed her that her tap water could be ignited. "The flame from the jug of water was this high," she said, indicating about 20 inches, "and that's what my kids and our family have been drinking."

The companies consider the chemical composition of their fracking fluids proprietary. The composition of fracking fluid has been unregulated since the oil and gas industry won exemptions in 2005 from federal environmental laws including the Clean Water Act and the Safe Water Drinking Act. According to the Endocrine Disruption Exchange reports, about a third of fracking chemicals may cause cancer; half could damage the brain and nervous system, and almost 90 percent have the potential to harm skin, eyes, and other sensory organs.

In late 2009, New York's Department of Environmental Conservation analyzed thirteen samples of wastewater from drilling in the Marcellus Shale and found that they contain levels of radioactive radium-226 as high as 267 times the limit safe for discharge into the environment and thousands of times the safe limit for people to drink. Radium is a potent carcinogen that gives off radon gas, accumulates in edible plants, and takes 1,600 years to decay. The distribution of radon in wastewater from the widespread Marcellus Shale is unknown.

It should be noted that the pursuit of money is not restricted to the natural gas industry. Wayne Smith, a local farmer, said he made about $1 million in royalties over three years from gas taken from under his 105 acres. But he now wishes he never signed the lease and wonders whether tainted water is responsible for the recent deaths of four of his beef cattle and his own elevated blood-iron level.[21]

Coal bed methane forms 21 percent of the unconventional gas supply and is simply methane found in coal seams. It is formed from either a microbiological process or a thermal process stimulated by depth of burial. Several venture capital start-ups are experimenting with using microbes in the laboratory to turn coal into natural gas, as microbes do in coal mines. The microbes naturally metabolize coal. The theory is that if you can isolate and breed the very best methanogens (as they are called) and then inject them into coal seams and abandoned mines, you should get a stream of profitable gas back in return. It is another attempt to turn the ugly duckling of coal into the relatively white swan of natural gas, an energy alchemist's dream. But the realities of mining-by-microbe are no doubt a long way off, if ever, and coal pollution needs to be stopped as quickly as possible.

Natural Gas Hydrates

Perhaps the most exciting natural gas reserve on the production horizon are natural gas hydrates, deposits of methane trapped in ice-like crystals that are abundant both on land in Arctic permafrost regions at depths of a few tens of feet and at depths of 1,600 to 2,000 feet below the ocean floor. These methane hydrate accumulations look like normal ice but burn if touched by a flame. U.S. deposits are believed to contain about 150 times as much natural gas as known conventional gas reserves and about 900 times our current annual natural gas consumption.[22] There is no commercial production yet, but there probably will be in ten to fifteen years.[23]

About one-quarter of the world's hydrates are located within the borders of the United States, in places such as the North Slope of Alaska, the Gulf of Mexico, and off the coast of South Carolina.[24] The production of gas hydrates becomes economical when natural gas is priced at $4.00 to $6.00 per thousand cubic feet or when oil prices are at least $54.00 a barrel. According to the Department of Energy, if only 1 percent of the methane stored in the world's hydrates could be recovered, the world would be awash in natural gas.

Coal: The Worst of the Worst

Christmas tradition threatens a lump of coal in the stocking for those who were naughty. This could easily be done in the United States, where 28.6 percent of the world's supply of coal is located (table 6.7 and figure 6.6). We have about 250 billion tons of it, the equivalent of 800 billion barrels of oil, more than three times the oil reserves in Saudi Arabia. Coal supplies 22 percent of the energy needs of the United States.

Most coal is mined in Wyoming (38.4 percent), West Virginia (13.1 percent), and Kentucky (10.4 percent). We mined only 1.35 billion tons of our available 250 billion in 2009, so our reserves of coal are adequate for perhaps 150 years, depending on future production requirements. Coal supplies 23 percent of America's energy and 45 percent of its electric power (down from 57 percent in 1988) from more than 600 power

Table 6.7
Distribution of world coal reserves

Country	Percentage of world reserves
United States	28.6
Russia	18.5
China	13.5
Australia	9.0
India	6.7
South Africa	5.7
Ukraine	4.0
Kazakhstan	3.7
Poland	0.9
Canada	0.8
Brazil	0.8
Colombia	0.8
Germany	0.8
Others	6.2

Source: British Petroleum (BP Statistical Review of Energy, 2008).

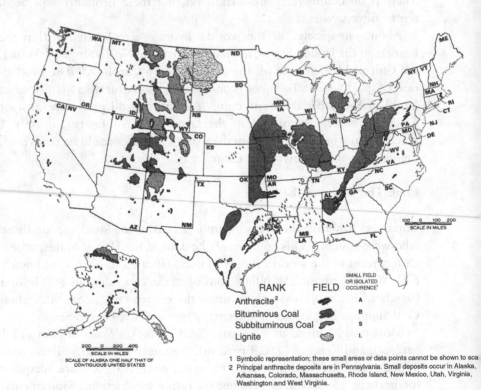

RANK FIELD SMALL FIELD OR ISOLATED OCCURENCE[1]

RANK	FIELD	SMALL FIELD OR ISOLATED OCCURENCE[1]
Anthracite[2]		A
Bituminous Coal		B
Subbituminous Coal		S
Lignite		L

1 Symbolic representation; these small areas or data points cannot be shown to sca
2 Principal anthracite deposits are in Pennsylvania. Small deposits occur in Alaska, Arkansas, Colorado, Massachusetts, Rhode Island, New Mexico, Utah, Virginia, Washington and West Virginia.

Figure 6.6
Coal in the United States, U. S. Geological Survey, 1996.

plants. A large coal-fired power plant can generate about 1 gigawatt of electricity if operated continuously, that is, with no shutdowns for maintenance—enough to power 1,200 homes in the Northeast but fewer homes in the South, where air conditioning is common.

It currently is the lowest-cost source of energy, at just 3 cents per kilowatt-hour. Coal is literally cheaper than dirt, about a penny or two a pound. Topsoil costs more than that. But if coal's heavy environmental costs such as pollution and health losses are added, the cost rises to 10 or 15 cents per kilowatt-hour, making it one of the most expensive sources of power.[25]

Mining

Thirty-one percent of America's coal is mined at the surface and 9 percent underground. Because of the extensive use of computer-regulated automation, coal mining today is more productive than in the past, and mining personnel must be highly skilled and well trained in the use of complex, state-of-the-art instruments and equipment. Many jobs require four-year college degrees. The increase in technology has significantly decreased the mining workforce from 335,000 coal miners working at 7,200 mines fifty years ago to 105,000 miners working in fewer than 2,000 mines today. The decrease in mine numbers and miners does not indicate a declining industry: coal production doubled between 1965 and 2005.[26] The Department of Energy forecasts that coal use will increase at the expense of oil during the next two decades.[27]

Surface Mining   Coal is mined at the surface if the seam is less than about 180 feet below the surface. However, if the coal seam is very thick, economical surface mining may extend to a depth of 300 feet. A particularly noxious surface mining practice in Appalachia is mountaintop mining, a process used when near-surface coal seams are thin but numerous. (About 14 percent of the coal used to generate electricity is obtained by mountaintop mining.) Enormous bulldozers and draglines scrape away mountain tops to expose the coal seams, and the coal is then trucked away. The leftover dirt and rock are dumped into adjacent valleys and streams. This method leaves ridge and hilltops as flattened plateaus and is highly controversial because it covers streams and disrupts ecosystems. Mountaintop mining has clipped the tops off 500 mountains and buried 2,000 miles of streams (figure 6.7). Environmentalists regard this as domestic terrorism by homegrown terrorists.

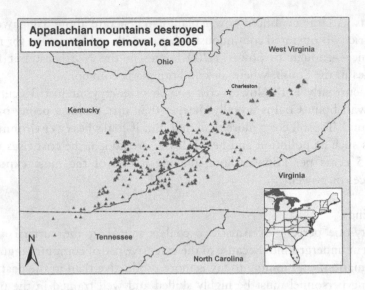

Figure 6.7
Locations of Appalachian mountains destroyed by mountaintop removal, 2005. (http://ilovemountains.org/resources)

Current federal law does not classify the rock waste dumped in the stream bed as a pollutant under the Clean Water Act. A court decision in 2009 gave the green light to many dozens of mountaintop mining projects that could destroy more than 200 miles of valleys and streams on top of the 1,200 miles that have already been obliterated. The federal government has always caved in to pressure from the coal mining industry and from the late Robert Byrd, West Virginia's senior senator.

Getting remedial legislation through Congress is not the only problem facing those who want mountaintop mining stopped, as Maria Gunnoe in West Virginia found out.[28] Her property has flooded seven times in the past nine years. In 2004 a spring rain turned a small creek in her back yard into a raging torrent 60 feet wide and 20 feet high. She blames the 1,200-acre mountaintop removal mine that has been leveling the ridge above her home.[29]

In 2007 the coalition she founded sued the Army Corps of Engineers to stop new mining at a mountaintop site near her home. Gunnoe gathered twenty local residents to join her in testifying against the site. But more than 60 miners showed up at the community hall to harass them; Gunnoe was the only resident willing to speak up.

The federal court ruled in her favor. Soon after, the mining company announced potential job losses at the site, and Gunnoe found her photograph on local "Wanted" posters, labeled "Job Hater." Her daughter's dog was shot. Friends heard rumors that Gunnoe too would be shot and her home burned with her children inside. Some nights she stays awake as a protective measure. She refuses to relocate. She feels she is right and refuses to give in.

During a supportive visit to a proposed mountaintop removal site, climate expert James Hansen said, "I am not a politician; I am a scientist and a citizen. . . . Politicians may have to advocate for halfway measures if they choose. But it is our responsibility to make sure our representatives feel the full force of citizens who speak for what is right, not what is politically expedient. Mountaintop removal, providing only a small fraction of our energy, should be abolished."[30] President Obama has chosen not to follow Hansen's advice and has approved forty-two of forty-nine planned removals of mountaintops by coal companies.

Subsurface Mining Underground coal mining in the early part of the twentieth century was very dangerous because of inadequate mine safety requirements (mine collapses, rock falls, gas explosions) and the health hazard known as black lung disease, in which coal dust penetrates deep into the miner's lungs, eventually causing emphysema, cancer, and often death. Coal mines today are required by law to be much safer, ventilated, and are heavily automated. But coal mining remains a dangerous occupation, and there are several dozen deaths each year.

Coal Rank
The abundance of domestic coal is not totally meaningful without consideration of its rank; higher ranks of coal generate more heat per volume. The rank of coal is its position in the series lignite, subbituminous coal, bituminous coal, and anthracite coal. America's reserves are 7 percent lignite (4,000–8,300 BTUs per pound), 46 percent subbituminous coal (8,300–13,000 BTUs per pound), 47 percent bituminous coal (10,000–15,500 BTUs per pound), and only a trace of anthracite coal (less than 15,000 BTUs per pound).[31] Production of each rank is about the same as its relative volume.

Western coal is of lower rank than eastern coal, and is therefore a less effective source of energy than eastern coal. Wyoming coal is mostly lignite and subbituminous and has, on average, only 8,600 BTUs of energy per pound. Eastern coal is mostly bituminous and generates over 12,000 BTUs per pound, meaning that power plants need to burn 50 percent more western coal to match the power output from eastern coal.

The advantage of western coal is that it contains less than a quarter of the amount of pollutant sulfur that eastern coal has. Lower ranks of coal, in addition to being less effective producers of heat, also leave more ash residue when they burn.

Coal Use

Ninety percent of coal use is to generate electric power. The United States uses 3.7 trillion kilowatt-hours of electricity per year, about 23 percent of total world use.[32] A small percentage of coal is used as a basic energy source in industries such as steel, cement, and paper. Only 7 to 8 percent of America's coal was exported in 2008, but the percentage is growing rapidly because world coal consumption is growing by more than 4 percent annually. It grew 30 percent between 2001 and 2007, twice as fast as any other energy source.[33]

There were 1,470 coal-fired power plants operating in the United States in 2007, and as of January 2009, 28 more were under construction, 7 additional ones had been approved and were about to start construction, and 13 more had applied for permits to build.[34] But power plants that burn coal are increasingly unpopular because coal is the world's dirtiest fuel. The 48 percent of power plants that are fueled by coal account for 93 percent of the emissions from the electric utility industry.[35]

About 8 percent of electricity generation in the United States comes from cogeneration, an integrated energy system that produces both electricity and heat (combined heat and power, CHP). The heat is recycled to provide another energy service. Most of America's CHP capacity is in industry rather than in power plants because of the relative constancy of industrial power needs compared to those of power plants. The United States has the potential to produce between 110,000 and 150,000 megawatts of electricity with CHP systems. Thermal heat is lost rapidly when transported, so CHP plants must be located near the point of use to be most effective.

Coal emits twice as much carbon dioxide as natural gas for the same amount of energy produced, ninety times the sulfur dioxide, and nine times the nitrogen oxides. Because of this, natural gas is currently the fuel of choice in new construction despite its much higher cost. Electricity produced using natural gas increased by 9 percent between 1997 and 2007. Two-thirds of the new generating capacity added in 2006 was fueled by natural gas. Using coal to generate electricity costs about $0.90 per 1 million BTUs, petroleum $7.00, and natural gas about a dime more than oil.[36] Economics rules.

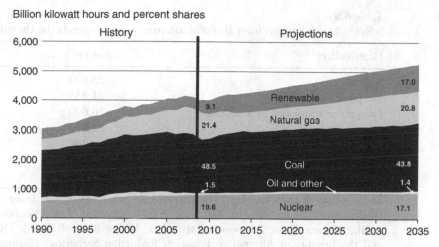

Figure 6.8
Predicted sources of electricity to 2030 (Energy Information Administration, 2009).

However, according to the Department of Energy the rapid increase in natural gas prices and in the volume of imports in recent years will soon cause a return to coal-burning plants (figure 6.8). Energy demand in the United States is growing at about 1.2 percent a year, and utilities must get energy from somewhere.[37] It is not clear that alternative and renewable sources such as cellulosic biofuel, wind, solar, nuclear, and geothermal can be brought online fast enough with enough generating power to accommodate this growth.

Environmental Cost of Coal

In order to promote the use of coal as an energy source, General Electric in 2006 started its "ecomagination" advertising campaign touting its new "clean coal" technology. One ad featured attractive scantily clad models (male and female) shoveling coal in a dark mine while Tennessee Ernie Ford's version of "16 Tons," played on the soundtrack. Near the end of the ad, a voice announced, "Harnessing the power of coal is looking more beautiful every day." The healthy and attractive models in the ad were not among the 23,600 Americans who die prematurely each year from black lung disease or from power plant pollution, most of it generated from burning coal (table 6.8).

When the Clean Air Act was passed in the 1970s, it included a grand-father clause that allowed older power plants to avoid meeting the modern pollution control standards that new facilities had to adapt. It

Table 6.8
Selected health impacts from air pollution from power plants in the United States

Health effect	Cases per year
Mortality	23,600
Hospital admissions	21,850
Emergency room visits for asthma	26,000
Heart attacks	38,200
Chronic bronchitis	16,200
Asthma attacks	554,000
Lost workdays	3,186,000

Source: Conrad G. Schneider, "The Dirty Secret Behind Dirty Air: Dirty Power" (Washington, DC: Clear the Air, June 2004), based on Abt Associates Inc., et al., "Power Plant Emissions: Particulate Matter-Related Health Damages and the Benefits of Alternative Emission Reduction Scenarios" (Boston: June 2004).

was anticipated that these older plants would soon be retired. However, because of the increasing need for electric power, many of these plants are still operating and have avoided installing modern pollution controls. As a result, most existing power plants today are between thirty and sixty years old and are up to ten times dirtier than new plants.[38] Nationwide, coal-fired power plants released an amount of carbon dioxide equal to the amount produced by 449 million cars (there are only 254 million cars in the U.S.) and accounted for 36 percent of carbon dioxide emissions in the United States in 2007.[39] Three-quarters of the emissions came from the plants built before 1980, about half of the nation's plants. Some are as old as seventy years. In addition to carbon dioxide, coal-fired power plants release 59 percent of total U.S. sulfur pollution, 50 percent of particle pollution, 40 percent of mercury emissions, and 18 percent of total nitrogen oxides every year.

Many of the coal-fired plants that have installed scrubbers in their chimneys in compliance with air pollution laws to clean the plant's emissions have simply transferred the noxious waste to the region's water supply. At an Allegheny Energy plant in Pennsylvania, every day since the scrubbing equipment was installed in June 2009, the company has dumped tens of thousands of gallons of wastewater containing chemicals from the scrubbing process into the Monongahela River, which provides drinking water to 350,000 people. No federal regulations govern the disposal of power plant discharges into waterways or landfills.

Even when power plant emissions are regulated by the Clean Water Act, plants that violate the law are not penalized. Since 2004, 90 percent of 313 coal-fired power plants that have violated the Clean Water Act were not fined or otherwise sanctioned by federal or state regulators.[40] Plants that were fined paid only trivial amounts.

Coal Ash

Coal plants generate 96 million tons a year of toxic fly ash, bottom ash, and scrubber sludge, combustion waste containing arsenic, mercury, and many other poisonous elements. Coal ash piles are so toxic and unstable that until June 2009, the Department of Homeland Security refused to divulge the location of the nation's forty-four high-hazard-potential coal sites. A high-hazard potential rating is related not to the stability of these impoundments but to the potential for harm should the impoundment fail. The department and the EPA feared that terrorists would find ways to spill the toxic substances.

The waste is pumped into active and abandoned mines for decades, into about 400 landfills, and 1,300 holding ponds at 156 coal-fired power plants in thirty-four states; only 26 percent of the ponds have liners at their base, so the waste can leak into streams, aquifers and drinking water. At least 137 sites in thirty-four states have poisoned surface or groundwater supplies from improper disposal of combustion waste (coal ash).

Spills from these ponds have killed or injured hundreds of people, most recently in Tennessee in December 2008 where 5.4 million cubic yards of coal ash accumulated since 1958 spilled from its earthen retention pond near Knoxville, destroying or damaging twenty-six houses and releasing more hazardous material than the *Exxon-Valdez* oil spill did. Three hundred acres of land were buried, roads and railroad tracks were covered, and sludge flooded into the nearby Emory River, attaining a depth of 30 feet in some places.[41] Residents have complained of health problems from the spill. The EPA called the spill one of the largest and most serious environmental releases in the nation's history. In the first four months of the cleanup by the Tennessee Valley Authority, very little of the ash was removed. The cost of the cleanup is approaching $1 billion, not counting lawsuits and penalties. The total cost will not be known for some time.

Long-term effects of coal ash, such as cancer from arsenic consumption or liver damage from exposure to cadmium, cobalt, lead, and other pollutants in coal ash are uncertain. EPA data show a disturbingly high cancer risk for up to one out of every fifty Americans living near wet ponds used to dispose of ash and scrubber sludge from coal-fired power plants in the United States. In 2000 the EPA said it wanted to set a national standard for these holding ponds but has since backed away from regulating the wastes because of fierce opposition from the coal industry and utilities.[42] As a result, coal ash ponds are subject to less regulation than landfills accepting household trash. This may soon change. The current EPA administrator planned to issue new coal ash regulations by the end of 2009, but they had not appeared by mid-2010.

An excellent example of the difficulty of passing needed legislation in Congress is provided by the public's expressed desire to decrease the use of highly polluting coal. California gets only 1 percent of its electricity from coal-fired plants. A congressional representative from the state proposed a climate change plan in 2009 that would put a price on carbon dioxide emissions, a concept of charging for pollution favored by the Obama administration in Washington. Coal-producing states and states that rely heavily on coal-fired power plants are opposed to the California bill because it would harm their economies. West Virginia, Indiana, Kentucky, and Wyoming are among the most dependent on coal, with more than 90 percent of their electricity generated in-state coming from coal-burning plants. They, the Midwest, and the Plains states will lead the opposition to the bill. As an executive at a large coal-producing company noted, California, the Northwest, and New England are the ones driving the legislation because they're the ones that aren't going to have to pay for it. As these three groups of states have many fewer than half the votes in the House of Representatives, the bill from the California congressman faces an uphill battle.

Legislating changes in entrenched but outmoded ways of doing things is always difficult, whether it is the American habit of eating meat and dairy products, driving gas-guzzling cars, eliminating the use of pesticides in agriculture, or decreasing the use of polluting energy sources. If voters do not make such changes the most important consideration in their voting choices for representation in their local, state and federal governments, change is unlikely to happen.

Mercury

Coal-burning plants are the largest source of mercury in the atmosphere.[43] Perhaps half of the mercury spreads considerable distances, while the rest is deposited locally, creating so-called hot spots. The primary route of human exposure is fish consumption. The mercury is deposited from the atmosphere and is converted by bacteria in sediments to methylmercury. This chemical is taken in by fish that are subsequently eaten by humans and birds. Nearly every state has issued fish consumption advisories, especially for pregnant women because fetuses are vulnerable. A survey by federal scientists in 2009 found mercury contamination in every fish sampled in 291 streams around the nation, and a quarter of those fish had mercury levels higher than those considered safe for human consumption. The Centers for Disease Control has found that 6 percent of American women have mercury in their blood at levels that would put a fetus at risk of neurological deficits such as brain damage, mental retardation, blindness, seizures, and inability to speak. American women face a 10 to 15 percent risk of bearing children with mercury levels high enough to slow their mental development.[44] Some 200,000 to 400,000 children born in the United States each year have been exposed to mercury levels in utero that are high enough to impair neurological development, according to data gathered by the National Health and Nutrition Examination Survey. In a 2009 study, UCLA researchers reported that deposits of mercury in the bodies of Americans are increasing at an alarming rate.

Nine of the top ten mercury-emitting power plants reported an increase in mercury emissions in 2007 compared to 2006. As of 2010 there are no federal rules restricting mercury from coal-burning power plants.

Newer coal plants capture most of the particulates that used to pour from the smokestacks of coal-burning power plants, but a significant percentage of these particles still spew into the air.[45] These particulates have been shown to increase respiratory and cardiac mortality, increase asthmatic distress, and have other negative effects on human lungs. Coal fly ash and sulfur and nitrogen oxides form a major portion of the particulate matter. Coal-fired plants produce nearly two-thirds of the sulfur dioxide emissions that are responsible for much of the acid rain in the eastern United States.

Coal also contains low levels of naturally occurring radioactive isotopes such as uranium and thorium. The elements are present in only trace amounts, but enough coal is burned for power that significant amounts of these substances are released (figure 6.9). When coal is burned, the uranium and thorium are concentrated by a factor of 10 in the resulting fly ash. The radioactive emission from a 1,000 megawatt

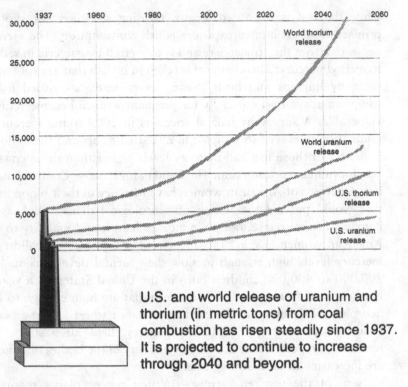

U.S. and world release of uranium and thorium (in metric tons) from coal combustion has risen steadily since 1937. It is projected to continue to increase through 2040 and beyond.

Figure 6.9
Releases of uranium and thorium from coal combustion in the United States and the world since 1937, with estimated 2040 figures based on the projected increase in the use of coal (Oak Ridge National Laboratory, 1993).

coal-burning power plant is 100 times greater than a comparable nuclear power plant with the same electrical output; including processing output, the radiation output from the coal power plant is more than three times greater.[46] Coal-burning power plants are not regulated for their radioactive emissions.

The energy content of nuclear fuel released in coal combustion is greater than that of the coal consumed. Americans living near coal-burning power plants are exposed to higher radiation doses than those living near nuclear power plants that meet government regulations.[47] If radiation emissions from coal plants were regulated, their costs would increase and coal-fired power would be less economically competitive.

Equally objectionable is the carbon dioxide that coal plants pump into the air—35 to 40 percent of all the carbon dioxide released in the United States, more than the emissions from all the cars and trucks on American roads.[48] Coal-burning power plants are the major cause of global warming and should be replaced as soon as possible. Coal power plants are the least carbon-efficient power stations in terms of the level of carbon dioxide produced per unit of electricity generated.

"Clean Coal"

Coal producers and industries that rely on coal have for some years been promoting a concept they call "clean coal." There is no such thing. Coal is inherently a dirty fuel, and there is no way to make it clean at any acceptable cost. As Al Gore put it in 2008, "Clean coal is like healthy cigarettes."

Clean coal is a theoretical concept whereby the carbon dioxide produced by the burning coal is captured and disposed of so it does not enter the atmosphere. Conventional pulverized coal plants burn coal in air, producing an effluent composed of 80 percent nitrogen, 12 percent carbon dioxide, and 8 percent water. The carbon dioxide can be chemically separated from the nitrogen, then extracted and compressed for injection into storage locations. This uses energy—roughly 30 percent of the energy from burning the coal in the first place—and may raise the cost of generating electricity from coal by 50 percent.[49] As of the end of 2009, there were no operating clean coal commercial power plants in existence, although a small pilot plant started operating in eastern Germany in September 2008.[50] The carbon dioxide will be captured, liquefied, and transported by truck 150 miles from the plant to be injected 10,000 feet underground into a depleted gas field. Ideally, in the future, the gas will be carried by pipeline to underground storage. Vaclav Smil of the University of Manitoba, Canada, has calculated that handling and transporting just 10 percent of today's carbon dioxide emissions would require more pipelines and other equipment than is now used worldwide to extract oil from the ground.[51]

The storage reservoir proposed by the coal enthusiasts is the subsurface. Simply pump the carbon dioxide below ground where, the coal companies believe, it will remain indefinitely. Unfortunately, subsurface injection has been used legally for many years to dispose of toxic waters produced by oil refineries and chemical plants, and the result has been aquifer contamination in at least twenty-five states. The geology of rocks in the subsurface can never be known well enough to ensure that leakage will not occur.

And then there is economics. On January 30, 2008, the administration of President George W. Bush announced it was terminating a joint project with thirteen utilities and coal companies to build a demonstration coal-fired power plant in Illinois with underground carbon sequestration because of massive cost overruns. The original cost of $950 million in 2003 had climbed beyond $1.5 billion by early 2008, with further rises on the horizon.

What began as a minor annoyance to coal producers has evolved into a near-national consensus against the use of coal. Opposition has come from environmental, health, farm, and community organizations and an increasing number of state governments. The public at large has turned against coal as well. In a September 2007 national poll by the Opinion Research Corporation about which electricity source people would prefer, only 3 percent chose coal.[52]

China

Because coal is cheap and abundant, it has become the fuel of choice in developing countries, particularly in China and India, the largest two nations on earth. China has the world's third-largest coal reserves, after the United States and Russia. China produces 80 percent of its electricity from coal and burns more of it than the United States, European Union, and Japan combined—39 percent of annual world coal consumption. Coal consumption in China doubled between 2000 and 2008.

China burns more than twice as much coal as any other country and intends to increase its dominance in this arena. At the start of 2007, about 550 additional coal-burning plants were under construction in China, and the expectation is that it will open one new coal-fired plant each week over the next five years, an incredible pace of construction.[53] The country plans to increase coal production by 30 percent by 2015. The polluting emissions from the increase will swamp any emission reductions elsewhere.

Experts say the least efficient plants in China today convert 27 to 36 percent of the energy in coal into electricity. The most efficient plants achieve efficiencies as high as 44 percent, meaning they can cut global warming emissions by more than a third compared with the weakest plants. In the United States, the most efficient plants achieve around 40 percent efficiency. But by continuing to rely heavily on coal, China ensures that it will keep emitting a lot of carbon dioxide; even an efficient coal-fired power plant emits twice the carbon dioxide of a natural gas-fired plant.

China is the world's largest emitter of mercury. Estimates are that half of the atmospheric mercury in the United States comes from overseas, especially China.[54] The trip across the Pacific takes five to ten days. Soot, toxic chemicals, and climate-changing gases from the smokestacks of China's coal-burning power plants are now one of that country's most significant exports. The country's emissions doubled between 1996 and 2006. China is now the largest emitter of carbon dioxide, emitting 24 percent of the world's total, surpassing the United States at 21 percent. The increase in global-warming gases from China will probably exceed that for all industrialized countries combined by 2030, surpassing by five times the reduction in such emissions that the Kyoto Protocol seeks.[55]

China is racing toward economic development on the back of coal, and the rest of the world is paying part of the price. The United States raced toward its present affluence for the past hundred years on the back primarily of oil and secondarily on coal at a time when pollution and other environmental concerns were not on America's, or the world's, agenda. In a sense, we were fortunate that no one cared about our environmental crimes. We were at the wrong place at the right time. China is at the wrong place at the wrong time.

China is not only increasing its use of coal, but is developing alternative energy as well. Renewable sources provide 8 percent of the country's energy today, but Beijing aims to increase that to 15 percent by 2020. In comparison, the United States hopes to generate 10 percent of its energy from renewables by 2012. The Chinese in mid-2009 said they plan to have 150 gigawatts of wind power capacity by 2020. In comparison, in the first three months of 2010 only 0.5 gigawatts were added in the United States. The 150 gigawatt figure is six times the target the government established only eighteen months earlier. Solar and nuclear power are also being expanded. China is developing its energy-generating potential at an incredible rate. It can do this partly because it does not have the kind of democratic complexities that Western countries do, and so it can move more quickly and without the kind of resistance from narrow economic interests that might make it more difficult.

It seems clear that world air pollution is going to rise for many decades to come, regardless of policy in the United States. And, as noted earlier, air circulates.

Summing Up: What Should We Do?

It is apparent that America's supply of fossil fuels can last long past the lifetimes of children not yet born, assuming the political will and the money are available to exploit unconventional sources. Both the supply of domestic coal and the reserves of oil in Western oil shales are projected to last for more than a century, and natural gas from methane hydrates may be correspondingly large, although there is not yet a reliable estimate of reserves. Producible fossil fuels are still abundant in the United States; in 2006 they supplied 70 percent of the nation's electricity.

The key issue is whether we want to use them. They are responsible for the air pollution in our cities and are thought by most scientists to be largely responsible for such climate changes as global warming, changes in precipitation patterns, the increasing severity of hurricanes along our Gulf Coast and Atlantic coastlines, and the rise in sea levels. Political realities and the time needed to develop the infrastructure for alternative energy sources mandate that fossil fuels will be our main source of energy for many decades to come.

Alternative energy sources are abundant but are of small importance at this time. Their research and development require time and considerable money from the government during their early stages. The Obama administration is well aware of these facts and is determined to replace fossil fuels with alternatives to the extent possible. But it will not happen overnight. The bottom line on fossil fuel use was well expressed by Senate Majority Leader Harry Reid in June 2008: "The one thing we fail to talk about is those costs that you don't see on the bottom line. That is, coal makes us sick, oil makes us sick. It's global warming. It's ruining our country. It's ruining our world. We've got to stop using fossil fuel. We have for generations taken carbon out of the earth and put it in the atmosphere, and it's making us all sick. It's changing our world."

Conservation: A Pollutant Saved Is a Pollutant Earned

As Steven Chu, Nobel Prize–winning physicist and current secretary of energy, believes that conservation and energy efficiency will remain for the next couple of decades the most important thing the world can do to get on a sustainable path. "If I were emperor of the world, I would put the pedal to the floor on energy efficiency and conservation."[56] It is a method that can be implemented immediately and have a noticeable effect, unlike the development of renewable energy sources that will take many decades to have a major impact, regardless of federal government policies.

Current government financial policies that foster conservation, such as increasing the use of renewable energy sources and the construction of energy-efficient buildings, are essential. But additional measures are needed, such as decoupling the profits of electric utility companies from the amount of power they sell by introducing a regulatory formula that rewards utilities for providing the best service at the least cost. California regulators have made this change, and as a result of this and other policies, Californians use less than half as much electricity per person as other Americans do; their per capita usage is comparable to that of Japan, Germany, Belgium, and Denmark.[57] California's per capita usage is matched by New York, Rhode Island, Oregon, Vermont, and Washington State. Many Americans, however, believe that if we reduce our carbon footprint, we will compromise our quality of life and end up drinking warm beer in a dark room. However, the per capita usage in these six states proves that it is possible to have a U.S. lifestyle with a European-sized carbon footprint. If all other states were to emulate these six, the nation would go a long way toward the nation's national emissions goals. The potential for energy efficiency (conservation) in the United States is greater than anywhere else in the world, according to a global consulting firm.[58]

Shower Heads   Probably the most easily implemented conservation measure is the installation of low-flow shower heads, which not only save freshwater but also reduce the need for heating water using fossil fuels. A new shower head costs only a few dollars and takes only a few minutes to install. An additional saving on heating water can be achieved by taking shorter showers—say, three minutes of running water instead of fifteen.

Compact Fluorescent Light Bulbs   Another inexpensive and easy way to conserve fossil fuel is by removing Thomas Edison's incandescent light bulbs and replacing them with energy-efficient compact fluorescent lamps (CFLs). Ninety percent of the energy that Edison's bulbs emit is in the form of heat, not light. The CFLs consume 70 to 80 percent less electricity and last six to ten times longer on average. A 20 watt CFL is equivalent to a 100 watt regular bulb and the 11 watt is equivalent to a 60 watt incandescent bulb. A CFL is more expensive to buy than an incandescent bulb, but because of its long life, it more than overcomes the extra purchase cost over the lifetime of the bulb. In addition, you save 70 to 80 percent of your electricity bill on light. CFLs are a sound financial investment in addition to reducing the need to pollute the air, water, and our bodies by burning coal, natural gas, and oil. Lighting accounts for 17 percent of global power consumption.[59]

The Department of Energy has estimated that if every household in the United States switched one of Edison's bulbs to a CFL, the country would save enough energy to light 2.5 million homes and reduce greenhouse gas emissions equal to 800,000 cars (0.3 percent of all the motor vehicles on American roads). Another way to envision the saving from CFLs is that replacing just one incandescent bulb with a CFL reduces the need for 500 pounds of coal.[60] Americans bought 21 million CFLs in 2000, 100 million in 2005, and 400 million in 2007.[61] There are 112 million households in the United States, so each household replaced more than three incandescent bulbs on average. The U.S. Congress has set new efficiency standards that will make today's incandescent bulbs obsolete by 2014.

CFL bulbs contain a small amount of mercury, much of which is released if the bulb breaks. Open windows to dissipate the mercury vapor. Then, wearing gloves, use sticky tape to pick up the small pieces and powdery residue from the bulb's interior. Place the tape and pieces of the bulb in a plastic bag. After vacuuming the area, place the vacuum bag inside doubly sealed plastic bags before discarding. Unbroken bulbs that have exceeded their lifespan should be recycled, not thrown into a landfill where they may be broken.

The next advance in lighting will be light-emitting diodes (LEDs), light generators that are twice as efficient as CFLs and last twenty years.[62] A fluorescent light bulb might last 3,000 hours while an LED fixture lasts more than 100,000 hours. The current downside to LEDs is their cost, about $100, as opposed to $7 for a regular bulb, so it would take five to ten years to recoup the money in energy savings at today's electricity prices. But the cost is decreasing rapidly, so the sticker shock will quickly become less intimidating. Studies suggest that a complete conversion to LEDs could decrease carbon dioxide emissions from electric power use for lighting by up to 50 percent in twenty years. Some cities and expensive hotels in the United States, as well as cities in China and Italy, and Buckingham Palace in England have installed LEDs as replacements for burned-out bulbs. Note that these municipalities and the palace do not have the price constraints that individuals do. Cities can raise taxes, hotels can increase their room prices, and the queen's budget is a bit larger than yours or mine.

Dwelling Modifications   Many modifications to houses or apartments will conserve energy and reduce both the use of fossil fuels and electric bills. Perhaps the least costly is to place weatherstripping along the sides of a doorway. Installing double-pane windows will generate additional savings, as will adding insulation. The latter two changes will probably cost hundreds of dollars, but with the ever-increasing cost of heating oil and natural gas, the money spent will be recouped in a fairly short time.

The biggest opportunity to improve the energy situation in the United States would be a major investment program to make homes and businesses more efficient. Homes would account for about 35 percent of the potential efficiency gains, while the industrial sector accounts for 40 percent and the commercial sector 25 percent.[63] Stricter building codes for energy conservation have been fought by the powerful builders' lobbies, which contend that they can add $2,000 or more to the cost of a house. Nevertheless, in a few cities and in California, tough new rules have been adopted.

Heating water in a typical single-family home accounts for about one-fourth of the home owner's total energy use. Up to 85 percent of the water heating cost can be saved by installing a solar heating system on the roof. In Israel they are required by law, and all dwellings have them. There is backup electrical heating for the winter months. In the United States, a solar heating system costs about $2,000 installed (only $1,000 in Israel, possibly because of mass production).

Transportation Motor vehicles are responsible for almost a quarter of the annual emissions of carbon dioxide in the United States, so doing something about how we move from place to place can make a significant impact. Americans are wedded to their cars, a tie that has a major negative impact on both their wallet and the environment. Car ownership accounts for 15 to 22 percent of all household expenditures, and vehicles powered by fossil fuels are essentially pollution-generating, global-warming factories on wheels. Two-thirds of the oil used in the United States is for transportation. Each day we use 350 million gallons of gasoline, spewing enormous amounts of pollutants into our air and our lungs (see chapter 10).

Changing the car you drive, however, is the most expensive green change you can make. As everyone is aware, smaller is better, and hybrid cars are better for the environment than those that run only on gasoline. Unfortunately, new cars are expensive, even with a government tax credit for the purchase of hybrids, and most people buy a new car only when necessary—about every fifteen to twenty years on average. Compounding the problem is the much higher cost of hybrid cars. Savings can be made with a gasoline-powered vehicle by proper maintenance, such as keeping the correct tire pressure, keeping the odometer at 50 to 60 mph on open roads, and keeping the car tuned up. In 2010 the government specified that by 2016 auto fuel efficiency must be 35 miles per gallon, a 40 percent increase over existing requirements.

Can the emissions from the tailpipes of private cars be stopped? Probably not completely, although the high cost of gasoline in 2008 and the weak economy in 2008 and 2009 did reduce the number of miles most Americans drove. A recent Harris poll showed that 92 percent of Americans believe that gasoline prices will trend upward in the future, perhaps indicating that the reduction in miles driven may be permanent.

But emissions certainly could be permanently reduced by an increase in the availability of public transportation. At present, only 5 percent of U.S. commuters take public transportation because most lack convenient access.[64] Fewer than 5 percent of Americans live within a half-mile of light-rail transit, probably as far as most of us would walk. Only half of Americans live within a quarter-mile of any public transportation. Nevertheless, the number of passengers on public transportation has increased steadily in recent years—25 percent between 1997 and 2007.[65] Ridership in 2007 and 2008 was the highest in fifty years, and 2008 saw a 4 percent increase over 2007.[66] More than 300 metropolitan areas experienced an increase in public transit between 2007 and 2008 as gasoline prices soared. If public transportation were more easily available, there is a strong possibility that ridership would increase. Bus and rail officials say support for public transport goes beyond the increase in ridership. On election day in 2008, voters around the country passed twenty-five of thirty-three ballot initiatives to increase local and state taxes for public transportation.[67] When people vote to increase their taxes, they are serious about an issue.

There is one form of public transportation that is not beneficial to the environment, commercial aviation, and air traffic is expected to double between 1997 and 2017. Jet fuel accounts for 10 percent of America's fuel consumption. A jumbo jet burns 3,250 gallons of this refined hydrocarbon per hour of flight, spewing large amounts of greenhouse gas pollutants into the atmosphere. And thousands of commercial aircraft are flying every day. Commercial aircraft currently generate 3.5 percent of greenhouse gas emissions.

Oceangoing ships are polluters as well. As international trade has exploded and shipping capacity has grown by 50 percent, cargo ships have become a major source of greenhouse gases. Passenger ships are far less abundant than cargo ships, but they are also polluters. The *Queen Mary II* burns 13,000 gallons of diesel fuel per hour. In comparison, a car burns about 2 gallons of gasoline per hour.

One form of public transportation that has been neglected until recently is streetcars, which were common in many cities in the first half of the twentieth century but have since almost disappeared. San Francisco is one of the few cities that never got rid of its streetcar system. Most Americans, in fact, have probably never seen a streetcar. They run on an overhead electrical wire and carry up to 130 passengers per car on rails that are flush with the pavement. Because they can pick up passengers on either side, they can make shorter stops than buses do. Portland, Oregon built the first major modern streetcar system in 2001 and has since added new lines interlaced with a growing light rail system. Residential building has boomed within two blocks of the streetcar line. Other modern streetcar systems have been built in Seattle and Tacoma, Washington, and Charlotte, North Carolina. Other cities are considering constructing them as well.

The United States may be at the start of a change in intracity transportation that was almost inconceivable only a few years ago, with the possibility that public transport may make a real dent in the use of personal automobiles.

Setting an Example: Leadership Is Needed

The federal and state governments in the United States should lead the way to alternative energy by adopting it themselves. All incandescent and fluorescent light bulbs should be removed in government buildings and replaced with compact fluorescent bulbs. Their buildings should be retrofitted with appropriate insulation, weather-stripping, and double-pane windows. Contractual arrangements for heating and cooling should be made only with utilities that use alternative energy sources to the extent possible. Solar panels should be mandatory on building roofs. All grassy lawns should be eliminated and planted with trees and perhaps organic gardens, as has been done with a small plot of land around the White House. All government vehicles should be hybrids or all-electric.

Secretary of Energy Steven Chu and President Obama have announced that photovoltaic solar panels and solar thermal apparatus will be installed on the White House roof by spring 2011.

7

Alternative Energy Sources: Energy for the Future

But times do change and move continually.
—Edmund Spenser, *The Faerie Queene*, 1589

Numerous sources of commercial energy other than fossil fuels are available, but only recently have they received more than scant attention from the government because of the ready availability and low cost of coal, natural gas, and oil, and the low cost per kilowatt-hour of electricity produced from power plants fueled by these fossil fuels. However, because of environmental concerns about coal and the rapidly rising prices for oil and natural gas, it is apparent that substitutes for these venerable fuels are needed. The Department of Energy forecasts that renewable alternatives will provide 21 percent of America's energy by 2030, with wind and biomass enjoying the greatest growth.[1] Most publicity has been focused on biofuels, wind power, and solar power, but other energy sources will compete with them in certain areas because of geological and geographical constraints.

In the United States as of June 2009, twenty-eight states and Washington, D.C., had established renewable energy portfolios and mandates, but not the federal government. Another thirteen states were considering joining the group. Meanwhile, Germany, Spain, and Japan have erected far more alternative energy installations, despite the fact that these countries are much smaller than the United States and have fewer options than we do. In 2008 more renewable energy than conventional energy was added in both the European Union and the United States.

Alternative Energy and Geography: Where Should We Put It?

Solar power is not the best energy source in Maine; wind will not be the mainstay of power in South Carolina; hydrothermal energy is not a good choice in Kansas; New Mexico will not depend on biomass; and geo-thermal energy is not the best bet in Ohio. But these sources of energy can be generated in other areas of the country and, assuming electrical transmission lines are available, can be transmitted to wherever they are needed, so geography is not an insurmountable barrier to the use of any of these alternative energy sources anywhere in the United States. But adequate transmission lines will not be available for many years to come.

However, it is not clear that coast-to-coast transmission lines are needed because renewable energy sources exist everywhere in the country. The West has abundant supplies of geothermal energy; the Southwest can develop solar power; the Northeast has strong winds offshore; the Midwest is rich in land-based wind power; several areas have abundant hydrothermal resources for power. In each area, developing these power sources and using them locally would be less expensive than piping in clean energy from thousands of miles away. The cost of building new power lines is estimated to be $2 million to $10 million per mile. And the farther electricity is transported, the more of it is dissipated, analo-gous to friction losses in objects moving along the ground. This "line loss" gobbles up an estimated 2 to 3 percent of electricity nationally.[2]

Current Development: Where Are We Now?

The tide of energy sources has begun to change across the country. Alternative and renewable resources made up 11 percent of U.S. energy sources in 2008 and are on a sharply upward trajectory, led by increases in wind power and biomass.[3] Renewables accounted for over 50 percent of new capacity in 2009. The federal government also forecasts that nonhydropower renewable power will meet one-third of total generation growth between 2007 and 2030, with natural gas from domestic uncon-ventional sources supplying most of the remaining growth (figure 7.2). Oil use is forecast to stagnate, and oil imports are forecast to decline, as will natural gas imports.

Clean energy is a major focus of President Obama's administration, the first administration since the Arab oil embargo in 1973 that has identified sustainable energy as an important goal. Although President Carter had declared oil independence the "moral equivalent of war," nothing was accomplished in his administration. In the stimulus bill passed in 2009, the administration inserted about $70 billion in grants, loans, and loan guarantees for Department of Energy head Steven Chu to award for high-tech research and commercial projects for renewable energy. This is nearly three times more than the baseline Energy Department budget and more than the annual budgets of the Labor and Interior departments combined. Energy's baseline budget received a 10 percent funding increase, solar energy research received $175 million, biofuels $217 million, hydropower $40 million, and additional funds were allocated for weatherization of buildings, energy efficiency, and vehicle technology.

Claims that it is not the government's place to support the development of new renewable energy technologies distort the country's history by denying the crucial role that government support has played in building the nation's current energy infrastructure. These claims are clearly motivated by the desire of the fossil fuel industries and their lobbyists to maintain their dominant position in the energy field.

The future looks bright for the development of alternative energy in the United States (figure 7.1) but the job of changing America's energy industry is immense. The energy industry, based on fossil fuels, is not only politically powerful, it is very large and hard to replace. As noted in chapters 2 and 6, the nation has about 500 coal plants that produce nearly half of the country's energy, thousands of miles of low-voltage electricity lines, more than 10 million old fashioned meters, and 354 million cars and trucks that run on diesel or gasoline with mediocre efficiency.

According to the Energy Information Administration, 71.7 percent of the electricity in the United States in 2009 was generated by power plants fueled by fossil fuels: coal, 44.9 percent; natural gas, 23.4 percent; and oil, 1.0 percent. Nuclear power (20.3 percent) is the dominant alternative energy source. Most of the rest is provided by hydroelectric plants. The residential cost of electricity at the end of 2008 averaged 10.9 cents per kilowatt-hour (12 percent higher than in 2004), but is variable among states because of the mix of energy sources used in different areas (table 7.1). New England is the most costly area, the northern midcontinent the cheapest.

Renewables in Other Countries

The United States is not alone in its increasing emphasis on renewable and nonpolluting energy development. At least eighty-three countries now have national targets for renewable energy.[4] At least sixty countries have policies to promote renewable electricity, and at least fifty-three countries, states, and provinces have biofuels mandates. The European Union planned to have 21 percent of its electricity generated by renewable energy by 2010. As of mid-2010, only seven of the twenty-eight EU

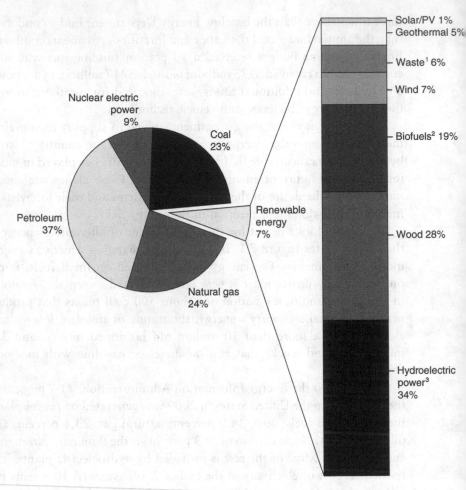

Figure 7.1
Forecast of growth in nonhydropower renewable energy. One-third of total growth in America's energy supply is forecast to come from these renewable sources between 2007 and 2030. (Energy Information Administration, 2008)

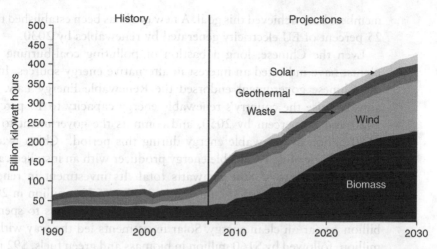

Figure 7.2
Nonhydropower renewable power meets 33 percent of total generation growth between 2007 and 2030 (Energy Information Administration, Annual Energy Outlook 2009 Reference Case Presentation December 17, 2008).

Table 7.1
Cost of electricity in the most expensive and least expensive states (cents per kilowatt-hour)

Most expensive		Least expensive	
Hawaii	29.3	Wyoming	7.9
Alaska	16.0	South Dakota	7.7
Connecticut	19.4	Oklahoma	7.6
Rhode Island	18.8	Washington	7.5
Massachusetts	18.2	Missouri	7.2
New York	17.1	Idaho	7.1
		Nebraska	6.9
		North Dakota	6.9

Source: Energy Information Administration (2009).

members have achieved this goal. A new goal has been established to have 25 percent of EU electricity generated by renewables by 2030.

Even the Chinese, long a bastion of polluting coal-burning power plants, have indicated an interest in alternative energy sources. In 2005 the Chinese government endorsed the Renewable Energy Law, which aims to raise the country's renewable energy capacity to 15 percent by 2020 and 30 percent by 2050, and commits the government to invest $180 billion in renewable energy during this period.[5] China is already the world's leading renewable energy producer with an installed capacity of 152 gigawatts of 800 gigawatts total. Its investment in renewable energy increased from $352 million in 2007 to $736 million in 2008 to $34.6 billion in 2009. The country has announced plans to spend $75 billion a year on clean energy. Solar investments led the way with $391 million, followed by $160 million in biomass and green fuels, $92 million in wind, and $39.3 million in technologies that increase energy efficiency. Attaining the goal of 15 percent renewable energy in China will be difficult because of the breakneck speed at which the country's energy consumption is growing.

A 2007 government edict requires that large power companies in China generate at least 3 percent of their electricity by the end of 2010 from renewable sources and 8 percent by the end of 2020. As a result, installed wind capacity increased from 2.3 gigawatts in 2005 to 13.8 gigawatts in 2009 and is expected to reach at least 30 gigawatts by 2020. A centralized government such as China's can accomplish changes quickly once decisions are made. The requirement for the increase in renewable energy excludes hydropower, which already accounts for 21 percent of Chinese power and 95 percent of its renewable energy portfolio, and nuclear power, which accounts for 1.1 percent. The government is pushing the development of alternative energy as fast as possible because of severe pollution from the country's coal-based manufacturing growth engine.

However, China suffers from the same lack of adequate transmission lines as the United States, a fact that will handicap plans for the development of alternative energy. Wind turbines are being built faster than the national grid can erect high-voltage transmission lines to carry the electricity elsewhere, the same problem facing the United States.

Employment
Some adjustments in employment will be required as the United States makes the transition to a cleaner energy future. As new jobs are created in new industries, others will be lost, requiring retraining of

displaced workers. Worker displacement is not a new phenomenon and has happened repeatedly in the nation's history because of new technologies. Whalers laid down their harpoons and found other professions during the transition to petroleum oil and then electricity for lighting. Blacksmiths lost most of their business when metal objects could be mass-produced and the automobile replaced the horse and wagon as the dominant method of transportation. Farmers were forced to retrain as their numbers dropped from a high of 29 million in 1900 to under 1 million today. Movie careers ended for many foreign-born stars with thick accents when "talkies" replaced silent films in 1929. Many radio celebrities were unable to adapt to the visual medium of television in the 1940s and 1950s and had to find new occupations. The occupational displacements resulting from the change from fossil fuels to alternative sources of energy is simply the latest in an unending stream of such displacements. These are the realities of employment in a vibrant and ever-changing economy.

In fact, the transition to alternative energy sources will have a net benefit in terms of job numbers.[6] According to estimates by the Lawrence Berkeley National Laboratory and the Renewable Energy Policy Project, solar energy generates thirty-two jobs per megawatt, and wind produces approximately six to nine jobs per megawatt, whereas natural gas and coal create only two to two-and-a-half jobs per megawatt. As of September 2009 about 85,000 people were employed in the wind energy industry, up from 50,000 only a year earlier. These employment numbers are in addition to the tens of thousands of jobs that will be created by renovation of the nation's electricity transmission grid. The nation has many reasons to welcome the era of alternative energy: cleaner air, cleaner water, cleaner soil, and an increased job market.

Energy from Plants: Biological Energy

Biomass currently provides 53 percent of the renewable energy in the United States, more than half of it directly from wood and the remainder from biofuels, municipal solid waste, and animal manure. Energy from plants, from algae to trees, is forecast to be the fastest-growing energy source between now and 2030, increasing from 40 billion kilowatt-hours to 230 billion kilowatt-hours. Biomass can be used for power production, fuels, and products that would otherwise be made from fossil fuels. The burning of biomass releases about the same amount of carbon dioxide as burning fossil fuels. However, fossil fuels release carbon dioxide captured by photosynthesis tens or hundreds of millions of years ago, inserting an essentially "new" greenhouse gas into the modern world. Existing biomass, however, releases carbon dioxide that is largely balanced by the carbon dioxide captured in its own growth.

Most biomass use in the United States today involves some form of incineration to produce electricity or fermentation to make ethanol and biodiesel fuels. Biofuels supply about 3 percent of the energy we consume, thanks primarily to waste biomass from the paper industry, forest industry residues, and municipal solid waste used to generate electricity.[7]

Energy from Manure

Livestock in the United States produce almost 2 trillion tons of manure annually, most of which is left to rot.[8] If it were collected and converted to methane in anaerobic digesters, it could generate a significant percentage of annual U.S. electricity demand. It is a stable and reliable source of fuel. A typical lactating cow produces about 100 pounds of "pies" and 50 pounds of urine each day. There are about a hundred methane-producing cattle farms today.[9] Typically a minimum herd of 300 dairy cows or 2,000 swine is needed to make such a system feasible.[10] It takes twenty to thirty days to turn the manure into heat and power.

Questions have been raised about the economic viability of such projects.[11] Calculations by several investigators indicate that large biodigester operations will need federal subsidies to be economically viable and that proposing to build them is therefore inadvisable. Particularly problematic is the site-specific nature of cattle feedlots, which produce four-fifths of the manure. The low energy density of manure makes transporting waste to centralized digesters difficult, and large cattle feedlots are often located in areas far from where the energy is needed.

However, as is usual with calculations that indicate alternative energy sources are uneconomical, no consideration is given to the enormous amounts of money given to the fossil fuel industries in the form of military protections and domestic subsidies and the undependable character of the nations that supply America's imported fossil fuel (see chapter 6). The federal government has provided many billions of dollars in subsidies for the fossil fuel industries, and subsidies in rich countries worldwide for fossil fuel energy were estimated to be about $73 billion a year in the late 1990s, with an additional $162 billion spent in other countries.[12] In an age of climate change and air pollution, these subsidies are like giving money to drug addicts to fuel their life-destroying habit. When these facts are inserted into the financial calculations, most alternative energy sources seem cheap.

Biofuels

Farmers traditionally grow three Fs: food, feed, and fiber. Now a fourth F has been added: fuel. It is hard to scan a newspaper, magazine, or popular science publication today without encountering an article about biofuels. Global production of ethanol more than doubled from 2004 to 2008, with the United States leading the way. Biofuel production in 2008 was 8 billion gallons. The federal government is pushing it, the agricultural lobby is in favor of it, and even the automobile industry wants it because it will decrease noxious emissions from motor vehicles. However, the generation of biofuels is not all benefit according to environmentalists.

Ethanol   The energy bill passed by Congress in 2005 mandated that at least 7.5 billion gallons of renewable fuel (ethanol and biodiesel) be produced annually by 2012 and 36 billion by 2022. By 2008, annual production was already almost 10 billion gallons (94 percent of it ethanol), having risen from 1 billion in 2005. The 10 billion gallons was 7 percent of America's annual gasoline use of 140 billion gallons. Ethanol plants have mushroomed across the nation like fungi in wet soil, especially in the Midwest, where most of the nation's corn, the foodstock currently used in the United States for ethanol, is grown. As of June 2009, 180 ethanol refineries were operating. This number is the equivalent of 150 operating oil refineries. When money is to be made, things move fast.

Corn planting increased from 76 million acres in 2001–2002 to 85 million acres in 2008–2009 even as yields per acre were increasing, with one-third of the 2009 crop scheduled for ethanol distilleries rather than as feed for animals and other traditional sources. Corn prices doubled between 2005 and 2009. The percentage of America's corn crop used for ethanol was 5 percent in 2001, 14 percent in 2006, and 31 percent in 2009–2010.[13] Dedicating all current corn production to ethanol would meet 12 percent of America's gasoline demand.

Many Americans wince when they become aware that 70 percent of the nation's corn is fed to cattle, swine, and chickens rather than humans, but at least the farm animals are food for humans. However, it appears that soon farm animals may be relegated to second place as corn consumers, and even less corn will be used for food. To avoid this possibility, some companies are building ethanol plants next to cattle feeding operations, so that corn kernels can go through the ethanol distillery, and then residues from the operation—basically corn minus the starch—can go straight to the cattle as feed. Ideally, the corn will pass through the cow, and the manure may return to the plant and go into an anaerobic digester to produce methane, natural gas.[14]

Corn is not the only feedstock chemically suitable for ethanol. Research is underway to derive ethanol from wheat, oats, and barley, and others are investigating the use of microbes genetically engineered to produce enzymes that will convert the cellulose in crop waste, wood chips, and other plants to ethanol. Cellulose, essentially any kind of plant fiber, is available all over the United States, not only in the Midwest where corn is grown and most ethanol plants are located. Ethanol plants using cellulose could be located near coastal areas, alleviating the problem of transporting ethanol from the Midwest. But ethanol is corrosive and absorbs water, making it difficult to transport in pipelines.

Cellulosic ethanol is much more energy efficient than ethanol from corn and has a higher productivity. Using cellulose from native rather than from cultivated plants ("grassoline") does not require fertilizer or the petroleum-using machinery to spread it. Using waste materials such as corn stover, wood pulp, or trash can be even more efficient and has the additional positive effect of putting waste materials to work. The energy bill that Congress passed in December 2007 requires that 3 percent of the nation's federally mandated ethanol be derived from cellulosic sources by 2012 and 44 percent by 2022. As of mid-2009, there was no economically competitive commercial production of cellulosic ethanol, but there is intensive research by many corporations, and technological barriers may soon be overcome.[15]

The amount of available cellulosic biomass in the United States could produce more than 100 billion gallons of grassoline a year, about half its current annual consumption of gasoline and diesel. Similar projections estimate that the global supply of cellulosic biomass has an energy content equivalent to between 34 billion and 160 billion barrels of oil a year, numbers that exceed the world's current annual consumption of 30 billion barrels of oil. Cellulosic biomass can also be converted to any type of fuel: ethanol, ordinary gasoline, diesel, even jet fuel.

Ethanol distilled from sugar cane is much cheaper to produce and generates eight times more energy per unit of input than corn does.[16] The United States does not use sugar cane to produce ethanol for several reasons that are related to politics rather than to science. The favors granted to the American sugar industry, such as a guaranteed share of 85 percent of the market, keep the price of domestic sugar so high that using it for ethanol is not cost-effective. And the tariffs and quotas for imported sugar mean that no one can afford to import foreign sugar and turn it into ethanol the way that oil refiners import crude oil from the Middle East to make gasoline. Moreover, Congress imposed a tariff of 54 cents per gallon on sugar-based ethanol in order to protect corn producers from competition. Analysis has determined that if the tariffs were removed, prices would fall by 14 percent and almost 300 million more gallons of ethanol would be used in the United States. Former President George Bush proposed eliminating the ethanol tariff in 2006, but Congress killed the proposal. President Obama voted to keep the tariff when he was a senator, and the sugar quotas seem to be sacrosanct, so it is unlikely that sugar-based ethanol will appear in the United States.

Corn-based ethanol contains only two-thirds as much energy as gasoline. So even if it costs the same at the pump, it will not take you as far on the road. In addition, low-ethanol blends such as 10 percent ethanol in gasoline increase smog levels because the addition of ethanol increases the ease of evaporation of the fuel, so more of the hydrocarbons evaporate into the air and create smog. But as is common in political hassling, scientific realities take a back seat to successful political lobbying.

Biodiesel About 23 percent of the fuel used in America's motor vehicles is diesel, almost all of it in commercial trucks, ships, and farm machinery rather than cars.[17] Because of its chemical composition and molecular structure, corn kernels are not suitable as a feedstock for the production of diesel fuel. Soybeans are. The relative percentages of farmland devoted to corn and soybeans fluctuate yearly, but both are increasing as farmland previously idled for environmental or other reasons is brought back into production. Biodiesel is easy to manufacture and has a high net energy balance; soybean-biodiesel produces a 93 percent energy gain versus 25 percent for corn-ethanol.[18] And biodiesel emits 78 percent less carbon dioxide emissions than petroleum diesel.[19]

Soybeans are not the only vegetable materials suitable as a feedstock for biodiesel. Practically any type of vegetable oil or animal fat can be used, a fact that has not been lost on the large meat corporations. Animal fats are cheap and plentiful. The largest chicken producers— Tyson Foods, Perdue Farms, and Smithfield Foods—have established renewable-energy divisions. Tyson Foods is the nation's biggest producer of leftover fat from chicken. Maybe one day you will go to McDonald's and order Chicken McNuggets, fries, apple pie, and 10 gallons of fuel.

In 2005 a Canadian company began producing fuel from bones, innards, and other inedible animal parts. Although the fuel is cheaper than crude oil, there is a strange side effect: an odor of popcorn or french fries.

Algae Perhaps the most exciting possibility as a feedstock for biofuel is algae, with an oil yield of up to 50 percent.[20] Algae are like microscopic factories using photosynthesis to transform carbon dioxide and sunlight into lipids, or oil. Some strains can double their weight in a few hours under the right conditions. They can grow in fresh, salty, or even contaminated water, and they do not compete with food crops. Scientists think algae might grow fatter and faster if they were force-fed extra carbon dioxide, which could also help alleviate the buildup of greenhouse gases. Smokestack emissions from power plants and other sources could be diverted directly into the ponds, feeding the algae while keeping greenhouse gases out of the atmosphere.

Algae have emerged as a promising feedstock for biofuels due to their high energy content, energy yield per acre, fast growth, and ability to grow in water of varying quality. Algae's potential, at least in theory, is remarkable. According to the Department of Energy, it may be able to produce more than 100 times more oil per acre than soybeans or any other terrestrial oil-producing crop. Soybeans generate about 50 gallons of biodiesel per acre per year, but algal species can produce up to 8,000 gallons per acre per year. Because of its high energy content, oil from algae can be refined into biodiesel, green gasoline, jet fuel, or ethanol, making it an all-purpose biofuel. And algae need only water, sunlight, and carbon dioxide to grow. And they grow rapidly. About 100,000 strains of algae are known, and some are more efficient producers of biofuel than others. Some researchers estimate that algae can compete economically with petroleum at $60 per barrel. In 2007 only 5 companies were experimenting with algae as biofuels; today there are at least 200.

There are still biological, technological, and economic problems to be overcome before algal biofuel can be commercially successful, but research is intense, and breakthroughs can be expected in a few years. At present the cost of producing biofuels from algae is exorbitant, and beyond the technical challenges, issues such as land and water use have to be addressed. Almost all commercial algae biomass production at present occurs in open ponds, which take up large areas of land and are susceptible to contamination by native algal species and evaporation and usually produce algal oil of lower density than in more controlled settings.

Environmental Concerns   Environmentalists have several concerns about the production of ethanol from corn and biodiesel from soybeans:

• Clearing grassland to plant corn or soybeans releases ninety-three times the amount of greenhouse gas that would be saved by the fuel made annually on the land.[21] So for the next ninety-three years, climate change is being made worse, not better, by biofuels. Using naturally occurring cellulosic perennial plants would eliminate this problem.
• Conventional corn farming can remove 30 to 50 percent of the carbon stored in the soil.[22] In contrast, cellulosic ethanol production does not deplete the soil because plants are mowed as they grow naturally, often on land that is already in conservation reserve.
• Corn is a thirsty crop, and increasing corn planting will cause considerable harm to the nation's water quality from increased nitrogen and phosphorus pollution. This adds to the problem of dead zones in the Gulf of Mexico and other water bodies like Chesapeake Bay. Cellulosic ethanol production does not have this problem.
• Corn is grown in an area where irrigation is required from the Ogallala aquifer, which provides irrigation for much of the southern Great Plains. The water table has already dropped significantly in these areas, threatening the water supply of agriculture in general. Cellulosic plants such as switchgrass do not require irrigation.
• As is true of all row crops, growing more corn means applying more pesticides and artificial fertilizers to the soil. These inputs have a large carbon footprint in addition to the poisons they add to the soil. Using naturally occurring plants does away with this problem.
• The economics of corn and ethanol production means that farmers will grow corn on acres they otherwise would have rotated with soybeans to restore the nitrogen in the soil. The soil becomes depleted in an essential nutrient. Using naturally occurring plants avoids such choices.
• It is immoral to use food, which is essential for human existence, to provide fuel for cars. Naturally occurring grasses are not food.
• The skyrocketing price for a bushel of corn caused by the ethanol craze puts poor people in the position of subsidizing the motor vehicles of wealthier people. Distilling ethanol from cellulosic plants does away with this objection.

Biofuels appear to have a bright future for cars and trucks in the United States if naturally occurring nonfood plants are brought online as feedstocks.

Bioelectricity

Recent research has added a new twist to the biofuels debate. Converting biomass to electricity for charging electric vehicles may be a more climate-friendly transportation option than using biomass to produce ethanol fuel.[23] An acre of switchgrass, for example, would allow vehicles to travel 81 percent farther on average if it were pressed into pellets and burned to generate electricity than if it were converted to liquid fuel. Converting switchgrass to electricity would also reduce greenhouse gases by 108 percent more than if the feedstock were converted to ethanol.

Energy from Moving Water: Our Major Source of Electricity

Hydroelectricity is second to biofuels as a source of the nation's alternative energy, but it is the major renewable energy source for electric power (table 7.2). Only 3 percent of the dams in the U.S. currently generate electricity, so there is considerable room for expansion. Generating electricity using water has several advantages, primarily its low cost once the dam is built. Plants are automated and have few personnel on site during normal operation. In addition, because there is no fossil fuel use,

Table 7.2
Generation of hydroelectric power in the United States in 2008 (thousands of megawatt-hours)

Washington	70,283
Oregon	30,747
New York	24,346
Idaho	9,055
Montana	8,858
Arizona	6,827
Tennessee	5,736
Alabama	5,371
Arkansas	4,362
Maine	3,551
North Carolina	3,039
TOTAL	172,175 = 71% of total generation

Source: Energy Information Administration (2009).

operational costs are low, there are no sulfur or nitrogen dioxide emissions or particulate air pollution, and thermal pollution is limited compared to nuclear plants. Hydroelectric plants also tend to have longer economic lives than fuel-fired generation; some plants still in service were built 50 to 100 years ago.

Because fairly large rivers are best suited for the generation of hydropower, geography is a major control on the location of large dams. With the exception of Arizona and the Colorado River, all of the states with the highest hydropower generation are in the rainy and mountainous northwestern states or are tapping the abundant moisture in the mountainous Appalachian region.

Water supply may not be the same year-around, so the electric energy produced can vary seasonally. In some installations the water flow rate can vary by a factor of 10:1 over the course of a year. This variation can be overcome with pumped storage. Low water flow and times of peak demand can be accommodated by moving water between reservoirs at different elevations. At times of low electrical demand, excess generation capacity is used to pump water into a higher reservoir. When demand is higher, water is released back into the lower reservoir through a turbine. Pumped storage schemes currently provide the only commercially important means of large-scale energy storage in hydropower facilities. Hydroelectric plants with no reservoir capacity are called run-of-the-river plants, since it is not possible to store water.

Hydroelectric plants are not without environmental drawbacks. Despite the fact that they do not burn fossil fuels, they produce global-warming carbon dioxide and methane gas, and in some cases this can have a climatic effect worse than that of plants running on fossil fuel.[24] Methane is twenty times more effective at trapping heat than carbon dioxide. Emissions vary from dam to dam, and scientists do not agree on the quantitative importance of the methane release. The gases are released from the trees and other plants that rot after they are carried into the reservoir behind the dam. After the first pulse of decay when the organic debris releases carbon dioxide, the plant matter settles to the floor of the reservoir and is buried, where it decomposes in the absence of oxygen and generates methane gas.

There are seasonal changes in water depth in most reservoirs, and in the dry season when the water level in the reservoir drops, plants colonize the area around the reservoir. Subsequently the plants are drowned when the water level rises in the rainy season. Low areas around a reservoir can extend over many hundreds of square miles, providing an abundant and inexhaustible supply of organic debris.

Dams can last for many decades, but their life span can be rapidly cut short by the accumulation of sediment behind the dam. Abundant sand and mud are continually deposited in the reservoir. More sediment in the reservoir means less room for water, and dredging reservoirs is expensive. As water depth decreases, the ability of the dam to generate electricity also decreases.

A more futuristic source of power and electricity from moving water involves harnessing the oceans that cover 71 percent of the earth's surface. There is nearly 900 times more energy in a cubic foot of water as in a cubic foot of air because of the difference in density. But there are significant technological obstacles to harnessing wave and tidal movements for human use. Some have been overcome, but difficulties remain. Past governments in the United States have not been supportive of developing electricity from ocean power, in contrast to Europe, Australia, and elsewhere. Fourteen countries now operate wave- or tidal-power stations, but most are tiny, experimental, and expensive.[25] It is difficult to build devices that can survive storm damage, seized bearings, broken welds, and snapped mooring lines. The amount of overbuilding required as a safety factor to overcome these dangers is uncertain, and overbuilding increases costs.

Wave Power

The best areas for harnessing wave energy to generate electricity are where the waves form far out at sea and are driven shoreward as the winds pump energy into them. The areas of strongest winds are between latitudes 40° and 60° in both the Northern and Southern Hemispheres on the eastern sides of the oceans. In the United States, the most favorable areas are along the Pacific coast.

No commercial wave power generators are operating in the United States. The world's first wave farm opened in 2008 in Portugal. It consists of three machines, and expansion to twenty-five machines is planned, but it generates only 2 megawatts of power, an order of magnitude too small for the farm to be considered commercial. Other farms are in the planning or development stage in Scotland, England, and Australia. Environmental dangers to the commercial development of wave energy include displacement of commercial fishers from productive fishing grounds, alterations in the pattern of beach sand nourishment, and possible hazards to safe navigation.

Tidal Power

In contrast to the lack of commercial development of wave power, a commercial tidal power plant that generates 240 megawatts of electricity has been operating since 1966 in the estuary of the Rance River in northwestern France. The plant produces 0.01 percent of France's electric power at 70 percent of the cost per kilowatt-hour of the country's nuclear power installations.[26] The design of the tidal station has been superseded by newer technologies that have a lesser environmental impact on aquatic life.

Electricity can be generated by water flowing in and out of a bay. As there are two high and two low tides each day, electrical generation from tidal power plants is characterized by periods of maximum generation every 12 hours, with no electricity generation at the six-hour mark in between. Alternatively, the turbines can be used as pumps to pump extra water behind the linear barrage during periods of low electricity demand. The water can then be released when demand on the system is greatest, thus allowing the tidal plant to function with some of the characteristics of a pumped storage hydroelectric facility.

Selection of location is critical for the tidal turbine. In order to produce commercial amounts of electricity a difference between high and low tides of at least 16 feet is required. Tidal systems need to be located in areas with fast currents where natural flows are concentrated between obstructions, such as the entrances to bays and rivers, around rocky points, headlands, or between islands or other land masses. Three possible sites have been suggested in the United States: Cook Inlet in Alaska, the Golden Gate in San Francisco Bay, and the East River in New York City. As of mid-2009, five tidal units of a planned 300 have been successfully placed in the East River. When the project is completed, it will generate 10 megawatts of electricity, barely commercial but enough to power 10,000 homes.

Ocean Currents

More futuristic is the possibility of obtaining energy from fast-moving ocean currents such as the Gulf Stream. The amount of water moving past Miami is more than fifty times the total of all the rivers in the world, and the surface velocity sometimes exceeds 8 feet per second.[27] The total power of the kinetic energy of the Gulf Stream near Florida is equivalent to approximately 65 gigawatts. How much of this energy can be captured economically is unknown.

Energy from Nuclear Fission: Splitting Atoms

Nuclear energy is obtained by bombarding an atom of uranium-235 with neutrons in a nuclear reactor and splitting it, thereby releasing an enormous amount of energy. One ounce of uranium-235 produces the same amount of heat as burning 17,000 gallons of oil or 170,000 pounds of high-grade coal.

Uranium is a relatively common metal that occurs in rocks and seawater; it is approximately as common as tin or zinc. Economic concentrations of it are not uncommon. Australia is the Saudi Arabia of uranium ore with nearly a quarter of the world's known reserves (table 7.3). Canada, however, has the richest deposits. As is true of any other commodity obtained by mining, the amount of reserves depends in large part on the economics of extraction of the metal from the rock, processing, and importance of the metal. Changes in costs or prices or further exploration may significantly alter the amount of reserves. Since 2004, Australia's percentage of the world's reserves has decreased from 28 percent to 23 percent and Canada's from 14 percent to 8 percent, but the percentage in the United States has increased from 3 percent to 6 percent.

Table 7.3
Uranium reserves in a known current total of 5,469,000 tons of ore

Country	Percentage of world ore
Australia	23
Kazakhstan	15
Russia	10
South Africa	8
Canada	8
United States	6
Brazil	5
Namibia	5
Niger	5
Ukraine	4
Jordan	2
Uzbekistan	2
India	1
China	1
Mongolia	1
Others	4

Source: World Nuclear Association (2008).

Uranium mining in the United States has been confined to relatively dry areas of low population density in Utah, New Mexico, Wyoming, and Nebraska, in part because of concerns about the possible spread of radioactive pollution caused by mining operations. Recently, however, a deposit of the metal in moist and populated Virginia that may be the largest in the United States is being investigated as a possible mining site.[28] Because of environmental regulations, it takes about ten years to construct an operational mine after a uranium source has been identified. The estimated 110 million pounds of uranium in the Virginia deposit could supply all of the nation's nuclear power plants for about two years.

The world never runs out of any natural product, but there may always be a limit to what people are willing to pay; when this is exceeded, we say the supply is exhausted and replacements must be found. The amount of reserves of a metal in rocks depends not only on costs but also on the intensity of past exploration efforts and is basically a statement about what is known rather than what is present in the earth's crust. At ten times the current price or with the development of new technology, seawater might become an economically profitable source of uranium. The estimated uranium reserves are about 6 million tons, and current use is 71,000 tons per year. So known reserves are adequate for about eighty-five years.

Very little exploration for uranium took place between 1995 and 2005, but exploration has increased since then because of a renewed interest in nuclear energy, and this has led to a significant increase in known reserves. In just two years, 2005 and 2006, the world's known uranium reserves increased by 15 percent. On the basis of current geological knowledge, further exploration by geologists and higher prices will certainly yield additional reserves.

In addition, the increasing efficiency of reactors has decreased the need for uranium. Between 1975 and 1993, the electricity generated by nuclear energy increased 5.5-fold, while the amount of uranium used increased only 3-fold.[29] On the basis of analogies with other metallic minerals, a doubling of prices from today's levels could be expected to create about a tenfold increase in reserves. Uranium is not a renewable resource, but there is no danger of running out of it in the foreseeable future.

Ninety-two percent of the uranium used in the United States for nuclear power plants is imported—about half from Russia and half from Canada; Australia, Namibia, and Kazakhstan are minor suppliers. The imports that come from Russia do so under a deal that is converting about 20,000 Russian nuclear weapons into fuel for U.S. nuclear power plants. That deal ends in 2013, but in 2008, a new contract was signed that permits Russia to supply 20 percent of U.S. reactor fuel until 2020 and to supply the fuel for new reactors quota free. According to the Nuclear Energy Institute, an industry trade association, former nuclear bomb material from Russia accounts for 45 percent of the fuel in American nuclear reactors today; another 5 percent comes from American bombs.

The price for uranium was $7 per pound in 2003 when it began a slow rise to $25 in 2005. It then skyrocketed to a high of $139 in 2007 before joining the sharp decreases in oil and natural gas prices and falling to $40 in 2009. The current relatively low price for all three of these commodities is certain to increase as the world recovers from the 2008–2010 economic decline. However, unlike oil or natural gas, the cost of the uranium fuel is only a small part of the cost of the electricity produced by the power plant, so even large fluctuations in the price of uranium do not have a major effect on the price of the electricity produced.

All existing full-power operating licenses for nuclear power plants in the United States were issued between 1957 and 1996. Of the 132 plants that were granted licenses, 28 have been permanently shut down, so 104 nuclear power plants are now in service; the most recent came online in 1996 in Tennessee. They operate at 92 percent of capacity rather than 100 percent because of shutdowns for maintenance and repair. This is a much higher percentage than for coal (73 percent), natural gas (16–38 percent), or oil-fueled (30 percent) plants.[30] Although no nuclear plants have been built in the United States in recent years, existing facilities have substantially improved their performance and lowered operating costs.

Most nuclear plants whose operating licenses were approaching their end have requested and been granted twenty-year extensions. And because of heightened concerns about global warming and carbon dioxide emissions from coal-fired plants, the Nuclear Regulatory Commission (NRC) received twenty-one applications in 2007 and 2008 for permission to build thirty-four new nuclear plants at various sites around the country.[31] These are the first such applications in thirty years. A large percentage of proposed nuclear power plants whose applications are approved are never built, so the future of nuclear energy in the United States is uncertain despite the flood of applications. Most of the applications were submitted by utilities in the South. The NRC estimates it will need two and a half years to review each application and an additional year to conduct hearings on its conclusions. As a further setback to the proponents of nuclear energy, the NRC in 2008 delayed approval indefinitely on all plants under construction because of needed design changes.

Nuclear energy currently generates about 100 billion watts of electricity,[32] 9 percent of America's energy needs and 20 percent of our electricity. These percentages have varied little since 1988 despite a continual increase in the amount of energy used. Vermont, New Jersey, Illinois, New Hampshire, and South Carolina rely on nuclear energy for more than half of their electricity.[33]

The possible increase in the number of nuclear-fired power plants is as controversial as the possible construction of additional coal-fired power plants. Some people believe there is no other choice if the United States intends to continue its economic development. However, the environmental consequences of burning coal are well known and understood, and no group in the United States other than the coal industry is enthusiastic about building more coal-fired plants. In contrast, the social, economic, political, and environmental consequences of nuclear power plants are less well defined, and the issues are more complex. They are discussed in some detail in the next chapter.

Energy from the Earth's Interior: Heat beneath Our Feet

Nuclear power plants are designed to boil water to drive steam turbines without producing massive amounts of pollutants. However, these plants are superfluous. The earth comes to us with a safe and clean nuclear power plant built in that can provide all of the energy we need.[34] Almost all of the earth's volume, 99.9 percent, is hot enough to boil water. We might skip all the effort involved in the generation of nuclear energy and replace it with energy from the hot rocks under our feet. This is what hydrothermal geothermal energy is about. Currently geothermal plants produce just over 3,000 megawatts of electricity, compared to 9,183 megawatts of solar power and 28,200 megawatts of installed wind capacity. In 2009 six new geothermal plants came online, generating a 6 percent growth in geothermal power capacity, 91 percent of it in California and Nevada. Another 188 plants were under development, and the outlook is for even stronger growth in this industry.

In October 2009, the Department of Energy awarded $338 million in stimulus funding to geothermal energy research and development projects, the largest injection into new geothermal technology development in more than twenty-five years. The funds will support 123 projects in thirty-nine states.

Geothermal heat is inexhaustible and releases no greenhouse gases or radiation. The many tens of thousands of holes drilled in the United States in search of oil and natural gas have revealed the variation in depth temperature over the United States (figure 7.3). The data show possibilities for harnessing the earth's heat directly almost everywhere in the country. In many places in the West, it is suitably hot at depths of only a few hundred feet, but in mid-America, sufficiently high geothermal temperatures occur only at depths of 10,000 to 20,000 feet. In the East, suitable temperatures are present only at even greater depths, and it is not clear that this deep heat can be tapped into economically. However, Google in 2008 invested $11 million in new deep drilling technology that can drill through hard rock five times faster than current methods. If this development succeeds, geothermal power will be practical almost anywhere.

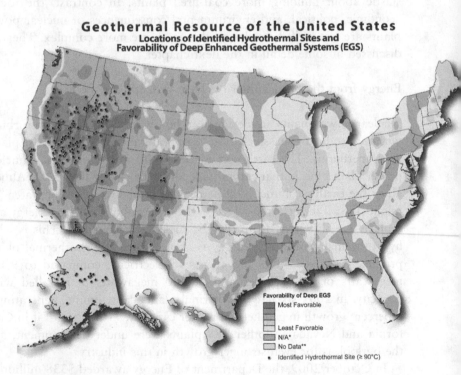

Geothermal Resource of the United States
Locations of Identified Hydrothermal Sites and
Favorability of Deep Enhanced Geothermal Systems (EGS)

Favorability of Deep EGS
- Most Favorable
- Least Favorable
- N/A*
- No Data**
- • Identified Hydrothermal Site (≥ 90°C)

Figure 7.3
Temperatures in the forty-eight contiguous states at a depth of 20,000 feet. In some western areas, temperatures exceed 300°F. Although Hawaii and Alaska are not shown on the map, Hawaii has active volcanoes, as does coastal Alaska, indicating both states have geothermal resources at shallow depths. (Department of Energy) 2004

The subsurface heat is generated by the decay of radioactive elements in the crust itself and at greater depths from molten rock within the earth's mantle and core. It is transported upward from these areas by convection and conduction. Sometimes molten rock is very near the surface, and we see evidence of its heat as volcanic eruptions or, less dangerous, as geysers. The fluid shooting into the air at a geyser is liquid water and water vapor emanating from molten rock at shallow depths, combined with normal groundwater that has been heated by the hot rock. The water in geysers is dangerously hot at the surface and tends to be at least 300°F to 400°F at depths less than 300 feet.

Geothermal capacity in the United States in 2009 was 3,100 megawatts.[35] In 2009, a report by the Geothermal Energy Association identified 144 projects underway in fourteen states, from Alaska to Florida and from Hawaii to Mississippi. When developed, these projects could provide nearly 7,000 megawatts of new electric power—enough electricity to meet the needs of roughly 7 million homes. In large areas of the western United States, temperatures are hot at depths of just a few hundred feet, and this heat has been tapped into for many decades. In Oregon, Utah, and Idaho, geothermal heat is used for power generation and space heating, providing about 600 megawatts of heat, roughly enough energy to heat and cool more than 400,000 homes. Nevada has fifteen geothermal power plants that generate enough electricity for 73,000 homes, and two dozen new plants are under development. One company in 2008 in Nevada announced plans to deliver 1,000 megawatts of geothermal power to Los Angeles and Las Vegas from massive natural steam zones in northwestern Nevada. In California, hot waters from 350 wells at the Geysers, 72 miles north of San Francisco, supply 5 percent of the state's power needs. (The billows of steam at the locality prompted the first European visitor in 1847 to believe he had discovered the gates of hell.) On the volcanic island of Hawaii, 25 percent of the electrical supply comes from geothermal energy.

The Bureau of Land Management says that a dozen western states could generate 5,500 megawatts of geothermal energy from 110 plants by 2015, and that number could rise by another 6,000 megawatts by 2025. Geothermal power plants such as the Geysers have system availabilities of at least 95 percent, much higher than plants fueled by coal, oil, natural gas, or uranium. And unlike wind and solar systems, a geothermal plant works night and day. However, their efficiencies are only 8 to 15 percent, less than half that of coal plants. High construction expenses and the low efficiencies make the cost of hydrothermal electricity about double that of coal, which sells for around 5 cents per kilowatt-hour.

Heat Pumps

Even without a nearby hot spot, the ground can be tapped to reduce the need for fossil fuels. All of the soil in the United States is suitable for a technology known as geothermal heat pumps, which transfer heat from the ground to homes in the winter and reverse direction to provide cooling in the summer. Heat pumps have the potential to provide nearly all households with heat and hot water.

Installing a heat pump has a higher initial cost than a natural gas furnace—about $7,500 for an average-size home. But monthly savings could total between 25 and 50 percent of current utility bills, with the system paying for itself in two to ten years. In 2004 a heat pump system was built into a new building in Washington for the Department of Agriculture. The building is heated and cooled for less than ten cents per square foot. About 1 million heat pumps have been installed, and roughly 50,000 to 60,000 more are being added annually.[36]

Engineered Geothermal Systems

Geothermal systems do not need to be restricted to areas where shallow groundwaters are naturally heated. An MIT study in 2007 concluded that converting geothermal heat into electricity by pouring water onto hot rocks underground and using the steam generated to turn turbines is arguably the most promising alternative and renewable source of green energy on the planet. The process is called *engineered geothermal systems* (EGS). The researchers calculated that there is more than enough extractable hydrothermal energy available to generate the entire 27 trillion kilowatt-hours of energy consumed in the United States in 2005. In fact, a conservative estimate of the energy extractable from hot rocks less than 30,000 feet beneath American soil suggests that this almost completely untapped energy resource could support U.S. energy consumption at its present rate for more than two thousand years. The resource is technically available for exploitation right now. Simply pump cold water into one borehole, and when it hits the hot rocks, it creates a network of fractures, which allow the water to travel horizontally to a second hole, where steam is allowed to escape and drive a turbine generator. The spent steam is condensed and recycled back down to the hot rocks again, making the water consumption insignificant. Once a geothermal plant is built, there are no fuel costs, so production cost is less than for a coal or nuclear plant. The main cost of geothermal energy is the initial cost of drilling the wells. Geothermal energy could be an ultimate answer to the world's growing energy needs.

Energy from the Wind: Hold On to Your Hat

The idea of harnessing energy from the wind goes back to the earliest recorded history. Along with the waterwheel, windmills are the oldest tools to capture and use nature's power for human purposes. Today's wind turbines can harness enough energy from the wind to supply thousands or even millions of homes with electricity, and in many European countries wind power is an essential and growing part of the energy supply. More than eighty nations now tap the wind to produce commercial electricity. The wind generates more than 1.5 percent of the world's electricity, up from 0.1 percent in 1997.

In the United States, wind supplies 1 percent of our electricity needs, powering the equivalent of 4.5 million homes. Capacity increased by 60 percent in 2009 to 35,000 megawatts. Wind power is growing rapidly, with Texas leading the way, 9,708 megawatts. It accounted for 42 percent of new capacity additions in the United States in 2008, second only to natural gas for the fourth year in a row.

The United States now leads the rest of the world in cumulative capacity and electricity generation. Because of the extensive flat Midcontinent area, we are well supplied with strong winds (figure 7.4). In such a flat area, over 90 percent of the land area may be favorably exposed to the wind, in contrast to areas of rugged topography where only the ridge crests may be suitably exposed. Iowa, second only to Texas in wind power capacity, surged past California in 2008. Fourteen percent of Iowa's electricity now comes from wind power. Kansas is also rapidly increasing its wind-generating capacity. Wind in the Midcontinent may turn out to be a more valuable commodity than crops in some areas. Wind power is now less expensive than natural gas and is tied with geothermal power as the least expensive form of alternative, renewable, and nonpolluting energy supplied by nature. It is the fastest-growing renewable energy source on the planet.

Energy Potential

Based on data from 2400 locations, researchers at Stanford University believe that there is ample wind in the United States to supply all the country's electricity.[37] This may be overly optimistic but it suggests that

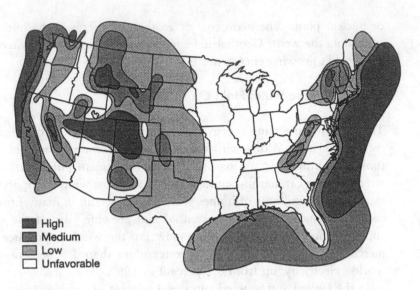

Figure 7.4
Wind resources in the United States. (Department of Energy, 1987)

the current rapid growth of wind energy will continue. Some experts believe a 20 percent contribution by wind energy is a realistic goal within a decade.

The United States led the world in new wind turbine installations for the fourth year in a row in 2008, with a record-shattering 8,358 megawatts of wind capacity added, increasing cumulative installed capacity to 19,549 megawatts—enough to power 6.5 million homes and offsetting the carbon dioxide emissions of twelve average-size coal-fired plants.[38] Texas and Iowa have the most installed capacity. California, once the location of practically all the wind energy activity in the United States, is now in third place. Iowa and Minnesota now get approximately 7.5 percent of their electricity from the wind, Colorado 6.1 percent, South Dakota 6 percent, and five other states at least 3 percent.[39]

Wind power was the largest new resource added to the U.S. electrical grid in 2008, ahead of new natural gas plants and far ahead of new coal-fired installations. The surge in installations was driven by the federal production tax credit and by renewable energy mandates in twenty-five states and the District of Columbia. The Department of Energy forecasts that installed capacity will increase rapidly between now and 2020.[40] Even the economic downturn of 2008 did not slow the growth of wind power. Although the U.S. economy hemorrhaged 2.6 million jobs in 2008, the wind industry enjoyed 70 percent job growth

(35,000 new jobs), which supported a 60 percent surge in new wind capacity installation over 2007.[41] The increase was 8,358 megawatts, 5,000 new turbines, which accounted for 40 percent of all new electricity capacity. Wind capacity in the United States now totals over 25,000 megawatts.

The rapid increase in wind power installations has affected the economics of wind power. Between 1992 and 2007, the costs of wind-generated electricity dropped by 50 percent as efficiency, reliability, and power rating experienced significant improvements. But costs have increased in recent years due to a shortage of turbines and parts, rising costs of materials, and increasing manufacturing profitability. Despite the higher costs, wind power remains competitive with new natural gas plants, and it will become increasingly competitive with coal if Congress puts a price on carbon emissions, which seems likely.

Four main variables determine how much electricity a turbine can produce: wind speed, blade radius, tower height, and air density. For a turbine to function, wind speed must be at least 8 mph, and energy can be produced economically at 15 mph; maximum electricity is generated at velocities between 25 and 55 mph. Most turbines are programmed to shut down at about 55 mph because they can be damaged by stronger winds.

The energy carried by the wind increases as the cube of the wind speed, so a wind speed of 25 feet per second yields 15,625 (25^3) watts (15.6 megawatts) for every square foot exposed to the wind, and at 50 feet per second, energy production is 125,000 (50^3) watts (125 megawatts) per square foot. This relationship puts a premium on sites having the strongest winds and explains why the wind industry strives to increase the scale of the equipment it installs. This relationship also explains why the area swept by the turbine blades is so important. The most efficient turbines have increased in size over the years, with each of the three large blades longer than a football field. In 1980 turbine blades were about 50 feet in diameter; today they can exceed 400 feet. Electrical output has increased correspondingly, from 50 kilowatts to a possible 10 megawatts, 200 times as much. The average generating capacity of existing wind turbines in 2008 was 1.67 megawatts. The blades spin a shaft that connects to a generator that produces the electricity. Wind velocity is lowest at ground level because of friction between the moving air and the ground, and the velocity increases with height. Hence, turbines are placed high above the ground. The effect of friction is zero above about 300 feet.

An emerging market for wind turbines is the recent development of small wind turbines that can be attached to a rooftop and can generate electricity at home in winds as low as 2 miles per hour.[42] In urban and suburban settings, the high density of nearby buildings and trees tends to block the higher wind speeds needed by conventional turbines that must be mounted on the ground. Ninety percent of the wind sources in North America are lower than 10 miles per hour. Each small turbine generates 2 kilowatts of power (much more than is needed by an average household) and sells for $4,500, with an additional installation cost of $1,500.

Onshore versus Offshore

Winds are generally stronger and more consistent offshore than onshore because there are no topographically high areas in the sea to block the moving air. A report by the Interior Department in 2006 said that wind energy within 50 miles of the U.S. coast has the potential to provide 900 gigawatts of power, close to the total currently installed electrical capacity in the United States. And 78 percent of the nation's electricity is used in areas touching oceans or the Great Lakes. By comparison, the best wind sites on land tend to be in rural areas in the middle of the country.

Locating an energy source near the coast would relieve some of the burden on long-distance transmission lines. A study by the University of Delaware and Stanford University concluded that the wind resource off the mid-Atlantic coast alone could supply the energy needs of nine states from Massachusetts to North Carolina, with enough left over to support a 50 percent increase in future energy demand.[43] Despite these assessments, there are no offshore wind farms along the coastline of the United States, although the Department of the Interior has approved construction of a large wind farm offshore of Cape Cod, Massachusetts. There are at least eleven other offshore wind projects in development. Offshore wind installations are common in Europe. They occur in offshore Britain, Ireland, Sweden, Denmark, and the Netherlands.[44] Total offshore capacity in Europe grew from 40 megawatts in 2000 to more than 1,200 megawatts in 2007.

All current offshore wind farms are in shallow water, but facilities farther offshore to better wind locations are possible because of technological innovations by the oil industry drilling in deep waters in the Gulf of Mexico.[45] Floating offshore wind turbines may be erected in the future, although they are not economical at present.

The reasons the United States has lagged behind in this area of renewable energy are primarily objections by those who live on the coast and do not want their ocean views spoiled and, secondarily, the cost. Offshore turbines cost 8 to 12 cents per kilowatt-hour compared to 5 to 8 cents for onshore wind.[46]

Energy from the Sun: It's Hot Out There

Most of the power generated by people originates from the sun. Millions of years ago, sunlight nurtured life that became today's oil, gas, and coal. It is the solar heating of the earth's atmosphere and oceans that fuels wave power, wind farms, and the rainfall that permits hydroelectric schemes. But using the sun's power directly to generate power is rare. Solar cells account for less than 1 percent of the world's electricity production, with Germany and Japan leading the way, and only one-hundredth of 1 percent in the United States. But annual global production of photovoltaic cells has risen exponentially from about 250 megawatts in 2000 to 7,000 megawatts in 2008, largely because of growth in Europe, particularly in Spain and Germany. Rapid growth is likely to continue.

Solar energy, like geothermal energy, is inexhaustible, but it has proven difficult to turn enough of this energy into electricity for its cost to be competitive with other energy sources. At present, it is 30 cents per kilowatt-hour, but it is expected to drop dramatically soon because of China's entry into the production market.[47] The record conversion efficiency for a solar array is 31.25 percent, set in January 2008 in New Mexico.[48] The previous record was 29.4 percent, set fourteen years earlier. Solar cells in the laboratory achieved 42.8 percent in 2007, breaking the previous record of 40.7 percent set in 2006. Much higher efficiencies from arrays of solar panels are required for solar electricity to be commercially competitive with either other alternative energy sources or fossil fuels, which are below 10 cents per kilowatt-hour. Only in places unconnected to an electrical grid, such as rural Africa and rural Asia, are solar cells commercially viable today.

Although sunny California currently has two-thirds of the nation's grid-connected photovoltaic installations, the largest solar power plant in the United States is located on Nellis Air Force Base in Nevada. It is a collection of 72,000 solar panels built on 140 acres, including part of an old landfill. The plant, a public-private venture that took six months to build, cost $100 million and generates about a quarter of the electricity used on the base, where 12,000 people live and work. It will save the federal government nearly $1 million a year and reduce carbon pollution by 24,000 tons a year, the equivalent of removing 24,000 cars from American roads.

Photovoltaic cells function by separating electrons from their parent atoms and accelerating them across a one-way electrostatic barrier formed by the junction between two different kinds of semiconductors. Photovoltaic cells are already widely used in small items such as calculators, watches, toys, and a variety of other consumer products, but getting high conversion efficiencies from large cells for electricity generation is a challenging problem. Traditional solar cells are made of crystalline silicon, which is expensive and requires a lot of energy to manufacture. It takes about three years for a conventional photovoltaic panel and the equipment associated with it (the rigid frame used to mount it and the power-conditioning electronics that attach it to the grid) to produce the amount of electrical energy required to manufacture this equipment in the first place.[49]

Experiments with dye-sensitive cells made of titanium dioxide have shown promise in reducing costs—they cost one-tenth the price of pure silicon—but they have low conversion efficiencies. Experiments using plastics have shown that they are poor conductors of electrons; cells using them have efficiencies of only 3 or 4 percent. However, because of the low cost of plastic, an efficiency of even 7 percent would make such cells economically competitive with silicon.[50] What is needed is a way to boost the efficiency of cells made from cheap materials.

One emerging technology that is promising is thin-film technology, which does not require silicon.[51] It is cheaper to produce but has lower efficiencies than standard photovoltaic panels. Its global market share doubled from 7 percent in 2006 to 14 percent in 2008 and is projected to reach 31 percent by 2013. Thin films are composed of very thin layers of photosensitive materials and require less energy and materials to make than conventional silicon-based solar cells. They can be integrated into roof shingles, siding, and the windows of buildings. Conversion efficiency for commercial uses is only 10 percent, but efficiencies have reached 20 percent in the laboratory.

Because of the cost and conversion efficiency problems associated with solar cells made of silicon, solar thermal technology, which has been operating in California's Mojave Desert since 1985, has undergone a rebirth in the United States. The technology is called concentrated solar power (CSP) or solar thermal electricity (STE). CSP systems capture and focus the sun's rays, using mirrors, to heat a water, oil, or molten salt to about 750°F and use it to drive a turbine to generate electricity. As a source of large-scale power, CSP is less expensive and more practical than photovoltaic systems, not least because the technology can deliver power for hours after the sun sets or the day is cloudy by keeping the hot fluid in insulated tanks.[52]

America's southwestern deserts are an abundant source of sunshine that could meet the country's power needs several times over if it were full of CSP equipment (figure 7.5).[53] CSP is still expensive at 12 cents per kilowatt-hour but the cost is falling as the technology improves and cost-reduction strategies are implemented, and is now half the price of electricity produced by photovoltaic cells. In 2008, the Bureau of Land Management received more than thirty planning requests to develop large-scale CSP plants across the United States. In 2009 an Israeli company signed an agreement with a California company to build the world's largest solar thermal plant in California. The plant will generate 1,300 megawatts of electricity and generate 286,000 megawatt-hours per year by 2013.

Although the cost of solar power appears to be much greater than that of fossil fuels or nuclear energy, the numbers are deceptive because of federal subsidies Congress has granted to the nuclear industry and the

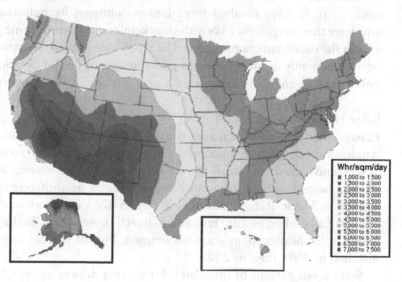

Figure 7.5
Average daily solar radiation, 1961–1990. (National Renewable Energy Laboratory)

enormous amounts of money given each year to fossil fuels. The military expenditures were noted in chapter 6, but the number of nonmilitary subsidies is equally mind-boggling. These subsidies are embedded throughout the generation and delivery value chain; ten of them have been enumerated by the Environmental and Energy Study Institute.[54]

Subsidies, incentives, and giveaways are applied throughout the energy life cycle, including the phases of research and development, extraction, transport, production, consumption, and decommissioning and are extremely difficult to uncover. But they are there, although not mentioned by supporters of fossil fuel use. When the external costs of environmental degradation and health costs of fossil fuel use are added to the subsidies, it is difficult to maintain the frequently quoted statistic that solar photovoltaic power is several times more expensive than fossil fuels and nuclear energy. It is likely that the reverse is true.

Federal subsidies are a way of life in the United States for both industries and families. Businesses get rapid depletion allowances on new equipment, farming conglomerates get agricultural subsidies, parents get tax deductions for children, poor families get both federal and state assistance, and so on. Examination of almost any industry or aspect of life reveals that Washington has legislated government assistance for most of the things Americans do, whether at work or in family life. Many of these aids were granted with the best intentions a long time ago, and some of them have resulted from intense lobbying by industries, but whatever their origin, they skew the marketplace in favor of some groups and to the disadvantage of others. The renewable energy industries have been consistently disadvantaged by government policies for decades. This policy may be changing because of a new attitude in Washington.

Solar Heating
Passive solar heating is increasingly used in building construction, where windows are located to maximize the greenhouse effect in rooms and offices and reduce the need for heating. Active solar heating systems pump a heat-absorbing medium such as air, water, or antifreeze through a small collector. Active collectors are generally located adjacent to or on top of a building and are required in Israeli construction. In the United States, solar heating panels are uncommon, but 70 percent more were installed in 2008 than in 2007.

Sixty-seven percent of installed solar heating devices are in China. In Europe, new installations roughly tripled in 2008 and demonstrated that an area does not have to be in the sun belt to benefit from solar panels.

Germany, at the same latitude as southern Alaska, often has cloudy skies and has 27 percent of the population of the United States. Yet the number of installed solar panels in Germany is more than three times that in the United States. Spain and Japan, also with much smaller populations, are also ahead of us.

Germany is one of the leaders not only in solar heating (second to Spain) but in wind power as well. Germany (and Spain as well) has instituted feed-in tariffs (FITs), as have eighteen other EU countries (including Spain) and forty-six countries worldwide.[55] FITs guarantee that anyone who generates electricity from a renewable energy source— whether a home owner, small business, or large electric utility—can sell that electricity into the grid and receive long-term payments for each kilowatt-hour produced. Payments are set by the government at preestablished rates, often higher than what the market would ordinarily pay, to ensure that developers earn profitable returns.

The push for solar power by the Spanish government generated a rapid production response from industry. Spain is now the world leader in solar photovoltaic installations; solar installations there quadrupled from 2007 to 2008. When political leaders establish policies, innovators create new industries or enlarge existing ones. It is noteworthy that Germany has a solar profile like that of Alaska; Spain's is like that of Idaho. Clearly the intensity of sunlight is not vital for solar power to be commercial.

Germany's Reichstag in Berlin is set to become the first parliamentary building in the world to be powered 100 percent by renewable energy. Soon the entire country will follow suit. The country is accelerating its efforts to become the world's first industrial power to use 100 percent renewable energy, perhaps by 2050.[56] Germans believe the switch will add 800,000 to 900,000 new jobs by 2030. In 2008 the percentage of renewables in Germany's primary energy consumption was only 7.3 percent, but the government expects that to increase to 33 percent by 2020 and 50 percent by 2030 as the country thunders ahead in its conversion program.

The United States has relied primarily on two state-level policies to promote renewable energy, net metering and renewable portfolio standards, but they have not been as effective at promoting renewable energy as the FIT system has. Net metering has been implemented in forty-two states and the District of Columbia. Energy produced by a utility customer can offset the power it consumes from the utility-fed grid, much like being able to obtain store credit by selling home-grown peaches to the grocer. Typically, however, the utility customer is limited in the amount of power that can be sold to the supplier; the most the customer can do is break even. Credit can be carried from month to month, but unused credit expires at the end of the year. The utility gets without cost whatever is left over at the end of the twelve-month billing cycle.

Renewable portfolio standards (RPSs) set minimum shares of total electricity generation that must be met with renewable energy. Twenty percent is a common target for a set date in states that have RPSs. The dates are typically ten to twenty years from the time the RPS is established. Under RPSs, a utility can meet its renewable energy obligation by securing tradable renewable energy credits in three ways: producing renewable energy itself, contracting for the long-term purchase of credits or renewable power, or buying credits on the spot market. In theory, competition to provide utilities with the requisite credits is supposed to lower electricity rates.

Job Creation

A widely quoted report from the University of California concluded that the nation can expect 86,370 new energy jobs if we continue with the energy mix current in 2007.[57] But if 20 percent of our energy were to come from renewable sources, 188,000 to 240,850 jobs could be created, depending on the relative proportions of wind, solar, and biomass energy. The American Council for an Energy-Efficient Economy estimates that 1.1 million jobs could be created through investments in energy efficiency technology.

Another study concluded that solar photovoltaic creates fifty-five to eighty times as many direct jobs as natural gas, and solar heating creates two to eight times more direct jobs than conventional power plants do.[58] Investing in clean energy technologies would both reduce the nation's trade deficit and reestablish the United States as a leader in the largest global industry today.

Fuel Cells: Splitting Water for Power

In his 1874 science fiction tale *The Mysterious Island*, Jules Verne predicted that "water will be the coal of the future." Early in the twenty-first century, Verne's vision is starting to come true in the form of a fuel cell—an electrochemical energy device that converts hydrogen and oxygen into water, producing electricity.[59] It has no moving parts. It is similar to batteries in its design but does not run down or need recharging. A fuel cell provides a direct current voltage that can be used to power motors, lights, and many other electrical appliances. It can be scaled and stacked until the desired power output is achieved.

The oxygen needed for a fuel cell to function is obtained from the air, but air contains no hydrogen. Available commercial fuel cells today rely on a fossil fuel as a source of hydrogen, but on the horizon are cells that split water to get the hydrogen, a process called electrolysis. The combination of a fuel cell using hydrogen with electricity generated from clean, renewable energy sources would create a truly sustainable energy system. Fuel cells release only water as a fluid by-product. They also produce heat that can be used to provide hot water or space heating for a home or office.

Criticisms of the technology and economics of generating, transporting, and storing hydrogen fuel are on the verge of being silenced.[60] Researchers at the Weizmann Institute of Science in Israel in 2005 developed a method for producing a continuous flow of hydrogen and steam under full pressure inside a car.[61] The method could also be used to produce hydrogen for fuel cells. A full-scale prototype is being constructed.

Summing Up: We Know What to Do. Will We Do It?

We already have the scientific, technical, and industrial understanding to solve energy, carbon pollution, and climate problems. Many methods are available. It has been estimated that a ninety-mile-square area (only 10 miles by 10 miles) of desert in one of our southwestern states could host enough solar power facilities to meet today's electricity needs.[62] It has also been estimated that there are sufficient wind resources off our coasts and in the Midwest to do the same job and that marine renewables and geothermal energy can make significant contributions to the mix. As a practical matter, no energy source is likely to do the entire job by itself, but the range of possibilities is large enough that different energy sources can be tailored to specific areas.

The main reason the United States has not progressed further in the search for replacements for fossil fuels is a shortage of money from the federal government for research and development. Because of the political clout of the fossil fuel industry and its lobbyists, most government funding related to energy in past decades has gone to the fossil fuel industry rather than to alternative energy sources. This may be changing with the election of a president who has made the development of alternative energy a top priority. In an address to Congress in February 2009, he said, "The nation that leads the world in creating new sources of clean energy will be the nation that leads the twenty-first century global economy."[63]

The passage of the American Recovery and Reinvestment Act of 2009 provided more than $70 billion in direct spending and tax credits for clean energy and transportation programs, including $11 billion toward developing a smart grid to transport electricity; $4.5 billion to make federal buildings more energy efficient; $2 billion in grants for advanced batteries for electric vehicles; $8.4 billion for mass transit; and $20 billion in tax incentives and credits for renewable energy, plug-in hybrids and energy efficiency. Added to these funds is an eight-year extension of the investment tax credit for solar energy, a three-year extension of the production tax credit for wind, new rules that allow utilities for the first time to participate in investment tax credits, and a new provision that allows renewable energy developers to receive up to a 30 percent government grant instead of a tax credit. And perhaps in either 2010 or 2011, additional supports will be put in place, such as a national renewable portfolio standard and perhaps a cap-and-trade system for carbon emissions.

The transition from fossil fuels to renewable and nonpolluting sources will require many decades, no matter what the government and private companies do. But the future is bright for wind, solar, geothermal, and hydrogen as energy sources. Of these four, only wind is now in widespread use.

Developments such as fuel cells for automobiles are attacking a giant. The internal combustion engine has been around for more than a hundred years to reach the limits of performance and reliability that current users expect and is supported by a solid refueling and repair infrastructure. But the future for alternative and renewable sources of energy is bright. The grandchildren of today's teenagers will wonder what took their grandparents so long to wise up.

8

The Nuclear Energy Controversy: Radiation for Everyone

Man has lost the capacity to foresee and to forestall. He will end by destroying the earth.

—Albert Schweitzer

The energy that can be released from nuclear reactions was introduced to the public in 1945 when two atomic bombs were dropped on Japan to end World War II. Nine years later, the first nuclear plant that produced electricity for commercial use started operating in Russia. The number and generating power of commercial nuclear facilities grew rapidly during the 1970s, from a generating capacity of 16 gigawatts at the start of the decade to 135 gigawatts in 1980. As of March 15, 2009, there were 436 nuclear power plants in thirty-one countries with a net installed electric capacity of about 370 gigawatts (table 8.1). These power plants produced 2,600 billion kilowatt-hours, about 15 percent of the world's electricity. On average, the plants operate at 90 percent of their capacity. Ninety percent of existing reactors were built before 1987, and there has been no net change in the number of reactors since 1997. However, fifty-four plants are under construction, most of them in China, the Russian Federation, India, and South Korea.[1] The number of plants being built is double the total of just five years ago. China alone is preparing to build three times as many nuclear power plants in the coming decade as the rest of the world combined, up to ten each year.

Many nations are heavily dependent on nuclear power for their electricity. The United States is the world leader in both number of reactors and megawatts generated, with 104 reactors generating 100,582 megawatts, 20 percent of our needs; France is second, with 59 reactors producing 63,260 megawatts, 78 percent of that nation's electric needs. Japan is third, with 53 plants generating 45,967 megawatts, nearly one-third of its electricity. No other country generates more than 20,000

Table 8.1
Nuclear power plants worldwide, in operation and under construction, March 2009

Country	In operation		Under construction	
	Number	Electrical net output (in megawatts)	Number	Electrical net output (in megawatts)
Argentina	2	935	1	692
Armenia	1	376		
Belgium	7	5,824		
Brazil	2	1,766		
Bulgaria	2	1,906	2	1,906
Canada	18	12,577		
China	11	8,438	11	10,120
Czech Republic	6	3,634		
Finland	4	2,696	1	1,600
France	59	63,260	1	1,600
Germany	17	20,470		
Hungary	4	1,859		
India	17	3,782	6	2,910
Iran	—	—	1	915
Japan	53	45,957	2	2,191
Korea, Republic of	20	17,647	5	5,180
Lithuania	1	1,185		
Mexico	2	1,300		
Netherlands	1	482		
Pakistan	2	425	1	300
Romania	2	1,300	—	
Russian Federation	31	21,743	8	5,809
Slovakian Republic	4	1,711		
Slovenia	1	666		
South Africa	2	1,800		
Spain	8	7,450		
Sweden	10	8,958		
Switzerland	5	3,238		
Taiwan	6	4,949	2	2,600
Ukraine	15	13,107	2	1,900
United Kingdom	19	10,097		
United States	104	100,582	1	1,165
Total	436	370,120	44	38,888

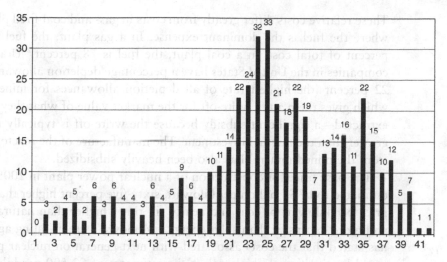

Figure 8.1
Number of operating reactors by age. (European Nuclear Society, 2009) *Note:* Reactor age is determined by its first grid connection.

megawatts. In the European Union, 30 percent of the electricity is generated by nuclear power.

Once completed, a reactor will supply electricity for approximately forty years. Most of today's reactors are at least twenty-five years old (figure 8.1). For nuclear power to play a significant role in satisfying future energy needs, countries must build new reactors to replace those ending their service life and expand significantly the number of commercial reactors in service.

Is Nuclear Energy Renewable?

Nuclear energy is an alternative energy source that in principle could help relieve America's dependence on highly polluting fossil fuels. Nuclear fuel occurs as an element (uranium) in a mineral mined from rocks and is therefore not renewable. But it is sufficiently abundant that neither the United States nor the rest of the world will run out of uranium during the twenty-first century regardless of how many new nuclear power plants may be built.

How Much Does Nuclear Energy Cost?

Operation and maintenance costs in a nuclear plant are 74 percent of the cost of the electricity generated; the fuel itself is only 26 percent.

These relative costs differ greatly from costs in gas- and coal-fired plants, where the fuel is the dominant expense. In a gas plant, the fuel is 94 percent of total cost; in a coal plant, the fuel is 78 percent.[2] Uranium companies in the United States have a percentage depletion allowance of 22 percent (the highest rate of all depletion allowances for minerals), which gives them a tax write-off for the market value of what they have extracted—a significant subsidy because the write-off is typically much greater than their actual investment. The manufacture of the reactor fuel from the mined mineral has also been heavily subsidized.

The base price for construction of a nuclear power plant in 2009 was estimated to be $7 billion per kilowatt, sixty-five percent higher than the price per kilowatt of coal and nearly six times higher than natural gas and more expensive than solar energy.[3] Moody's, a credit rating agency, has stated that the costs associated with next-generation nuclear plants could be significantly higher than the estimates of $3,500 per kilowatt cited by the industry.[4] It is noteworthy that two next-generation plants under construction in Finland and France are billions of dollars over budget. Moody's cautioned that nuclear investment could have a negative effect on a corporation's credit ratings, a concern that had a chilling effect on corporate financial officers. Bolstering this concern, new nuclear plant cost estimates for plants designed by Westinghouse had more than doubled to $12 billion to $18 billion by the end of 2007.

Banks are not willing to loan money for the construction of nuclear power plants, considering them to be too risky, so the federal government has stepped in to prop up the industry. An energy task force established by Vice President Dick Cheney in 2001 endorsed the construction of nuclear power plants, and the administration of President Bush did what it could to stimulate new construction and licensing. The Energy Policy Act enacted in 2005 contained large amounts of money for nuclear tax breaks and loan guarantees. The U.S. government is now offering incentives worth more than $13 billion as seed money for new nuclear plant construction. According to an analysis by the nonprofit group Public Citizen, the Energy Policy Act includes $2.9 billion for R&D, at least $3.5 billion in construction subsidies, more than $5.7 billion for operating subsidies, and $1.3 billion for shutdown subsidies. The package also includes $2 billion for risk insurance, which allows builders of the first six reactors to collect for any delays in construction or licensing, including challenges by the public on grounds of safety. It includes production tax credits of 1.8 cents per kilowatt-hour for eight years, an estimated $5.7 billion to $7.0 billion that would otherwise go to the U.S.

Treasury. There are also provisions for taxpayer-backed loan guarantees for up to 80 percent of the cost of a reactor. Also granted was protection against a nuclear catastrophe. The operator's liability was capped at $10 billion.

But even these subsidies have failed to entice investment from private sources. Thomas Capps, former CEO of Dominion Resources, one of the groups seeking a license for a new reactor, said in 2005 that if his company announced it was going to build a nuclear plant, debt-rating agencies such as Standard & Poor's and Moody's "would have a heart attack, and my chief financial officer would, too."[5]

A reactor meltdown, which is unlikely but possible, could cost much more than $10 billion. The General Accounting Office estimated in 1987 that a major nuclear accident could cost as much as $15 billion, and that was more than twenty years ago. Sandia National Laboratory, working for the Nuclear Regulatory Commission (NRC), estimated that a meltdown would cost $56 billion to $314 billion, not including the cost of losing the facility. (The NRC estimated a 45 percent chance of having a meltdown by 2013.)[6] Taxpayers may have an enormous liability bill to pay.

Polls indicate that the American public supports the construction of new nuclear plants, although no one wants them built in their neighborhood. Public Citizen concluded that the public could subsidize as much as 80 percent of new nuclear costs. Based on public objections to bailing out banks, insurance companies, and other financial institutions in 2009, it is not at all obvious that the public favors the idea of having the government shoulder the financial risk for private enterprises.

The owner of the infamous Three Mile Island nuclear plant in Pennsylvania said in 2007 that if his company wanted to start a nuclear plant today, the licensing, design, planning, and building requirements are so extensive it would not open for at least eight years. And even if the company got all the approvals, it could not start building "because the cost of capital for a nuclear plant today is prohibitive."[7] As of 2008, all nuclear plant construction in the world was running years late and billions of dollars over budget. In the United States, the NRC has delayed approval on all plants under construction indefinitely because of needed design changes. The earliest delivery date for the prototype plant is 2012, so other new plants will not be built for at least a decade.

The nuclear industry in 2006 quoted the cost of nuclear electricity at 1.68 cents per kilowatt-hour (wind 0.2 cents, hydroelectric 0.5 cents, coal 1.9 cents, solar 2.48 cents, natural gas 5.87 cents),[8] but its cost estimate for nuclear electricity does not take account of the soaring costs of plant construction, managing pollution, insuring power stations, protecting them from the twenty-first-century plague of terrorists, decommission risks, and almost-certain repeated litigation.

The current projected cost for new nuclear power plants is about 20 cents per kilowatt-hour, seven times more expensive than coal.[9] Nuclear utilities depend on government bailouts, insurance, and subsidies. From a strictly financial point of view, nuclear power plants should not be the wave of the future in the United States. From a business perspective, nuclear is the highest-risk form of power generation. The electricity these plants generate has moved from the 1960s estimate of "too cheap to meter" to the 2010 estimate of "too costly to matter."

Reactor Construction in Other Countries

Several dozen nuclear power plants are under construction in Europe and Asia, financed at least in part by national governments rather than completely by private enterprise. The construction frenzy has been generated largely by concerns about the rising tide of carbon dioxide emissions worldwide and its effect on climate change. Other important factors are concerns about national energy security and the rapidly increasing cost of oil and natural gas in nations that lack coal.

Many of the fifty-four reactors being built around the world will be operational within a decade. The growth in reactor construction is exemplified by China, whose rapid industrial growth and economic success have been underwritten by greatly increased energy consumption. Most of the energy has been supplied by China's vast coal reserves, and the country has suffered worldwide criticism because of its rapidly increasing emissions of carbon dioxide. China views its increasing use of nuclear power as one way to reduce the country's dependence on coal, a partial response to international criticism of its increasing rate of production of greenhouse gases. Under plans already announced, China plans to build 32 nuclear plants by 2020. Some analysts say the country will build 300 more by the middle of the century, almost as many as the generating power of all the nuclear plants in the world today.[10] The speed of the construction program has raised safety concerns, and China has asked for international help in training a force of nuclear inspectors. The country must maintain nuclear safeguards in a national business culture where quality and safety sometimes take a back seat to cost cutting, profits, and corruption, as shown by scandals in the food, pharmaceutical, and toy industries and by the shoddy construction of schools that collapsed in Sichuan Province in 2008. At present, China derives only 2.7 percent of its electricity from nuclear power.

China is not relying only on nuclear energy as an alternative source. The government is also increasing its use of other alternative energy sources such as hydropower, wind, and solar. Industrial growth depends on growth in energy supplies. However, the planned enormous growth of nuclear power in the coming decade will still generate only 9.7 percent of the country's power because of the rapid growth of electrical demand, according to the Chinese government. It will reduce emissions of global warming gases by only about 5 percent, compared to the emissions that would be produced by burning coal to generate the power. China's economy is growing so fast that even with all the nuclear plants it expects to build in the next ten years, total emissions of greenhouse gases are forecast to rise 72 to 88 percent by 2020.

The United Kingdom is contemplating resuming its long-dormant nuclear construction programs, but many regulatory hurdles remain before any new plants can be started. The proposed new plants will have only a minimal effect on its emissions of carbon dioxide but are fueled by a concern over a need for additional supplies of energy for economic growth.

A number of European countries have banned or restricted nuclear power over the past twenty years. Italy closed all its reactors, and Germany and Belgium have long prohibited the building of new reactors, although those that were open were allowed to operate for their natural life span. However, concerns about energy security are now prompting a reassessment of nuclear power in many countries. Switzerland failed to renew its nuclear ban in a 2003 referendum, and Italy, Germany, and Sweden are among those reconsidering earlier bans. Finland and politically uncertain Iran have nuclear plants under construction, and other politically unstable nations such as Pakistan and Egypt hope to build nuclear power plants. As of 2010, it appears that nuclear power plants are going to become more widespread during the next few decades.

Will Nuclear Power Plants Slow Global Warming?

One of the most cited reasons for the use of nuclear power is its effect on emissions of carbon dioxide gas, thought by most scientists to be the major, if not the only, cause of global climate change (chapter 9). Unlike fossil fuels, no carbon dioxide is generated during the operation of nuclear plants. Unfortunately, calculations indicate that for nuclear power to be a significant part of a mitigation plan for greenhouse gases, it would require that the world build 14 new nuclear plants and replace 7.4 retired ones every year for the next fifty years to have a significant effect.[11] That's 21.4 new plants every year—1,070 over the fifty-year span. It seems doubtful that even the most enthusiastic supporters of nuclear power believe this is possible. Nuclear power plants are a slow-growing (maybe) technology that cannot address global climate change rapidly enough to make a difference.

NASA scientist James Hansen said a few years ago that we had a ten-year window before global warming reaches its tipping point and major ecological and societal damage becomes unavoidable.[12] Recall that even if a nuclear energy project were given federal government approval today and no lawsuits were filed to stop construction, it would take about ten years for the plant to begin delivering electricity. The idea that nuclear power plants are going to stop or even delay climate change is as unfounded as the claim by the nuclear industry that it is cost-effective.

Accidents and Explosions at Nuclear Power Plants

Everyone has heard of the nuclear accident at Three Mile Island, Pennsylvania, in 1979 and the Chernobyl disaster in Ukraine in 1986, but there have been many more less publicized nuclear power plant accidents. Some occurred because of human negligence, and others because of equipment malfunctions; neither cause can be totally prevented. Neither humans nor the things they build are infallible.

Thousands of mishaps have taken place in reactors around the world—30,000 in the United States alone.[13] Adding to the concern raised by these events are falsified safety reports at operating reactors. In 2002, General Electric International, which built and maintains many reactors in Japan, admitted that it had falsified safety records at thirty-seven locations.[14]

Equipment failures are illustrated by the cracks that have occurred in some Japanese reactors and in the nozzles of reactors in the United States. Between September 2000 and April 2001, there were at least eight forced shutdowns in reactors in the United States due to equipment failures caused by aging. It is not comforting that many of America's reactors have been granted twenty-year extensions because new plants have not been built and the government is concerned about a national shortage of electricity. As a former policy advisor at the Department of Energy said, nuclear reactors "are just like old machines, but they are ultra-hazardous." By pushing their operating span to sixty years, he says, "disaster is being invited."[15] He did not mention that most commercial nuclear power plants are located in densely populated areas, where the electricity they generate is most needed.

In 2002, forty-eight of fifty-nine nozzles in the reactor vessel head of a reactor in Virginia were found to be cracked. Leaks have been discovered at the bottom of a reactor's pressure vessel in Texas. In 2006, 148 gallons of a highly enriched uranium solution leaked into a laboratory in Tennessee, causing a seven-month shutdown of the plant.

In 2002, workers discovered a foot-long cavity eaten into the head of a reactor vessel at a plant in Ohio. In one part of the lid, less than 0.4 inch of metal remained to contain over 80,000 gallons of highly pressurized radioactive water. The defect had not been discovered earlier because of falsified reactor vessel logs. The company that operates the plant acknowledged that it had resisted an inspection because that would have forced the temporary closing of the reactor. After reviewing the Ohio event, the NRC inspector general found that "the fact that [the licensee] sought and the [NRC] staff allowed Davis-Besse to operate past December 31, 2001, without performing these inspections was driven in part by the desire to lessen the financial impact on [the licensee] that would result in an early shutdown."[16] The fact that concerns about profitability can trump concerns about public health and safety in a plant using nuclear fuel is both shocking and horrifying.

Seventeen nuclear plants in Japan have been closed because of safety concerns. Nations in eastern Europe and the countries of the former Soviet Union house dozens of operating nuclear reactors that are not safe enough to be licensed in the United States, but these countries do not have the money to make repairs or upgrades or alternative energy sources that would allow them to shut down the reactors. In 2003 partially spent fuel rods ruptured and spilled fuel pellets at a nuclear power plant in Hungary. Boric acid was used to prevent the loose fuel pellets from achieving criticality, the point at which a chemical reaction goes nuclear. In 2005, 20 tons of uranium and 350 pounds of plutonium that were dissolved in nitric acid leaked for several months from a cracked pipe into a sump chamber at a nuclear fuel reprocessing plant in England.

One of the near-misses of catastrophe in nuclear reactor operation occurred in Sweden in 2006.[17] A former director of the reactor said, "It was pure luck there wasn't a meltdown." The reactor was out of control for twenty-two minutes when backup generators failed to fire during a power cut. When such a failure occurs, the operator loses instrumentation and control over the reactor, leading to an inability to cool the core. Failures such as this have also been reported in the United States and Germany. Core meltdown was narrowly averted when an engineer disobeyed rules and overruled safety systems to source power from other parts of the facility.

Reactors in Japan

There have been more than 12,000 cumulative reactor-years of operations worldwide since the 1950s with only two major events resulting from human error—one in the United States in 1979, the other in Ukraine in 1986. Both were widely publicized. Less publicized was the event in 1999 at Tokaimura, Japan, a town 75 miles from downtown Tokyo. Workers accidentally mixed together enough uranium-235 to trigger a runaway chain reaction that burned uncontrolled for twenty hours. The accident unleashed radiation 20,000 times the normal level and injured at least forty-nine people, some critically. Local people were exposed to radiation levels estimated to be one hundred times the annual safe limit. Radioactive rain fell in the surrounding area. Investigators subsequently found higher-than-expected uranium concentrations in the environment around the reactor site, possibly indicating prior unreported accidents.

Japan should be particularly concerned about nuclear power plants. The land area of the Japanese archipelago is the size of Montana, has fifty-five reactors, and is located in an area where several continental and oceanic plates meet. This is the cause of the many earthquakes that rock the islands. There have been ten with magnitudes between 6.6 and 7.9 since 2004. The most recent major one was a 6.5 magnitude quake in July 2009 that created some fifty problems at a large nuclear plant. In addition, the tremors tipped over several hundred barrels of radioactive waste, opening the lids on dozens of the barrels. And 317 gallons of radioactive water flowed into the Sea of Japan. Japanese nuclear power plants are disasters waiting to happen. Building nuclear reactors in Japan is about the same as scattering bread crumbs for flea-infected rats to eat during the fourteenth-century bubonic plague in Europe.

The inspection system in Japanese nuclear plants is lax, and there is a culture of secrecy. For the most part, the nuclear power companies are expected to monitor themselves. Perhaps a bigger problem according to industry analysts is that the power companies employ many workers who are either unqualified for their jobs or badly trained. In Tokaimura, for example, the workers who inadvertently caused the uranium they were handling to ignite in a blue flash did not know what criticality is. According to a scholar at Tsuda College in Tokyo, 90 percent of the workers in the nuclear power industry are "temporary employees who work at plants for one to three months at a time. These people are mostly farmers, fisherman, or day laborers seeking to supplement their incomes. Some of them are homeless."[18]

Chernobyl

The Chernobyl event spread radioactivity from northern Ukraine over nearly all of Europe in 1986, an expanse that is about 5 percent of the world's land area. Heavy doses of radiation spread from 200 miles east of Moscow westward to Scotland, and from northern Norway and Sweden in the north to Greece in the south. Greenpeace has estimated that 270,000 cases of cancer will eventually be attributable to Chernobyl radiation and that 93,000 of them are likely to be fatal. Other investigators have estimated deaths at only a few thousand. There have been unusually high numbers of physical abnormalities and birth defects in people living in the vicinity of Chernobyl. The true cost of the disaster will not be known for many decades, if ever. That disaster was a clear illustration that a nuclear accident anywhere can be a nuclear accident everywhere, unlike accidents with other forms of alternative energy. The collapse of a wind turbine, the fracture of solar panels, or the corrosion of a wave energy apparatus does not endanger humanity's survival.

Statistical Safety

Clearly safety is a major concern in the operation of a nuclear power plant. Major malfunctions or accidents are rare, but how rare is "rare"? In the United States, there have been no immediate radiological injuries or deaths among the public attributable to accidents involving nuclear power reactors, although there may be injuries or deaths from latent cancers in people who live near operating reactors. A team of experts concluded in their 2007 report, "A severe accident at a nuclear power plant is both physically and statistically possible for existing plants, for plant designs under consideration in the near term, and for advanced designs."[19]

How does one evaluate the probability of a rare but possible event?

High-probability events such as automobile accidents or cancer occurrences can be analyzed using actuarial data, as insurance companies do.

In the case of very low-probability events, such as major nuclear plant accidents, actuarial data are either too limited to be meaningful or are nonexistent. Statistical models must be established, and their quality and reliability depend on the assumptions that go into them. The modeling permits a quantitative assessment of what can go wrong, the likelihood that it can go wrong, and the consequences if it does go wrong.

The scenarios are arguably uncertain due to the difficulty of quantifying human responses, and model quality varies from one reactor to the next. Many experts believe that quantitative risk calculations cannot reliably measure the absolute risk of low-probability catastrophic events but can be reliably used to compare relative safety. That is, the models can be used to say that one reactor or facility is safer than another, but we cannot know how safe either one is. This is not very reassuring. Moreover, there are no assessments for the multitude of reactors outside the United States, which is three-quarters of the world's operating reactors.

Irradiating Employees and Their Neighbors

Assume you are working in a nuclear power plant and live nearby. You have read what the experts say and are satisfied that the risk of a catastrophic equipment failure that would irradiate you is small enough that you have no hesitation about going to work. Furthermore, you feel that there are enough fail-safe procedures in place that human error is unlikely to be able to cause a massive failure in the functioning of the plant. Should you be concerned about leakage of radiation in and around the plant site?

The answer is yes. Nuclear facilities emit low-level radiation during normal operations, and many studies have documented adverse effects on animals and people living around the plant.[20] Radioactive ants, roaches, rats, gnats, flies, worms, and pigeons have been found near nuclear plants. A study in 2003 determined that cancer rates for children living within thirty miles of each of fourteen nuclear plants exceeded the national average. One in nine of these cancers was linked to radioactive emissions. A 2004 study reported a 40 percent increase in childhood cancers in two counties closest to reactors in St. Lucie, Florida. A study in the United Kingdom in 2006 found cancer rates among young women living near a closed nuclear power station to be fifteen times higher than the average in the United Kingdom. Another study the same year found the death rate from cancer in neighborhoods close to a U.K. nuclear plant to be as much as 70 percent higher than in other parts of the country. A 2007 study in Germany documented a childhood leukemia cluster surrounding two nuclear establishments near Hamburg.

Sealing the lid on the positive relationship between radiation from the operation of nuclear power plants and cancer in children is a report published in 2002 in the *Archives of Environmental Health*.[21] The researchers found that local infant deaths and childhood cancer rates dropped dramatically following the closure of eight U.S. nuclear plants. Infant mortality fell 17.4 percent in counties lying up to forty miles downwind of nuclear reactors in the two years following the closure of the reactors. Over the same period, the national decline was only 6.4 percent.[22]

There have been studies of the rate of cancers among workers in nuclear power plants in the United Kingdom and at nuclear weapons facilities across the United States.[23] Workers at the Sellafield nuclear plant in the United Kingdom and their children were found to have twice the normal risk of leukemia and lymphoma than others in the region. Among the children, the incidence of these cancers was fifteen times greater in a small village next to the nuclear plant. The risk for the children rose in line with the radiation dose received by their fathers. The U.S. government study found a high incidence of twenty-two different types of cancer at fourteen nuclear weapons facilities. And there is evidence that working in such plants affects both the body cells and sperm of male employees.

Some studies have failed to find increases in cancers in those living near nuclear plants. There is no known explanation for these results. The Nuclear Energy Institute, which promotes the use of nuclear energy to generate electricity, says, "Even if you lived right next door to a nuclear power plant, you would receive less radiation each year than you would receive in just one round-trip flight from New York to Los Angeles. You would have to live near a nuclear power plant for over 2,000 years to get the same amount of radiation exposure that you get from a single diagnostic medical X-ray."[24] The institute has not commented on the radiation-induced cancers I have described.

Fanatical Terrorists and Nuclear Power

Safety and security are interconnected because the systems, processes, and procedures that protect a plant and the public from accidents during normal operations are the same ones that are used to prevent a release of radiation in the event of a terrorist attack or sabotage. It is possible that terrorists who might succeed in breaching plant security could disable safety systems and trigger a fuel meltdown and the release of radioactive materials that would have a major impact on public health and safety and on the environment. Shortly following the 9/11 attack on the World Trade Center, it was reported that the Indian Point Nuclear Station thirty-five miles north of New York City had been mentioned in documents confiscated from terrorism suspects. Energy Secretary Spencer Abraham said there was evidence that terrorists may have specifically targeted the Palo Verde nuclear power station in Arizona, the largest nuclear plant in the United States.[25]

Because of security concerns, it is difficult for outside groups to assess whether today's security measures are adequate. However, it is significant that in January 2003, nineteen Greenpeace activists entered the Sizewell nuclear plant in the United Kingdom, avoiding the guarded entrances and scaling the reactor without resistance. Their goal was to expose the plant's vulnerability. If the intruders had been actual terrorists, the result would have been catastrophic.

The safety and security implications of the global nuclear expansion will depend at least in part on where the expansion occurs. As of 2010, most nuclear plant expansion is expected to occur in developing countries, many of which have weaknesses in legal structures (rule of law); construction practices; operating, safety, and security cultures; and regulatory oversight. The relationship between wealth and security in the construction and operation of nuclear installations is analogous to that of food imports: we expect the food grown in the United States to be cleaner, on average, than food from Third World countries. And it is.

Over the next two to three decades, the fleet of U.S. reactors will be dominated by existing reactors, not by new reactors with improved designs, because most existing reactors are receiving twenty-year operating license extensions. Thus, the safety and security issues surrounding nuclear power plants in 2010 will persist through the 2020s and 2030s. The safety issues may well become worse because of the increasing age of plants that were not designed to last sixty years. There is also the sheer law of numbers. The more plants that are operating and the longer they run, the greater the statistical chance there is of a significant core damage event, either accidental or malicious in origin.

Tearing Down Old Nuclear Reactors

When the first nuclear power plants were built in the 1950s, they were given thirty-year licenses by the government; in 1982 this was increased to forty years. And now most of the reactors have applied for and been granted additional extensions to sixty years.

Thirty-three states currently house America's 104 functioning nuclear reactors (figure 8.2). When they reach the end of their useful life, they must be decommissioned and dismantled, a lengthy and costly process that normally takes three to five years. The NRC allows up to sixty years for completion; the cost at present is about $450 million per reactor, an increase from the 1991 estimate of $120 million. The rules for decommissioning are set by the NRC, and a prime concern is safety. Residual radioactivity must be reduced to a level that permits release of the property for other uses, a process that involves cleaning up of radioactively contaminated plant systems and structures and removal of the radioactive fuel.

It takes three years to remove the 60,000 fuel rods that were in the reactor core during operation. This high level waste has to cool

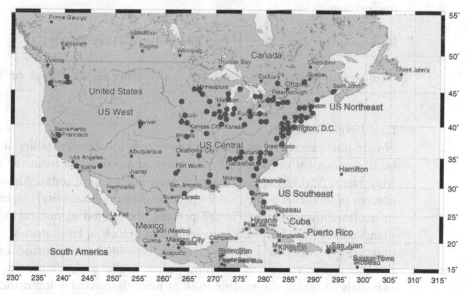

Figure 8.2
Location of nuclear power reactors in the United States. Many sites have more than one reactor. There are no nuclear plants in Alaska or Hawaii. (Nuclear Regulatory Commission)

for at least six months before it can enter interim storage, before being prepared for final disposal. Simultaneously work begins on removing other primary hazardous materials from the cooling systems. Both types of waste must be put in casks and sent for permanent disposal.

The facility must then be decontaminated, a process in which residual radioactive and chemical waste is removed from the buildings and equipment by washing, heating, chemical action, and mechanical cleaning. This creates a cocktail of chemical and radioactive particles that may evaporate into the atmosphere or leak out into the ground.

The reactor itself is sealed for perhaps 100 years, during which time the site is policed and monitored. Then the reactor core can be unsealed and disassembled, although it still contains dangerous radioactive waste. This intermediate waste must be cut into small pieces to be packed in containers for final disposal somewhere.

Finally the containment buildings are demolished. The rubble may be removed or buried on-site. The site is then decommissioned and the land made available for alternative uses. Decommissioning has not been completely achieved for any reactor because of the youthfulness of the nuclear age.

Radioactivity Never Dies; It Just Fades Away—Very Slowly

Radioactive waste is classified for convenience as low level or high level.

Low-Level Waste

Ninety-nine percent of all radioactive waste is low level, meaning it will be dangerous for only 300 to 500 years, approximately from the time King Henry VIII of England married Anne Boleyn until today. About 63 percent of low-level waste (but 94 percent of the radioactivity) originates in nuclear power plants. The 37 percent not related to nuclear power plants comes from hospitals, university and industrial laboratories, manufacturing plants, and military facilities. Low-level waste includes contaminated items such as clothing, hand tools, and, eventually, the materials of which the reactor itself is built. Radioactive items include fabric, metal, plastic, glass, paper, and wood.

Low-level wastes are usually sealed in metal drums, commonly after being burned in special incinerators to reduce their volume, and they are either stored above ground in vast holding pens, near the place where they are generated, or buried in shallow trenches beneath about 3 feet of soil. For decades, these storage areas have been described as temporary sites, to be emptied and transported to a permanent site somewhere. At these designated temporary shallow burial sites, the metal drums corrode as the years roll on, and stress is placed on them by the continuous radiation from the material in the drums. How long they will remain coherent is unknown. Many drums at some sites have leaked radioactive liquids into the soil and groundwater.[26]

Another disposal method for low-level waste, now banned, was to cast it into concrete, encase it in steel drums, and dump it into the deep ocean. Tens of thousands of tons of these drums were dumped prior to 1993 at an internationally agreed site in the Atlantic Ocean 1,300 miles southwest of England, where they sank 15,000 feet to the ocean floor. They are supposed to remain safe and undisturbed there for at least hundreds of years. However, if the drums eventually leak, the concrete in them, which is composed of calcium carbonate (limestone), will dissolve in the undersaturated ocean bottom water and release the drum's contents. Those who favor the sub-sea method of disposal believe that any leakage over time will be diluted into insignificance by the vast mass of ocean water and dispersed harmlessly into the surroundings. The flaw in this scenario is that fish that swim near the leaking drums become irradiated. When the smaller fish are eaten by larger carnivorous fish, the irradiation is spread in an ever-widening area whose limit is unknown. In other words, the barrel dumpers are conducting an experiment whose outcome is unknown. Although deep-sea dumping no longer occurs, radioactive waste is still pumped into nearshore waters off the coast of England.

Recently an Israeli company developed a process for dealing with low-level waste using plasma gasification melting technology.[27] First, plasma torches break down the waste; carbon leftovers are gasified and inorganic components are converted to solid waste. The gas is purified and is used to operate turbines to generate electricity. The remaining vitrified material is inert and can be cast into molds to produce tiles, blocks, or plates for the construction industry. The processing facility can convert up to a ton of waste per hour and produces energy, 70 percent of which goes back to power the reactor with a 30 percent excess, which can be sold. Processing costs $3,000 per ton and reduces the waste to 1 percent of its original volume. Considering that 99 percent of radioactive waste is low level, the plasma gasification process holds great promise.

High-Level Waste

High-level radioactive waste is a by-product of both nuclear weapons production and commercial reactors, although the relative amounts from these two sources have not been revealed. The waste consists of spent fuel rods and liquid materials involved in reprocessing spent fuel to recover highly radioactive plutonium that was produced at more than 100 former nuclear weapons sites. There are now more than 100 million gallons of high-level liquid waste in storage at 158 sites in forty-two states (figure 8.3).

The United States stopped civilian reprocessing of spent fuel in 1981 as part of its nonproliferation policy. This ban was put in place because plutonium, an element that does not occur naturally but is produced as

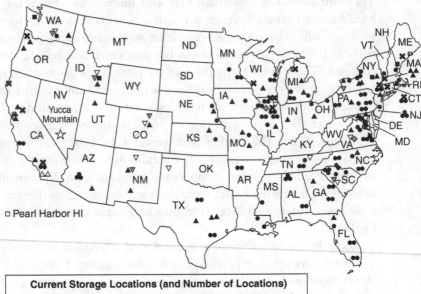

Current Storage Locations (and Number of Locations)
- ● Operating commercial reactors (104, at 72 sites in 33 states)
- ✖ Shutdown commercial reactors with SNF on site (14)
- ◆ Commercial SNF (not at reactor) (2)
- ■ Naval reactor fuel (7)
- ▲ Non-DOE research reactors (45)
- △ Shutdown non-DOE research reactors with SNF on site (2)
- ▽ DOE-owned SNF and HLW (10)
- ▼ Surplus plutonium (6)

Figure 8.3
Approximate locations of surface storage sites for spent nuclear fuel from commercial reactors and for other high-level waste and radioactive materials. (Department of Energy, 2009)

a by-product in nuclear reactions, is recovered during reprocessing and can be used to make nuclear weapons. The philosophy was that the less plutonium there is in the world, the lower the danger is that the element will fall into the wrong hands. Reprocessing is now allowed in the United States, but there are no operating reprocessing facilities.

Reprocessing can potentially recover up to 95 percent of the plutonium and remaining uranium in spent nuclear fuel, putting it into new mixed-oxide fuel. Segregating the uranium and plutonium in this way produces a reduction in long-term radioactivity within the remaining waste because it is largely short-lived fission products. It also reduces the volume by more than 90 percent.

Reprocessing of civilian fuel from nuclear reactors is currently done on a large scale in Britain, France, and Russia and an expanding scale in Japan; soon it will be done in China and perhaps India. France is the most successful reprocessor, recycling 28 percent (by mass) of its yearly fuel use—7 percent within France and 21 percent in Russia.

Spent fuel rods are stored temporarily in shielded basins of water, usually on-site. The water provides both cooling for the still-decaying fission products and shielding from the continuing radioactivity. After many decades, the fuel rods may be moved to a dry-storage temporary facility or dry cask storage, where the rods are stored in steel and concrete containers until their radioactivity decreases (decays) to levels safe enough for further processing. This interim storage stage spans years or decades or millennia, depending on the type of fuel. This means that a power plant site, even when it has reached the end of its electricity-generating life, must remain staffed to operate the cooling ponds, with full security to prevent site intrusions and theft of the spent fuel rods.

Who is going to pay for this? Consider the cost of staffing for perhaps 100 years when there is no revenue stream from electrical generation to cover this and no customers to carry the financial burden. This problem is only now starting to be realized because so few reactors have reached the end of life in the United States. The major electrical utilities that own and operate nuclear power plants have not considered this cost in their balance sheets.

Much more daunting is the fact that the material will remain radioactive for perhaps 100,000 years. No civilization has lasted for more than one thousand years. No spoken or written language (to mark the site or provide maintenance instructions, for example) has lasted more than a couple of thousand years. No objects, metal tools, or vessels created by humans have survived corrosion and structural failure for more than a couple of thousand years. Is it reasonable to believe we can build fuel rod containers that will last 100,000 years? Nuclear power plants are monsters that may come dangerously close to devouring us in the name of "more." What have we done?

Fernald Nuclear Site

An interesting view of the handling of nuclear waste at a "temporary" storage site is provided by the Feed Materials Production Center in Fernald, Ohio.[28] From the time it opened in 1951 until it closed in 1989, the facility enriched 500 million pounds of uranium, two-thirds of all the uranium used in the nation's cold war nuclear weapons program. The center also generated 1.5 billion pounds of radioactive waste. There appeared to be no problems until 1985, when neighbors discovered that the plant's waste had polluted their air, soil, and drinking water. The neighbors sued, and the resulting publicity prompted similar revelations at nuclear facilities around the country.

The site originally included a leaky silo filled with highly radioactive uranium sludge that emitted enormous amounts of poisonous radon gas. Officials at the Fernald center dumped radioactive waste into pits only twenty yards from a creek that sits on top of the Great Miami Aquifer, one of the largest aquifers east of the Mississippi. Rainwater carried uranium into the creek, where it sank and contaminated 225 acres of the aquifer.

When the Department of Energy (DOE) ran out of room to bury waste at the 1,050-acre Fernald site, it was packed into 100,000 metal drums, which were left outside, exposed to rain, snow, and wind. Accidental releases from the drums covered 11 square miles of surrounding farmland in radioactive dust. After lawsuits and much negotiating with the people living near Fernald, the federal government agreed to move 1.3 million tons of the most contaminated waste to other "temporary" storage sites in Texas, Nevada, and Arizona. The transfer required 197 trains, each with 60 railcars carrying 5,800 tons of contaminated soil. The final trainload of radioactive waste left Fernald in October 2006. Residents in the Fernald area agreed to place the rest, 4.7 million tons, in a landfill at Fernald.

The landfill is a winding trench 30 feet deep that is filled with uranium-laced soil piled to a height of 65 feet above the landfill. It is three-quarters of a mile in length. The landfill is bordered by a wall of rock, plastic, and clay 9 feet thick whose base is 30 feet above the aquifer. The DOE expects the wall to last at least 1,000 years. The government spent

$216 million on buildings to clean the site, and when the buildings were no longer needed, each had to be demolished, decontaminated, and placed in the landfill. The DOE also built a pumping system to suck contaminated water out of the aquifer and purify it. The cleansing process is expected to be completed in about 2023. The Fernald site is now open as a park, but the landfill is off-limits to the public. The cleanup at Fernald took thirteen years and cost $4.4 billion.

There are 104 operating reactors at sixty-five sites in the United States, and many more reactors exist at many more sites that are no longer operating. Fernald is an example of the problems, time, and cost of cleanups.

What Should We Do with Old Nuclear Waste?

In April 2005, the National Academy of Sciences (NAS) released a report assessing the dangers of keeping the spent fuel in cooling ponds at sixty-four power plants across the United States. The NAS report noted that choking off the water that cools the pools could trigger a radioactive fire that some scientists believe could cause as much death and disease as a reactor meltdown.

However, long before the NAS report, it was obvious to all who considered the problem that a large and permanent storage facility for high-level radioactive waste was needed, and in 1982 Congress ordered the federal government to find a suitable site. Such a site would need to have several characteristics.

• Geological stability. The repository should be stable for perhaps 1 million years, meaning no earthquakes, crustal deformation, faulting, or heat flow. Volcanic activity would be disastrous to the integrity of the storage cavity.
• Low groundwater content and flow at repository depths, with a projected stability of at least tens of thousands of years.
• Good engineering properties that readily allow construction of the repository as well as operation for periods of at least decades.

Congressional representatives in the 1980s lobbied intensely against locating the site in their districts. Who wants tens of thousands of tons of waste that is highly radioactive and will remain so forever in terms of human life span located near where they live? The 1980s was a good time for a state to have powerful representatives in Washington. Most states were clearly not suitable, and their representatives could rest easy.

For example, areas with abundant rainfall are likely to have high water tables that might feed corrosive water into the repository, so Florida and Louisiana were out of contention; California was eliminated because of frequent earthquakes; Hawaii and Washington have active volcanoes and so escaped serious consideration. Most states that might be suitable as possible sites found grounds (some technical but usually political) to have themselves ruled out of the running.

Ultimately Nevada, a state with relatively few people and no powerful representatives in Washington in the 1980s, drew the short straw. The location was to be at Yucca Mountain, 87 miles northwest of Las Vegas. Movement of high-level waste was set to begin in 1985. The state energized its legal apparatus immediately and has launched repeated lawsuits against the government. Gambling casinos in Las Vegas mobilized against the choice, fearing the repository would scare away the flow of customers to casinos. In addition, Las Vegas is one of the fastest-growing cities in the country, a growth that might be seriously slowed by the prospect of trains carrying dangerous materials passing through the city for decades on the way to Yucca Mountain. As a result, the planned opening in 1985 was pushed back to 1989, then 1998, 2003, 2010, and most recently to 2017.

Objections to locating the storage facility at Yucca Mountain are scientific as well as political. Although southern Nevada is dry and the water table is located 1,000 feet below the 1,000-foot depth of the repository, climates change over time, with or without global warming. Fifteen thousand years ago, much of the western United States was rainy, had high water tables, and contained many lakes (figure 8.4). The largest, of which the Great Salt Lake is a remnant, was 1,000 feet deep and was as large as Lake Michigan. Much of the Sahara Desert was fertile only 5,000 years ago. Climate regimes cannot be predicted thousands of years into the future.

Another worrying factor is volcanic eruptions. Eight volcanoes have erupted within 30 miles of Yucca Mountain in the past 1 million years. And there are concerns about earthquakes as well. Although the areas most likely to have earthquakes are well known, their exact locations cannot be predicted. A magnitude 5.2 earthquake struck near Yucca Mountain in 1992, and hundreds of smaller quakes have occurred within a 50-mile radius of the site during the decades that such records have been kept. And keep in mind that we are talking about a cavity in the ground that needs to be stable for at least tens of thousands, if not hundreds of thousands, of years.

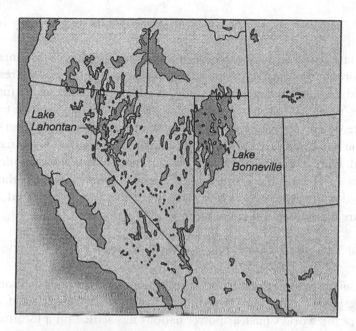

Figure 8.4
Location of large lakes in the western United States 15,000 years ago. The Great Salt Lake in Utah is the only large remnant still present.

In the meantime, high-level waste continues to accumulate at the nation's reactors. Companies claim that their storage areas are full or nearly full and demand immediate action from the federal government. The rising threat of terrorism has increased the urgency of their demands. Transporting the high-level waste to Yucca Mountain from around the country would require daily shipments for twenty years. According to the NAS, the material could be moved in 55,000 truckloads or 9,600 dedicated trainloads and 1,000 truckloads.[29] They were concerned, however, about the safety of the casks containing the spent fuel in the event of a fierce, sustained fire. Such fires do happen. In at least two cases, trains carrying petroleum-filled tankers burned for days before firefighters got them under control. The only way to minimize that risk, the panel concluded, is to make sure such petroleum trains go nowhere near trains carrying nuclear waste. The panel did not assess the dangers posed by terrorist attacks.

As of 2010, the 77,000-ton planned capacity of the Yucca Mountain repository is oversubscribed. And the estimate does not include storage space for 2,500 tons of glassified radioactive liquid wastes stored "temporarily" at 130 nuclear power plants and military installations.

So far, about $8 billion has been spent studying Yucca Mountain in the more than twenty years since the project was started under President Ronald Reagan. But in March 2009, President Obama cut off funding for the Yucca Mountain repository, fulfilling a campaign promise to shut the site down. The president is not a fan of increasing the nation's reliance on nuclear power, and Nevada is no longer powerless in the nation's capital. Harry Reid (D-Nevada) is the current Senate majority leader and objects to placing the radioactive fuel rods in his state. Whether the planned storage site is permanently dead or might be resurrected under a future administration is not clear.

Should There Be a Global Nuclear Waste Repository?

The United States is not alone in its search for a suitable million-year resting place for its high-level radioactive nuclear waste. As of 2010, none of the world's nuclear power nations has settled on a location and built a repository for their high-level waste. But a potential savior arrived in 2001: President Putin of Russia offered an area in a remote and sparsely settled area near Krasnoyarsk in south-central Siberia as a world repository.[30] Russia has immense nuclear experience and technical know-how, plenty of sparsely populated territory, and a government quite willing to negotiate terms for the acceptance of the world's more than 200,000 tons of waste. Experts estimate that the amount of spent nuclear fuel increases by about 5 percent a year.

Nevertheless, many nations have serious concerns. Given Moscow's notoriously poor record on the management of nuclear materials over the past five decades, the extensive and continuing nuclear contamination of Russian territory, a poor security record in controlling access to nuclear materials since the collapse of the USSR, continuing secrecy and corruption within the nuclear sector, and a tendency to ignore or squash unwanted public opinions in Putin's ever-more centralized Russia, the international community has thus far not taken up the offer.

Even after a decade of international efforts to promote better security, nuclear sites in Russia continue to be plagued by severe security shortcomings that would not be tolerated in Western European countries: security alarms routinely ignored or deactivated, doors left open and unguarded, fences in need of repair, widespread corruption and accounting irregularities, and, most alarming of all, rampant theft of valuable materials, including radioactive isotopes, typically by or with the collusion of insiders. The correction of these problems has been further hampered by the fact that the government agency responsible for monitoring the safe operation of nuclear facilities in Russia has long been underfunded and understaffed and has demonstrated little ability to exert independent oversight over either the civilian or military components of the Russian nuclear complex.

A country hosting an international repository must also be willing to subject its operation to international oversight, and its credibility in providing full information on the status of nuclear materials must be exceptionally high. Russia's record of continued secrecy and deception on key nuclear issues does not inspire confidence and trust. The Russian government has provided quite a bit of information on contaminated sites, but it continues to prosecute whistle-blowers and to fall back on old habits of secrecy and denial whenever convenient. Independent surveys reveal that 90 percent of Russians polled opposed importing nuclear waste. The government chose to ignore this information, a damning testament to the weakness of democracy in Russia today. Speculation about the security and viability of the Russian solution to the world's spent fuel dilemma cannot be divorced from speculation about Russia's future political trajectory.

What's the Bottom Line?

The evidence is overwhelming that nuclear plants should not be built anywhere in the world. They are more expensive to build and operate than other types of alternative renewable energy, and they leak radiation that many investigators have shown cause cancers of several types in both plant workers and people living nearby. Children are particularly affected. The possibility of reactor malfunctions and accidents caused by human error can never be eliminated, terrorist threats are always present, and there is no politically or scientifically acceptable way to dispose of nuclear waste.

Perhaps nuclear energy's biggest disadvantage is international security. The level of confidence in the Nuclear Non-Proliferation Treaty and nuclear safeguards is low. North Korea is the latest country to show how easy it is to divert nuclear fuel to make weapons, and nuclear sites are likely to remain targets for terrorists. Such concerns should limit where plants can be built, but they do not. Politically unstable countries are increasingly likely to build nuclear plants.

Presidential administrations in the United States have been more concerned about electricity shortages and resulting public outcry than about the long-range safety implications of nuclear reactors (and fossil fuels). But the chairman of the Federal Energy Regulatory Commission, an appointee of President Obama, has indicated a new direction for America's energy policy. He said that no new nuclear or coal plants may ever be needed in the United States and has emphasized the development of alternative energy sources.[31]

Scenarios illustrating the adequacy of nonnuclear alternative technologies have been compiled repeatedly by the Union of Concerned Scientists in the United States and organizations of scientists in Europe. While none of these analyses has ever been refuted, they are ignored by nuclear power advocates.[32] Alternative and nonpolluting sources of electrical power such as wind, solar, biomass, and geothermal can be brought online as rapidly as new nuclear plants can be built. Nuclear plants are unnecessary and extraordinarily dangerous. Nuclear reactors should be sent to a technology museum as an example of a bad idea. Future generations will look back at our current folly and wonder why their ancestors were so blind.

9

Climate Change: What Have We Done?

The climate system is an angry beast, and we are poking it with sticks.
—Wallace Broecker, oceanographer and climate scientist, 2008

After what seemed like interminable wrangling over the reality of climate change, the debate seems to have subsided. The evidence for global warming, the most publicized sign of the change, is now so overwhelming that even most of the hardened skeptics are onboard.

Arguments have raged for years about the adequacy and reliability of numerical data and computer modeling, but biological evidence cannot be refuted or doubted. The behavior of both plants and animals, which cannot lie, cannot be deceptive, and is unaffected by emotionally or politically based points of view, is totally convincing (table 9.1). Recognizing this, the hardiness zones determined by the Department of Agriculture—regional maps by which gardeners evaluate what can live in the ground through the winter—have been redrawn to reflect the northward march of plants into areas that were once inhospitable.

The nonbiological signs of warming have always been persuasive to most climate specialists but have been questioned by those who doubt the reality of climate change (table 9.2). Many of those who do not believe in climate change have this view because they confuse the issue of whether it is getting warmer with the question of whether humans are responsible. Polls reveal that Republicans, who typically are more protective of the fossil fuel industry, are more likely to doubt climate change than are Democrats. As discussed later in this chapter, it certainly is possible that astronomical factors are the main cause of the increase in the carbon dioxide content of the troposphere and that human additions have a relatively small effect.

Table 9.1
Biologic indicators of climate change

- Plants now bloom an average of 5.2 days earlier each decade.
- The density of vegetation has increased in the United States because of both warming and increased rainfall.
- As permafrost has melted, arctic shrubs in Alaska have invaded previously shrub-free areas. In Antarctica, mosses now colonize formerly bare ground.
- The ranges of migratory plants and animals have shifted an average of 3.6 miles per decade toward the poles (2,000 feet per year).
- The tree line in Siberia is moving north. Conifers now grow where no living tree has grown in 1,000 years. Tree lines in Europe and New Zealand have climbed to higher altitudes.
- Plants in Europe and North America now unfold their leaves and flower up to three days earlier than in previous decades. Trees are leafing earlier throughout Europe, and fall colors appear later.
- Marine mammals are being pushed northward as warmth-loving fish invade their feeding grounds in the northern Bering Sea.
- Warm-water fish are flourishing, but cold-adapted species are in decline.
- Over the past four decades, the traditional mid-February to April maple sugaring season in northern New England has slowly gotten shorter. It now starts a week early and ends ten days early. Maple trees are moving northward to Canada, being replaced by oaks.
- Mosquito-borne diseases have moved northward in Asia and Latin America over the past decade.
- Thirty-nine butterfly species in North America have shifted their range northward by 124 miles since 1975 (4.6 miles per year).
- The Audubon Society has reported a striking northward movement by most species of birds. The wild turkey has moved 400 miles north, the boreal chickadee 280 miles.
- Robins are building nests 240 miles north of the Arctic Circle.
- Barn owls, finches, dolphins, and hornets are arriving in Arctic villages where the natives have no words in their language to name these creatures—and no barns.
- Red foxes have moved north in Canada as Arctic foxes have retreated.
- In the Yukon, squirrels are mating and giving birth eighteen days earlier than their great-grandmothers because the amount of their main food (spruce cones) is increasing because of lengthened summers.

Table 9.2
Nonbiological indicators of climate change

- The twelve warmest years since 1900 have occurred in the last 25 years; the eight warmest have been since 1998.
- In 2005 the global average temperature was the warmest since scientists began compiling records in the late 1800s.
- Most of the world's glaciers and ice caps are melting.
- Arctic sea ice has disappeared at an average rate of 7.8 percent per decade since 1953. Commercial shipping may be possible by 2015.
- The Great Lakes are thawing two days sooner each ten years.
- Average global sea levels have risen about 9 inches since 1901; more than 10 percent of this rise occurred over the past decade.
- On the Antarctic Peninsula, the average temperature has risen by about 4.5°F since 1950, and 212 of the 244 glaciers on the peninsula are melting.
- The Arctic's winter temperature is now 3 to 4 degrees Fahrenheit higher than it was a hundred years earlier.
- The area of permanently frozen ground (permafrost) in Canada has decreased by 30 percent since 1900. Potentially arable soil is appearing.
- The Great Lakes, the earth's largest concentration of freshwater, are thawing earlier each spring, based on records dating back to 1846.

How Hot Is It?

Meteorological stations around the world have been recording daily temperatures near the ground for hundreds of years, so the database to determine an average earth surface temperature is very large. The data show that the average temperature has been increasing for more than a hundred years—not in a straight line, but the trend is clearly upward (figure 9.1). The ups and downs over short intervals result from interactions among the many feedback loops in the factors that influence climate. Studies indicate that positive loops, which reinforce higher temperatures, are more abundant than negative loops, which mitigate them.

It is noteworthy that the temperature has been increasing more rapidly in recent years than in the past. The rate of change in the last twenty-five years is triple the rate (in Fahrenheit) of the previous 100 years. This rate of change surpasses the worst-case scenarios predicted only a few years ago.

The average earth surface temperature in 2008 was 57.9°F, an increase of about 0.9°F, or 1.6 percent, above the twentieth-century average. This

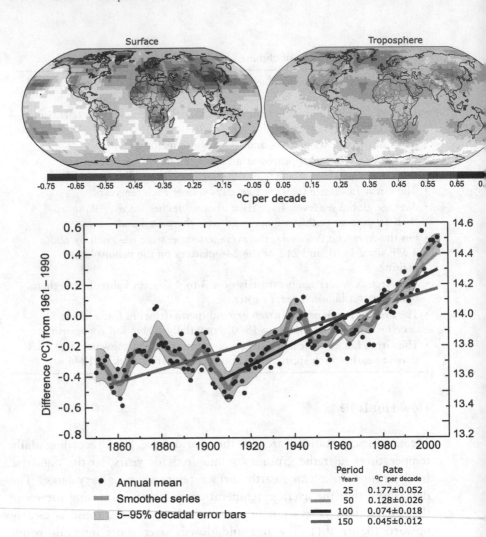

Figure 9.1
Global surface temperatures, 1980–2005. (°F = 1.8 °C + 32) (*Climate Change 2007: The Physical Science Basis*. Working Group 1. Contribution to the Fourth Assessment Report of the Intergovernmental Panel on Climate Change, Figure TS.6. Cambridge University Press).

increase seems quite small when compared to the regular occurrence of daily temperature variations of 10 to 20 degrees that everyone experiences, but the number of observed changes in the inorganic world and the migrations of plants and animals noted earlier testify to the exquisite sensitivity of both the organic and inorganic world to even small changes in average yearly temperature. And the ecological changes are not only in the range of mobile species. There is evidence that the genetic makeup of species such as fruit flies and birds is changing to adapt to the warmer temperatures.[1] The fact that a change in average temperature by less than 1 degree in a century, one-hundredth of 1 degree per year, can change so many biologically controlled things on the earth's surface is hard to believe. But it cannot be denied.

The scientific consensus is that the world can probably expect centuries of gradually increasing temperatures and attendant disrupting effects. Even if humans stop burning coal and oil tomorrow, we have already spewed enough greenhouse gases into the atmosphere to cause temperatures to warm and sea levels to rise for at least another century. This results from thermal inertia. Even if no more greenhouse gases are added to the atmosphere, an impossibility in the world's existing fossil fuel based economies, global air surface temperatures will rise about another 1°F and global sea levels will rise at least 4.3 inches by 2100.[2] As we continue to pump greenhouse gases into the atmosphere, the undesirable effects get worse—probably several degrees more of temperature increase and many more inches (perhaps feet) of sea level rise. Our ability to minimize these effects depends on how fast action is taken by people and their governments, but the longer we wait before implementing changes in our energy use and food consumption patterns, the smaller the effect we can have on them in this century.

Who's Going to Suffer?

What do we have to look forward to as the temperature climbs?

We can expect current trends to continue and probably accelerate—continuing migrations of plants, animals, and perhaps humans; spreading of malaria and other heat-dependant diseases; more frequent and severe heat waves; increased average rainfall occurring in heavier downpours; increased river flooding; water stress in lowland areas that depend on winter mountain snowpack for summer moisture; more severe hurricanes and storm surges; loss of most of the world's glaciers; rapidly rising sea level; increased coastal erosion; some acidification of the oceans; and possible changes in major ocean currents that affect coastal and national climates.

The effects will differ by region. The tropics and high-latitude regions will see more rainfall, while temperate regions, many of them already short of water, will get drier. High latitudes will warm faster than lower latitudes. On a more local scale, the changes become more complex and hard to predict.

Effects on Living Organisms

The poleward march of the living has already begun. This will affect American agriculture in terms of the areas best suited for growing crops. Economic disruptions are certain to occur as states previously too cool for certain crops become able to grow them, and states that dominated the production of certain crops lose their dominance. If needed, genetic modifications of crops will likely be engineered by the agrobiotech companies to mitigate the adverse effects of increased temperature. Weeds and crop pests will increase with warmer temperatures, but they can be controlled. Pollinating insects will increase. Overall, agricultural productivity in the United States as a whole will probably not be seriously affected.

Many plants and animals are already responding in subtle ways to local climatic changes. In the 1970s, mosquitoes became one of the first organisms in which scientists observed genetic changes that might be attributed to global warming. More recently, scientists reported that the genetic makeup of organisms ranging from fruit flies to birds might also be responding to climate trends. For example, in 1970, California white-crowned sparrows sang fast machine-gun trills. Just a few decades later, the sparrows sang noticeably slower songs because the bird's habitat has gotten scrubbier, and their melodies have evolved to better penetrate the thickets.[3] Slow songs and low-pitched sounds transmit better through dense vegetation, whereas high notes carry farther in open environments. Overall, grassland birds do have faster, shriller songs than those from leafy surroundings have. Adaptation apparently required no more than a few decades. Another example is the effect that increased carbon dioxide in the air has had on the quaking aspen, which has had a 50 percent acceleration in growth in suitably moist climates over the past fifty years.

According to the International Union for Conservation of Nature and Natural Resources, 20 percent of mammals, 30 percent of amphibians, and 12.5 percent of birds are threatened or endangered. But the Intergovernmental Panel on Climate Change (IPCC) appraised published studies of biological systems. Of 59 plants, 47 invertebrates, 29 amphibians and reptiles, 388 birds, and 10 mammal species, approximately 80 percent showed changes in the biological parameters measured, including start and end of breeding season, shifts in migration patterns, shifts in plant and animal distributions to higher elevations and toward the poles, and changes in body size and population numbers.[4] Living creatures are adaptable. The biological modifications the IPCC found are consistent with global warming predictions.

The fate of plants in a warming world concerns botanists, but their worries are likely unnecessary. Biologists from University Centre in Svalbard, Norway, studied the DNA of plants on the island. They found that over the past 10,000 years, different species of plants have had little trouble making the hop from the Norwegian mainland to the island 600 miles away. Plants are more adaptable to changing their habitable ranges than we might think. Those in the business community may justifiably be concerned as the things in the natural world that they depend on change locations as the climate changes, but for those interested only in the survival of life forms on planet Earth, there is nothing to worry about.

Climate change can have adverse or beneficial effects on species. For example, climate change could benefit certain plant or insect species by increasing their ranges. The resulting impacts, however, could be positive or negative depending on whether these species were invasive and undesirable (e.g., weeds, crop pests, tree-killing beetles, ticks, or mosquitoes) or valuable to humans (e.g., food crops, enemies of crop pests, or pollinating insects).

The observed changes are compelling examples of how rising temperatures can affect the natural world and raise questions of how the most vulnerable populations will adapt to future increases in temperatures and other climatic changes. For example, many fish species have begun moving into now-warmer arctic waters. Between 1982 and 2006, pollack moved their northward extent 30 miles northward; halibut 35 miles; rock sole 48 miles; and snow crab 55 miles.[5] The risk of extinction could increase for some species, especially those that are already endangered or at risk due to isolation by geography or human development, low population numbers, or a narrow temperature tolerance range. The IPCC estimates that 20 to 30 percent of plant and animal species are at increasing risk of extinction if the global average temperature increases by another 2.2 to 4.0 degrees.

The American pika, a small rodent that lives on the slopes of mountains in the western United States, can overheat when temperatures hit 80 degrees. Over the past century, these creatures have kept climbing, reaching new ranges that can be 1,300 feet upslope. When they reach the highest point in the landscape, it will be the end for them. Probably few people other than zoologists who specialize in rodents will lose much sleep over the possible decline of the pika, but humans may see some of their most beloved species decline or disappear.

Polar Bears   One warm, fuzzy, and charismatic mammal whose fate is causing concern among many people is the polar bear. Polar bears spend much of their life on the Arctic ice and use it as a hunting ground for fat and meaty seals. When ice on Canada's western Hudson Bay started to break up earlier than usual—three weeks earlier in 2004 than in 1974—the bear population fell by 21 percent in seventeen years.[6] Shrinking ice has also been blamed for cannibalism among polar bears off Alaska, something not seen before 2004. The seals may be pleased, but many humans are not.

Although data are very fragmentary, scientists say that the polar bear population has more than doubled since the 1960s and that 20,000 to 25,000 of them can be found across the Arctic region from Alaska to Greenland.[7] Nevertheless, in September 2007, the U.S. Geological Survey predicted that two-thirds of the world's polar bears could die out in fifty years.[8] The National Resources Defense Council cautioned that "birth rates among the polar bears are falling, fewer cubs are surviving, and more bears are drowning."[9] The World Wildlife Fund believes that polar bears could become extinct by the end of this century, and the Polar Bear Specialist Group agrees.[10]

Are these urgent calls to action justified? What do we know about the living habits and survival potential of polar bear populations? Contrary to media portrayals, they are not fragile "canary in the coal mine" animals, but robust creatures that have survived past periods of extensive deglaciation. The oldest polar bear fossil found so far is 130,000 years old, suggesting that the species is at least 200,000 years old and has persisted through past interglacial periods when global and Arctic temperatures were at least several degrees warmer than today (and sea level was 15 feet higher than today's level) and when levels of summertime Arctic ice were likely considerably lower than today.[11] The last warm span between ice ages peaked about 125,000 years ago. Although many of the bears may have drowned or starved during previous interglacial periods (no humans were around to save them), they have survived over at least two glacial cycles, in which the amplitude of environmental change in the Arctic was quite large, indicating they have the ability to adapt to and survive such changes. Earth's creatures are adaptable. If seals are not available, the bears turn to other food sources. Some polar bears in 2007 were seen swimming and eating fish in a river estuary packed with char, a relative of salmon, that were migrating upstream.[12] Panic about a possible total loss of the polar bear population due to melting Arctic ice does not seem justified based on existing knowledge.

Perhaps of more danger to polar bears (and other Arctic creatures as well) are the pesticides from agricultural lands far to the south of the bears' habitat. They are transported northward by rivers, ocean currents, and air and are consumed by the bears. Scientists report that polar bears have some of the highest concentrations of chemicals of any Arctic mammals. The bears already exhibit immune, endocrine, and reproductive effects. In December 2009, the Center for Biologic Diversity filed a lawsuit against the Environmental Protection Agency (EPA) for not considering the impact of Arctic pesticide pollution on the polar bear population.

Charismatic Megafauna Those concerned about global warming have long used charismatic megafauna that invoke powerful emotional responses in humans to raise awareness of potential ecological problems. Among the many examples are the giant pandas ("teddy bears" from our childhood), whales (lords of the seas), gorillas (they look like us), penguins (cute little guys in tuxedos), and eagles (our national bird). Penguins even got their own movie, *March of the Penguins*, in 2005. Polar bears are cute as cubs and majestic as adults, and were featured in Al Gore's film *An Inconvenient Truth* in 2006. The use of these "poster children" has been shown to increase financial contributions from concerned citizens. Warm and cuddly creatures open wallets as well as hearts.

When it began compiling lists of threatened and endangered animals and plants more than thirty-five years ago, the U.S. government gave itself the same mandate as Noah's ark: Save everything. But in practice, the effort has often worked more like a velvet-rope nightclub: Glamour rules.[13] Pretty things are more desirable than ugly things. The furry, the feathered, the famous, and the edible have dominated government funding for protected species, to the point where one subpopulation of salmon gets more money than 956 other plants and animals combined. The top fifty best-funded species include salmon, trout, sea turtles, eagles, bears—and just one insect and no plants.

Only fifteen U.S. species have officially been declared "recovered." They are three plants, two obscure tropical birds, and ten animals that would look good on a T-shirt. These include gray wolves, bald eagles, brown pelicans, and the Yellowstone subpopulation of grizzly bears. There clearly has been a very heavy bias toward charismatic megafauna. All other classes of fauna, and all flora, have gotten extremely short shrift.

Is the Environment Threatened or Resilient?   Environmentalists who focus on possible harm to plants and animals that might result from climate change invariably describe ecological environments as "delicate," "sensitive," "threatened," and "endangered." They never describe such environments as resilient, adaptable, mobile, robust, or accommodating. However, as the report by the IPCC illustrates, environments that harbor life expand, contract, or move their ranges as conditions change; they do not disappear. It is analogous to beaches. As sea level rises, beaches do not vanish. They move inland. There is always a beach where the land meets the sea.

There is simply no evidence that a significant amount of life on earth will be destroyed by whatever types of climate change may result from global warming. Some changes will occur, but their importance will be confined to academics and economics. Maine's cold-water lobsters and New England's maple syrup trees will probably partially relocate to Canada. The region's syrup production has been declining precipitously for fifty years, a reflection of the warming climate. Tropical fruits such as mangos, bananas, and papayas may no longer be confined to small areas in southern California. The state's grape-growing region may shift northward. Some species of plants and animals may vanish, but others will take their place, and life will go on. Such changes have been occurring for at least hundreds of millions of years without human interference.

Humans migrate easily and will not be irreparably harmed, only inconvenienced, by climate change (figure 9.2). The migrations may cause political turmoil in some areas. Populations may decrease in the Gulf states as people find the increased heat too uncomfortable, air-conditioning bills too costly, and increasingly severe hurricanes too dangerous and expensive. Immigrants from the Gulf Coast will repopulate

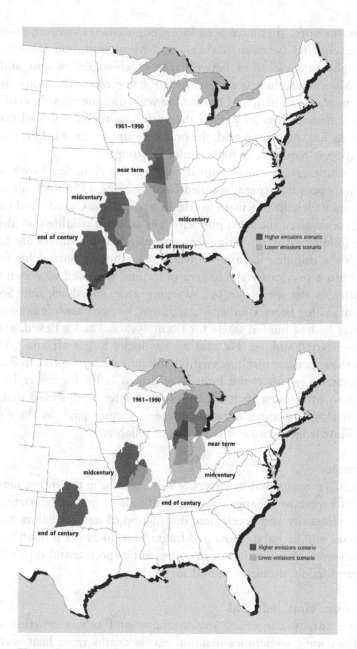

Figure 9.2
What the climate will feel like for people in Illinois and Michigan in the near term (2050)
and long term (2100). (Global Climate Change Impacts in the United States, Cambridge
University Press, 2009)

states to the north that have been losing population. However, given the range in climates between Alabama and Minnesota, and the range in topography and elevation between the Okefenokee Swamp and the Rocky Mountains, there is enough variation in ecological conditions that abundant plant, animal, and human life will continue regardless of temperature changes. It is all part of the overall nonlethal ebb and flow of history. As has often been said, the only constant in our world is change. Adaptation is required; change cannot be stopped.

The ethnic group most seriously affected by climatic change is the indigenous peoples generally called Eskimos. The 155,000 Inuits, who are spread widely from Russia to Alaska to northern Canada and Greenland, have received the most publicity. The fish and wildlife they depend on are following the retreating ice caps northward. Hunters are falling through the thinning ice and drowning. One hunter hauled his fishing shack onto a previously stable area of ice and then had to watch three days later as the ice broke up, sweeping away his shack and $6,000 worth of fishing gear. Another hunter drove his snowmobile onto ocean ice where he had hunted safely for twenty years. The ice flexed, and the machine started sinking. He said he was lucky to get off and grab his rifle as the expensive machine swiftly sank into the icy depths. In Russia's northernmost territory, the Inuit have drilled holes for water because there is so little snow to melt. One Inuit hunter commented that, this world is slowly disintegrating. Inuit elders cannot pass on their traditional knowledge because it is no longer reliable.

Biodiversity

The IPCC forecasts that biodiversity will decrease in the future independent of temperature change due to a multitude of pressures, particularly land use intensity and accelerated destruction of natural habitats. Most significant will be habitat loss and fragmentation as the human population increases; introduction of invasive exotic species; and direct effects on reproduction, dominance, and survival.

Death from Heat and Cold

Environmentally concerned organizations and others worried about global warming sometimes mention excess deaths from heat waves as another reason to worry about a warming climate. There certainly are deaths and illnesses due to heat waves, and they are always well reported by the media, particularly now that global warming is on everyone's mind. However, cold-related deaths are far more numerous than heat-related deaths in the United States, Europe, and almost all countries outside the tropics, and almost all of them are due to common illnesses that are increased by cold.[14] From 1979 to 1997, extreme cold killed twice as many Americans as heat waves, according to the U.S. Department of the Interior.[15] It is estimated that by 2050, global warming will cause almost 400,000 more heat-related deaths each year. But at the same time, 1.8 million fewer people will die from cold.[16]

As global temperatures increase, extremes of cold temperatures will be mitigated in two ways: global warming will increase maximum summer temperatures modestly and raise winter minimum temperatures significantly. Both factors should help reduce human death rates. An increase in average temperature should be a cause for rejoicing among the elderly, the infirm, and babies, the population subgroups most sensitive to extreme temperatures.

Changes in Precipitation

As the air temperature increases, evaporation from the oceans also increases, which leads to increased humidity and increased global precipitation. And because moisture in the air is the major absorber of the heat radiated from the earth's surface (more than half), a major positive feedback loop exists. Rising temperatures increase the amount of water vapor in the air, and the increased amount of water vapor increases the temperature still further. Data from more than 5,000 rain gauges on six continents indicate that average annual rainfall has increased by almost 1 inch during the past hundred years. Changes in the ratio of rainfall to snowfall have also been detected. Since the 1960s, the earth's snow cover has decreased by about 10 percent, and the thickness of the Antarctic ice cap has decreased by an astonishing 42 percent.

However, the increase in precipitation is not uniform over the globe. Over North America, the increase in precipitation has been more than an inch and a half, but wide variations exist among regions (figure 9.3). The rule of thumb is that wet places will become wetter and dry places will become drier. California, where the warm climate and irrigation with subsurface waters have made the state the heart of America's vegetable farming, was hit particularly hard by declining precipitation during the twentieth century, while the vast Midcontinental grain belt has seen substantial increases. The reason for these aerial variations is unknown.

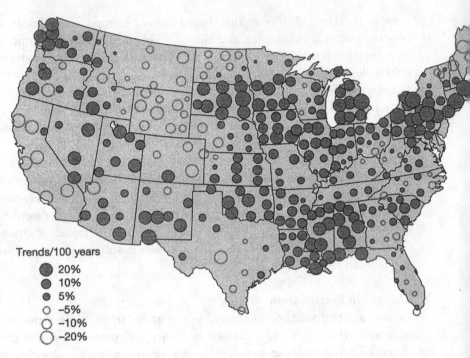

Trends/100 years

- 20%
- 10%
- 5%
- −5%
- −10%
- −20%

Figure 9.3
Trend in annual rainfall in conterminous United States since 1900 (T. R. Karl, R. W. Knight, D. R. Easterling, and R. G. Quayle, "Trends in U.S. Climate during the Twentieth Century," *Consequences* 1 [1995]: 5.)

Sea Level Rise

Sea levels are rising almost everywhere (8.7 inches since 1875), for two reasons: glacial melting and the expansion of water as its temperature increases. Thermal expansion is a less obvious process than melting ice because it cannot be seen, but scientists estimate it is responsible for half to two-thirds of measured sea level rise. About half of the amount that results from melting ice comes from the Antarctic and Greenland ice caps. The other half comes from melting of the world's 150,000 to 200,000 mountain glaciers—tongues of ice that are retreating landward at an average rate of about 30 feet per year.[17] It should be noted that the melting of ice already floating in the ocean, such as in the Arctic, does not affect sea level. Only the melting of land-based ice or its calving into the ocean can cause sea level to rise.

There are many uncertainties concerning how fast ocean levels will rise over the next few decades and centuries. For example, how fast will humans decrease their use of fossil fuels? The industrialized nations are beginning to use small amounts of alternative energy sources, but it will be at least fifty years before they form the major part of their energy use. And it is apparent that nations wanting to industrialize rapidly, such as China and India, may continue to increase their use of fossil fuels even further into the future.

Many feedback loops affect global warming, and the strength of each of them is changing with time. There are as yet no quantitative data to determine the effect of each one on global warming. As noted earlier, increasing water vapor in the air creates a positive feedback loop, reinforcing global warming. The warmer the air is, the more water vapor it can hold. Given the amount of temperature increase that is likely to occur in the lower atmosphere, the air will be not be able to hold more than about 4 percent of water vapor.

The water vapor increases the rate at which ice is lost from the sea, and decreasing the amount of sea ice creates an additional positive feedback loop. Ice reflects away a large portion of incoming sunlight, whereas the darker open ocean absorbs 80 percent more radiation. This is the main reason that the influence of warming is more pronounced at high northern latitudes than in temperate or tropical regions.

But there is also a negative feedback to the loss of sea ice. Water absorbs much of the carbon dioxide gas humans emit, so exposing more water permits additional absorption of carbon dioxide and decreases the warming effect of carbon dioxide emissions. From a high of 5.7 million square miles in 1950, the annual average extent of arctic sea ice has decreased to 4.2 million square miles.[18] Arctic summer ice is currently receding by an average of approximately 8 percent per decade, a pace that has been steadily quickening since 2001. Ice-free summers in the Arctic Ocean may occur within ten to twenty years. Winter sea ice is also decreasing in aerial extent.

The mass of ice-free water over most of the globe absorbs at least half of the carbon dioxide humans produce. But recent studies have revealed that the ocean is absorbing less carbon dioxide than it was in 1997. This is to be expected because warm water cannot hold as much gas as cooler water can. So here is an additional positive feedback loop. Warmer air drives carbon dioxide from the ocean into the air, and additional carbon dioxide in the air increases global warming and causes carbon dioxide to escape from the water.

Clouds are made of water, and as water in the atmosphere is increasing, global cloudiness must also be increasing. Increased cloudiness generates a negative feedback loop because clouds reflect much of the incoming solar radiation.

Possible changes in ocean currents create another uncertainty in heat transfer. Quite small changes in the transport of heat or salt can have large effects on surface temperature, and ultimately on climate. Ice, for instance, seems to be moving at an increasing rate out of the Arctic Ocean and moving both east and west of Greenland. Reduced sea-ice cover in the Arctic might also increase the influx of warm Pacific waters through the Bering Strait between Russia and Alaska.

Greenland   In addition to the uncertainties created by feedback loops, climatologists have discovered that they are unable to determine accurately how fast the ice in Greenland and Antarctica will melt. The ice cap on Greenland contains 800 trillion gallons of water, about the same amount of water as the Gulf of Mexico, and it seems that cracks are forming in the ice as it moves downslope to the ocean.[19] Millions of gallons of water from the melting surface of the ice flow into the cracks and downward to the contact between the base of the ice and the rock beneath. This lubricates the contact, allowing the ice above to increase its rate of flow toward the ocean. The ice cap cannot be thought of as a single block of ice that is melting under the summer sun.

As a result of the growing cracks and lubrication, Greenland is losing 48 cubic miles of ice each year, double the rate of five years ago.[20] One glacial tongue had a surge toward the sea that was measured to be 3 miles in 90 minutes.[21] One Greenland glacier puts enough freshwater into the sea in one day to provide drinking water for a city the size of New York or London for a year.

And the rate may increase even more, and not only because of the vertical cracks in the ice. Most of Greenland's ice mass is actually below sea level. The weight of billions of tons of ice over tens of thousands of years has depressed the rock beneath it so that the ground beneath the ice is actually a big shallow bowl. It is now filled with a subglacial melt-water lake that is 1,500 feet deep.[22] Seawater may soon flow in under the ice, effectively "floating" off most of the glacier into the ocean, rapidly accelerating its collapse. The current estimate of sea level rise of 0.1 inch per year may be an extremely poor predictor of the rise in the near future, but even at today's rate of climate change, the UN estimates that 25 to 50 million people in low-lying areas will have been displaced by 2010.[23] According to the UN report, "Societies affected by climate change may find themselves locked into a downward spiral of ecological degradation, towards the bottom of which social safety nets collapse while tensions and violence rise." The submergence of fertile river flood-plain agricultural land will threaten food production in some of the world's most densely populated regions in Southeast Asia.

The Greenland ice sheet contains enough ice to raise sea level 21 feet.[24] A rise of only 3 feet would endanger 108 million people around the globe (table 9.3). A rise of 20 feet would put significant parts of America's eastern seaboard and Gulf coast states, most of the Netherlands, and much of Bangladesh underwater. Residents of low-lying Pacific islands such as Tuvalu, Kiribati, and the Maldives will probably become the world's first global warming refugees as their homelands become submerged.

Table 9.3
Total surface area inundated and population at risk at global and regional scales

Global		
Sea level rise (feet)	Inundated area (square miles)	Population affected (millions)
3.2	407,000	108
6.5	507,000	175
9.7	594,000	234
13.0	685,000	308
16.2	774,000	376
19.5	847,000	431
Southeastern United States		
3.2	24,040	2.6
6.5	40,340	5.5
9.7	53,017	8.7
13.0	63,597	13.2
16.2	73,579	17.1
19.5	81,214	19.3
Southeast Asia and Northern Australia		
3.2	134,204	46.7
6.5	164,529	86.1
9.7	186,550	114.0
13.0	207,498	153.1
16.5	227,149	183.4
19.0	243,369	209.1

Source: EOS, American Geophysical Union, February 27, 2007

Antarctica   Nearly all of Antarctica is covered by an ice sheet that is, on average, 8,000 feet thick and has a volume of about 1,900 cubic miles. About 90 percent of all the ice on the globe is on the Antarctic continent. If all of this were to melt, following the lead of Greenland, the ocean would rise by 260 feet (figure 9.4). Current ice loss is as much as 36 cubic miles per year, 2 percent of the total Antarctic ice, and is the main source of the ice that fuels global sea level rise.[25]

Most of the ice in Antarctica is in the East Antarctic Ice Sheet (EAIS), which had been thought to be stable, with ice accumulation from snowfall balancing the loss from melting. But since 2006, net melting has occurred at the fringes of the ice sheet. Complete melting of the EAIS would raise sea level by 211 feet, submerging a large part of the east coast of the United States.

The Greenland-sized West Antarctic Ice Sheet (WAIS) makes up about 20 percent of Antarctica's surface area and stores the frozen water equivalent of 26 feet of global sea level.[26] The temperature on the Antarctic Peninsula, the northwestern extension of West Antarctica, has risen by 5°F since the 1950s, among the largest warming signals on earth. The biggest of the WAIS glaciers, the Pine Island Glacier, is moving 40 percent faster than it was in the 1970s, discharging water and ice more

Figure 9.4
Location of the shoreline around conterminous United States 18,000 years ago at the height of the last glaciation and where the shoreline would be if all of the planet's ice melted. Many states along the Atlantic and Gulf coasts would disappear.

rapidly into the ocean. The Smith Glacier, also in West Antarctica, is moving 83 percent faster than in 1992.

WAIS rests on the sea floor; its base is below the ocean surface. Unlike the primarily land-based EAIS, the bottom of WAIS rests below sea level, and therefore ocean water is always in contact with the edges of WAIS. Studies of the responses of marine-based ice sheets to past climate fluctuations suggest that they can melt and retreat quickly when ocean temperatures rise because of increased melting at the ocean-ice interface. In addition, the ocean water may begin to work its way underneath the base of the ice, which can accelerate melting and lead to greater instability, raising the ice sheet so it can cause an extremely rapid rise in ocean level. Total melting of WAIS would raise sea level 3 to 5 feet.[27]

Another unevaluated factor concerning the stability of the Antarctic ice mass is the presence of at least 150 lakes at the base of the ice, thousands of feet below the ice surface.[28] Most of them are beneath the EAIS. Antarctic glaciologists suspect there may be thousands more. The meltwater is produced by geothermal heat and collects in hollows on the rock surface, and there appear to be connections between some of the lakes—rivers beneath the ice. If there are enough of these sub glacial lakes in Antarctica, they would add significantly to the instability of the ice sheet.

Sea Level Fall
Some coastal areas are experiencing decreases in sea level, despite the thermal expansion of ocean water and the continuing melting of glacial ice. Ocean levels are indeed rising in these areas, but the land is rising even faster so that the shoreline is receding. The areas of rising land are in Northern Hemisphere areas formerly covered by billions of tons of glacial ice a few tens of thousands of years ago—Alaska, much of Canada, Greenland, Scandinavia, and Siberia. The land is rising in response to removal of the ice during the interglacial period that began 20,000 years ago. Land is rising from the water, shifting property boundaries and creating land whose ownership is unclear.

In some areas around Juneau, Alaska, the land is rising almost 3 inches a year.[29] Relative to the sea, land around Juneau has risen as much as 10 feet in about 200 years. As global warming accelerates and more ice melts, the land will continue to rise—perhaps 3 feet more by 2100, scientists say. Adding to the glacial rebound is movement of the Pacific plate as it slides beneath the North American plate.

Ocean Acidity

Little attention has been paid to the fact that increasing the amount of carbon dioxide in the atmosphere causes water to become more acid. The percentage increase in acidity of the oceans so far has not been determined, and the increased temperature of ocean water counteracts the effect of increased carbon dioxide because heat drives off gas dissolved in water. The ocean currently absorbs at least half of the carbon dioxide humans produce, but a ten-year study by scientists published in 2007 concluded that the amount of carbon dioxide absorbed by the ocean is decreasing.[30]

Researchers have been concerned that increased ocean acidity will be harmful to most shelled creatures because increased acidity makes it harder for these creatures to grow their shells. However, recent research has revealed a previously unknown complexity in this equation.[31] It seems that some shell-building creatures, including crabs, shrimp, lobsters, urchins, limpets, and those red and green algae that have calcium carbonate skeletons, build more shell when there is a slight acidification of the water in which they live. Apparently the organisms are able to manipulate their organic biochemistry to counteract small changes in ocean acidity. How they do this is not known.

Ocean Currents

Global warming may have a significant effect on the behavior of the major currents in the world ocean. These currents are driven by three things: the earth's rotation, which cannot change; the temperature difference between the poles and the equator, which is changing as the earth warms; and salinity differences between one part of the ocean and another, which is changing as melted ice pours into the ocean. The effect these changes will have on oceanic circulation is uncertain, but climate scientists believe there is reason for concern.

Tipping Points

Tipping points are instances of very rapid change preceded by long periods of slow change. The rapid change occurs in the same way as the slowly increasing pressure of a finger eventually flips a switch and turns on a light. Once the switch has occurred, the new and hostile climatic event lasts for perhaps centuries or even millennia. Earth's climate system can jump within a few years from one stable operating mode to a completely different one.

A 2002 report by the National Academy of Sciences said, "Available evidence suggests that abrupt climate changes are not only possible but likely in the future, potentially with large impacts on ecosystems and societies."[32] This was echoed in 2008 by James Hansen, director of NASA's Goddard Institute of Space Studies who described the climate as nearing dangerous tipping points.[33] Examples of possible tipping points that cause sleepless nights for climatologists include the rapid disintegration of the Greenland and Antarctic ice sheets with an accompanying rise of perhaps tens of feet in sea level within a few years; a sharp decrease in rainfall and consequent dieback of the Amazon rain forest with loss of its absorption of carbon dioxide; rapid release of methane from shallow-buried methane hydrates causing a major increase in the rate of global climate change; disruption of the Indian summer monsoon and West African monsoon with harmful effects on agriculture; and an influx of cold glacial meltwater causing a collapse of the Atlantic thermohaline (conveyer belt) circulation that keeps Europe habitable.

According to most climatologists, the Gulf Stream and its extension, the North Atlantic Drift, supply Western Europe with an amount of free heat equal to the output of about a million power stations. It keeps Western Europe 10°F to 20°F warmer than it would normally be at its latitude (frigid Labrador is at the same latitude). Without the Gulf Stream, the average temperature in Europe might drop by as much as 10°F in ten years or less; it has happened in the past.

About 12,700 years ago, average temperatures in the North Atlantic Region abruptly plummeted about 9°F and remained that way for 1,300 years before rapidly warming again. A similar abrupt cooling occurred 8,200 years ago. An abrupt warming took place about 1,000 years ago, allowing the Norse to establish settlements in Greenland. The climate turned abruptly colder 700 years ago, forcing the Norse to abandon their Greenland settlements. Between 1300 and 1850, severe winters had profound agricultural, economic, and political impacts in Europe.

Total disruption of the Gulf Stream and North Atlantic Drift would produce winters twice as cold as the worst winters on record in eastern United States in the past century. Abrupt regional cooling may occur even as the earth's average temperature continues to warm. In addition, previous shutdowns have been linked with widespread droughts around the globe. The new conditions would likely last for decades to centuries until conditions reached another threshold, at which time the circulation might revert back to the current state.

In December 2005, researchers at the National Oceanography Centre in the United Kingdom reported that the Gulf Stream slowed by about 30 percent between 1957 and 2004.[34] This has had no known effect on temperatures in Western Europe. Early in 2006, a senior researcher at the Ocean Physics Laboratory in France said that the changing climate could slow the Gulf Stream by 25 percent by the year 2100.[35] Clearly we are in uncharted territory when considering modifications of ocean currents.

Permafrost

Permafrost is defined as ground perennially frozen for more than two years. It is present at high latitudes and, to a lesser degree, at high elevations. Permafrost covers an area slightly smaller than South America and extends across a quarter of the Northern Hemisphere to depths of 1,000 feet or more. Sixty percent of Russia and 42 percent of Canada are covered by permafrost. It is present beneath the Arctic Ocean as well as on land. Most of the world's permafrost has been in this condition for thousands of years.

The southern boundary of permafrost, now generally located between 55°N and 60°N latitudes in Canada, Alaska, and Siberia, has been migrating northward in response to global warming. As recent warming trends continue and perhaps accelerate, the area covered by permafrost will continue to decrease. Glaciologists estimate that it will take several centuries for all of it to disappear, but such estimates are as uncertain as the melting rates of Antarctica and Greenland.

Recent warming has degraded large sections of permafrost across central Alaska, with pockets of soil collapsing as the ice within it melts. There is damage to construction founded on ice-rich permafrost, such as buckled roads and bridges, airports, destabilized houses, damaged pipelines, and "drunken forests"—trees that lean at wild angles. In Siberia, some industrial facilities have reported significant damage. Thawing permafrost is sending large amounts of water to the oceans. Runoff to the Arctic Ocean has increased about 7 percent since the 1930s.[36]

There is concern about emissions of greenhouse gases from thawing soils. Permafrost may hold 30 percent or more of all the carbon stored in soils worldwide. Thawing permafrost could lead to large-scale emissions of carbon dioxide or methane beyond those produced by fossil fuels. Decomposition of organic matter in contact with atmospheric oxygen will produce carbon dioxide; organic matter in water-saturated soil, such as peat bogs, will produce methane as it decomposes. This could have a major effect on climate. The volume of methane that may be produced as the permafrost melts is about the same amount that is released annually from the world's wetlands and agriculture. It would double the level of the gas in the atmosphere, leading to a 10 to 25 percent increase in global warming, another positive feedback loop.[37]

In addition, as the permafrost thaws, it reveals bare ground, and bare ground warms up more quickly than reflective ice, accelerating the rate of thawing—a positive feedback loop. As noted earlier, northern climates warm faster than temperate ones even without the added effect of melting permafrost. It is noteworthy that western Siberia, an area with extensive permafrost, is heating up faster than anywhere else in the world, having experienced a rise of about 5°F in the past forty years.

Permafrost under the Arctic Ocean is also melting, and in many areas, methane is bubbling up from the ocean floor. The amount of methane stored there has been calculated to be greater than the total amount of carbon in global coal reserves.[38] A methane time bomb may be lurking beneath the Arctic Ocean that has not been factored into existing climate models.

The thawing of the world's permafrost has been described by Sergei Kirpotin, a Russian ecologist, as an ecological landslide that is probably irreversible. He says that the entire western Siberian sub-Arctic region has begun to melt, and this has all happened since 2000. A featureless expanse of frozen peat is turning into a watery landscape of lakes, some several thousand feet in diameter. Kirpotin suspects that a critical threshold has been crossed, triggering the rapid melting.[39]

American scientists have reported a major expansion of lakes on the North Slope fringing the Arctic Ocean. Many of the permafrost lakes eventually disappear, because as the depth of thawing increases, the water is able to drain away underground.

As permafrost thaws and porous soil takes its place, the velvety carpet of moss and lichen is replaced by shrub bushes and eventually forests. The trees and shrubs will then accelerate warming because dark vegetation absorbs more solar radiation—another positive feedback loop. Although these higher plants absorb carbon dioxide from the atmosphere, the warming effect of a forested landscape far outweighs the cooling from the carbon dioxide uptake.[40]

The Nasty Gases

Many of the gases humans add to the atmosphere, as well as the soot spewing from our fossil fuel-consuming inventions, are contributing to global warming. Thirty-two percent of the greenhouse gases are generated from transportation, 43 percent from electricity production, and 25 percent from other sources.

Carbon Dioxide

Since the Industrial Revolution, the global atmospheric concentration of carbon dioxide has increased by 36 percent and is now 388 parts per million (ppm), the highest concentration in the past 800,000 years. It is rising at about 2 ppm per year. About 6.725 billion tons of carbon dioxide were emitted in the United States in 2007, an increase of 20 percent since 1990 and 2.5 percent since 2000.[41] Total world emissions of carbon dioxide are over 30 billion tons a year at present.

Excluding water vapor, carbon dioxide is the major gaseous contributor to global warming (40 percent), not only because of its dominance volumetrically, but also because of its lifetime in the atmosphere. A large number of sources, sinks, and feedback loops are involved in the cycling of carbon dioxide, so the lifetime of the gas in the atmosphere cannot be calculated accurately. However, experts agree that it will require at least hundreds, and probably thousands, of years for all of the carbon dioxide humans have put in the atmosphere to be removed by natural processes.[42] Because of slow removal processes, atmospheric carbon dioxide will continue to increase in the long term even if its emission is substantially reduced from today's levels. We have created a monster that untold numbers of generations will have to contend with during their lifetimes.

There is no question that Western industrialized nations are responsible for most of the carbon dioxide that has been added to the atmosphere. However, on a volume basis, developing countries, particularly China, are racing to catch up. China's annual energy-related carbon dioxide emissions surpassed those of the United States in 2006, years ahead of published international and Chinese forecasts. But the size of China's enormous population virtually guarantees the country will never catch up to the United States on a per capita basis.

Almost all (94 percent) of the carbon dioxide humans spew into the atmosphere comes from the burning of fossil fuels: coal, oil, and natural gas.[43] (This has recently been disputed. The World Bank estimates that when the entire life cycle and supply chain of the livestock industry is taken into consideration, livestock are responsible for 51 percent of greenhouse gas emissions.)[44] Thirty-nine percent of the total that is emitted comes from electricity generation, and 31 percent is from transportation (cars and trucks 25 percent, ships 3 percent, planes 3 percent). Industry supplies 14 percent, agriculture 8 percent, residential use 6 percent, and other sources 2 percent. It is clear that to make major reductions in our output of carbon dioxide, the most effective strategy would be to emphasize the reduction in the use of fossil fuels in motor vehicles and power plants—away from cars (even electric ones with multiple passengers) and into mass transit and away from coal and into alternative energy sources. Both of these changes have started in recent years and are continuing despite decreases in gasoline prices.

As usual, the fastest and most successful way to get industry and individuals to change their behavior is to hit their bottom line. If the cost to companies or people is larger than they want to or are able to bear, their behavior will change. It is as certain as death and taxes.

Two ways have been suggested to put a price on carbon to cause a reduction in consumption and a lowering of polluting emissions. The more straightforward way is to tax carbon use. A set tax rate is placed on the consumption of carbon in any form—fossil fuel, electricity, gasoline—with the idea that raising the price will encourage industries and individuals to consume less. The effectiveness of an increased price of carbon on reducing use was well illustrated by the decrease in gasoline use in 2008 as the price for a gallon rose from $1.50 to $4.00 independent of a tax increase.

A carbon tax was proposed in Congress in 1997 but got nowhere. The burden of such a tax would fall most heavily on coal-burning utilities and industrial users and processors of petroleum in its various forms. It would be no fun for long-distance trucking companies and individual car owners either. Lobbyists for the fossil fuel industries went into high gear, and the proposal was defeated. Any legislation in the United States that is called a tax or is even perceived as a tax by the public has about the same chance of getting through Congress as a fighting pekingese has of defeating a pit bull. Nevertheless, in March 2009, a bill to tax carbon use directly was submitted in the House of Representatives. It went nowhere.

The method that may be on its way through Congress to reduce fossil fuel consumption is called cap-and-trade.[45] Under a cap-and-trade program, the government sets a limit on the amount of pollution allowed (carbon dioxide, methane, soot, and so forth), and polluters are sold or given permits by the government to pollute up to that limit. Each permit allows the holder to release the equivalent of 1 ton of carbon dioxide. Companies that hold their emissions below their cap can sell their remaining pollution allowance on an open carbon market to those struggling to keep their emissions under the cap. This ensures that the industry stays within the overall emission limit, which is planned to decline over time. But until the planned decline occurs, cap-and-trade will not reduce emissions. The cap-and-trade system in force in the European Union caused the traditionally low cost of coal generation to exceed that of natural gas for much of 2008 and 2009, a trend that has continued into 2010.

A method for reducing carbon emissions that is highly publicized by some celebrities is called buying offsets, trumpeted by wealthy Americans as an effort to display their environmental colors by becoming "carbon neutral," the New Oxford American Dictionary's Word of the Year in 2006.[46] Those with lots of money can appear superior to ordinary folks who are unable to use this technique.

The concept is simple. You keep track of the number of miles you have driven, airline miles flown, or coal used to generate your electricity, and use an online carbon calculator to determine how much carbon dioxide you have generated. If you were responsible for 10 tons of carbon dioxide, you would pay a carbon offset company for 10 tons of carbon dioxide offsets. The company (or nonprofit organization) would then invest your money in a project meant to reduce greenhouse gas emissions. Because climate change is a nonlocalized problem, it does not matter where the emissions are reduced. Air circulates. Some critics have likened offsetting to buying pardons from the Catholic church in sixteenth-century Europe.

Nature's way is best for controlling the gases responsible for climate change. Short of cutting emissions, the world needs better management of forests, more careful agricultural practices, and the restoration of peatlands that could soak up significant amounts of carbon dioxide. Millions of dollars are being invested in research aimed at capturing and burying carbon dioxide emissions from power stations, but investing in ecosystems could achieve cheaper results. It would also have the added effects of preserving biodiversity, improving water supplies, and boosting livelihoods. Cutting deforestation by half by midcentury and maintaining that rate would save the equivalent of five years of carbon emissions at today's level. Agriculture has the largest potential for storing carbon if farmers use better techniques, such as avoiding turning over the soil and using natural rather than chemical fertilizers.

Forests

A frequent major omission in the inventory of sources of carbon dioxide is the ongoing destruction of the world's rain forests, primarily for agricultural purposes in South America and Southeast Asia. The destruction of rain forests is an important cause of climate change. Trees are 50 percent carbon. When they are felled or burned, the carbon they store escapes into the air as carbon dioxide. Carbon emissions from deforestation outstrip damage caused by planes and cars and are second only to the energy sector as a source of greenhouse gases. This is a major world concern in terms of carbon dioxide emissions. Forests still cover about 30 percent of the world's land area, and tropical forests absorb about 18 percent of the carbon dioxide added by the burning of fossil fuels, more than the emissions generated by all of the world's cars and trucks.[47]

An estimated 53,000 square miles of rain forest disappear every year, much of it in Indonesia and the Amazon Basin. Because of this, Indonesia and Brazil are behind only China and the United States as greenhouse gas emitters, despite the fact that neither country is noted for heavy industry.[48] Destruction of rain forests allows billions of tons of carbon dioxide to remain in the atmosphere. The Amazon rain forest is the largest living reservoir of carbon dioxide on the land surface. Its trees and soil contain perhaps 150 billion tons of carbon, about 20 years' worth of humankind's emissions from burning fossil fuels. An acre of forest stores about 200 tons of carbon through photosynthesis, carbon that is released into the atmosphere as carbon dioxide when the tree rots. Stopping forest destruction is difficult in countries where poor farmers want more land to increase their agricultural productivity. However, if we lose our forests, slowing climate change will be much more difficult.

A related but overlooked major source of carbon dioxide is peatlands in forested areas. Peat, formed over thousands of years from decomposed trees, grass, and scrub, contains large amounts of carbon that used to stay locked in the ground. When forested ground is shorn of trees in Third World countries to create farmland, then drained by canals and burned, carbon dioxide gushes into the atmosphere.

Deforested and formerly swampy peatlands in Indonesia in 2006 released an amount of carbon dioxide equal to the combined emissions that year of Germany, Britain, and Canada, and more than U.S. emissions from road and air travel.

Methane

Since the Industrial Revolution, the global atmospheric concentration of methane has increased by 137 percent. About 645 million tons of methane were emitted by the United States in 2007, nearly the same amount as in 2000. Its concentration in the atmosphere is only 1.8 ppm, but its overall impact on climate is relatively large because methane is about twenty times more potent as a heat absorber than carbon dioxide is. It is removed by chemical processes in the atmosphere (oxidation) much faster than carbon dioxide; it has a lifetime of only about twelve years.

The abundance of methane in the atmosphere rose significantly between 2007 and 2008, probably because of the increasing and unstoppable thawing of permafrost and consequent production of wetlands, the increasing number of cattle being produced in the developing world, and the increasing cultivation of rice among the exploding populations of southern and southeastern Asia (table 9.4). Cows, sheep, and goats produced one-third of the methane released in the world in 2007—90 percent by silent burping and 10 percent by flatulence.

The Environmental Protection Agency estimates that 24 percent of methane emissions in the United States can be attributed directly to livestock. Natural outgassing of methane hydrates is estimated to account for 1 or 2 percent of atmospheric methane. Methane emissions will certainly increase if methane hydrate reservoirs are exploited for natural gas (chapter 6), but a more critical concern is a possible catastrophic release of methane to the atmosphere or ocean from shallow

Table 9.4
Global sources of anthropogenic methane, 2005

Source	Percentage
Domestic livestock	34
Oil and natural gas	19
Landfills	12
Rice cultivation	10
Waste water	9
Other agriculture	7
Coal	6
Biomass burning	3

Source: Environmental Protection Agency (2008).
Note: Note the importance of cattle and also that methane from thawing permafrost and from wetlands are not included in this list.

gas hydrate deposits as a result of extraction activities or climate change. Recent research revealed that there are more than 250 plumes of methane gas bubbles rising from an area of seabed in the Arctic, which scientists believe are being released from methane hydrates that become unstable as the Arctic Ocean warms.[49] For marine gas hydrates, some of the released methane would probably be oxidized in the sediments and water column; some might dissolve in the ocean; and some might reach the atmosphere either immediately or over time. For permafrost gas hydrates, the potential for direct outgassing of methane to the atmosphere is considerably greater.

Other Gases

Small contributions to global warming are made by several other gases, but they are present in only tiny amounts in the atmosphere and are not a major source of concern. Nitrous oxide increases are due mostly to the increased use of fertilizers. It has an atmospheric lifetime of 114 years. Chlorofluorocarbons, refrigerant fluids used in refrigerators and air conditioners, are destroyed in the atmosphere at about the same rate as methane. CFCs are being phased out by manufacturers.

Solid Particles

Two types of particles in the atmosphere affect global warming: sulfate aerosols and soot. Sulfate aerosols arise from the burning of sulfurous coal and oil, and they have had a measurable effect on reducing the enhanced greenhouse effect produced by greenhouse gases because they reflect some of the incoming sunlight. The decrease in enhanced greenhouse warming caused by aerosol emissions is a rare example of a beneficial effect of fossil fuel use. Unfortunately, the effect of the sulfate aerosols is too small to completely counteract the effect of all the greenhouse gases we have pumped into the atmosphere. Pollution controls are reducing the amount of sulfur emitted from coal- and oil-burning industries and cars, a benefit to our lungs (chapter 10), but as the amount of anthropogenic sulfur in the atmosphere decreases, the rate of global warming will increase.

Because of our burning of biomass, coal, and fuel in motor vehicles, the United States is a major producer of black soot. In developed countries, most of it comes from the incomplete combustion of diesel fuel. In less developed countries biomass burning is most significant (table 9.5). Once in the air, it is spread around the world by winds. In the Northern Hemisphere, including the Arctic, soot is twice as potent as carbon

Table 9.5
Combustion sources of black carbon

Source	Percentage
Open biomass (cooking and forest fires)	42
Residential biofuel	18
Road transport (cars and trucks)	14
Nonroad transport (planes and ships)	10
Industry and power generation	10
Residential, coal and other	6

Source: Worldwatch Institute (2009).

dioxide in changing global air temperatures.[50] This may not be true in Greenland because the island has little anthropogenic pollution.

Soot, or black carbon, is responsible for 18 percent of global warming, second only to carbon dioxide as a greenhouse pollutant. In the Alps, soot particles reduced the reflecting power of the snow by 10 to 20 percent. Soot is the aerosol most responsible for the haze in rapidly developing and heavily coal-burning countries such as China and India, and it may be reducing the reflecting ability of snow in the Himalayas as it does in the Alps. In addition to its effect as individual particles, the soot covers some of the sulfate aerosol particles, changing them from reflectors of sunlight into dark-colored absorbers of the long-wave radiation emitted from earth's surface.

Soot has a very short life in the atmosphere, settling back to the ground in a few days or weeks, so reducing its abundance would have a rapid effect on global warming compared to a reduction in greenhouse gases. NASA's study shows that the "strongest leverage" on reducing global warming would be achieved by reducing emissions from domestic fuel burning in developing countries, particularly in Asia, and by reducing surface transport emissions in North America, especially from diesel engines. California, commonly a national leader in environmental matters, enacted measures in December 2008 to force diesel-burning trucks to fit filters to reduce diesel soot emissions by 85 percent.[51]

In December 2009 the Environmental Protection Agency formally declared that carbon dioxide and other greenhouse gases were a threat to human health and welfare. The declaration empowers the agency to regulate these emissions and gives President Obama an important tool if Congress fails to pass legislation to reduce global warming emissions.

Let's Be Realistic

There is no question that climate change poses numerous inconveniences and perhaps dangers for the United States and the world community. Given a choice, most of us would prefer to keep the climate we are used to. But this is not possible. Starting with the Industrial Revolution in about 1750, we unknowingly took off on an unstoppable trip into the unknown. Given human nature and the way the political systems of the world operate, it is questionable whether the anthropogenic effect on climate change can even be slowed. Climate change is a tragedy of the commons unlike any other that humans have faced.

The atmosphere is egalitarian. It distributes itself equally over the globe. This means that if climate change is to be slowed, there needs to be worldwide cooperation. As of 2010, there is little sign that the world's governments can accept the need for immediate drastic action. The economic pain would be too great. Lots of international meetings and some arguably useless agreements have been reached, but there is apparently no way to enforce them, and the largest contributors of greenhouse gases, the United States and China, have not signed onto the documents.

Getting different nations to agree on anything is like trying to herd cats. Getting them to agree on what should be done about climate change and how urgent is the need for action is like teaching cats to sit, shake hands, roll over, and bring you the newspaper every morning. The lack of agreement and the reluctance to act by many nations are understandable given the inequalities of economic circumstances that exist among nations. Any mechanisms agreed to must be thought to be fair to all if they are to succeed, but this is incredibly difficult because there is no universally accepted definition of the word *fair*. What developed nations believe is fair seems manifestly unfair to less developed nations. An observer from another planet would have no trouble understanding and sympathizing with the point of view of the downtrodden of the world. They are the least responsible for climate change but are likely to be the ones who suffer the most from it because they lack the means to adapt.

Several principles can be invoked to decide who should do what:[52]

The Egalitarian Principle   states that every person in the world should have the same emission allowance. This would benefit the more populous countries such as China and India.

The Sovereignty Principle   agues that each nation should reduce its emissions by the same percentage amount. Major emitters such as the

United States and China would cut more than low emitters such as Mexico and India.

The Polluter Pays Principle asserts that climate-related economic burdens should be borne by nations according to their contribution of greenhouse gases over the years. The U.S. bill would be by far the largest.

The Ability to Pay Principle argues that the financial burden should be borne by nations according to their level of wealth. This would again be most costly for the United States, but all Western nations would suffer.

Whichever mix of principles may eventually be agreed to by the world's sovereign nations will be costly in terms of money and disruptions in established ways of doing things. But inaction could be more costly and disrupting. No one really knows because the changes that will be brought on by climate change are at best uncertain and probably mostly unknown.

An added consideration is that any changes that might be made seem trivial in terms of what they will accomplish in any time span that humans can relate to. We are required to think about times a few hundred years from now, an inconceivably long time in terms of human life span. However, our Revolutionary War ancestors worried about life in the twentieth or twenty-first century. That is the reason they wrote the Constitution. We should be able to do the same.

The Environmental Protection Agency has determined recently that a massive and very probably impossible 60 percent reduction in carbon dioxide emissions by 2050 would reduce global temperature by only 0.2°F by 2095.[53] Does anyone believe that emissions from the world's power plants, motor vehicles, and agriculture can or will be reduced by 60 percent forty years from now? As of 2010, the world's emissions are continuing to increase, as they do every year, and there is little likelihood of declines in the next few decades as the less developed nations race to industrialize, using mostly the readily available fossil fuels, predominantly coal, as their main energy source.

The Framework Convention on Climate Change was negotiated at Rio De Janiero in 1992. Since then, the world's emissions of carbon dioxide have increased by 28 percent.[54] The Kyoto Protocol was adopted by most of the world's governments in 1997, but since then, carbon dioxide emissions have increased by 22 percent. In 2008, they reached about 36.5 billion tons, and the trend is still upward. In the 1990s, carbon dioxide emissions grew at an average annual rate of 0.9 percent; since 2000, the increase has been 3.5 percent annually. In 2009, a widely hyped international climate meeting was held in Copenhagen, Denmark. However, as was the case with earlier climate meetings, there were great expectations but few achievements. Promises were made, but no legally binding and enforceable agreements were signed. It proved impossible to forge consensus among the disparate blocs of countries fighting over environmental guilt, future costs, and who should referee the results. However, the question of refereeing results may be solvable without on-site inspections by unwanted foreigners. It is becoming possible to monitor emissions from space.[55]

No part of the world had a decline in emissions from 2000 to 2008. In February 2009, scientists announced that the pace of global warming is likely to be much faster than recent predictions because industrial greenhouse gas emissions have increased more quickly than expected and higher temperatures are triggering self-reinforcing feedback mechanisms in global ecosystems. Sea levels are rising faster than expected. We are basically looking at a future climate that is beyond anything that has been considered in climate model simulations. Despite this forecast, an examination of 131 land use plans from Florida to Massachusetts reveals that 60 percent of the land less than 3 feet above sea level has or will soon be developed.[56]

Many industrialized nations are trying to switch at least partly to alternative and renewable sources of energy. This is certainly commendable and necessary. But energy needs grow, so that any decrease in greenhouse emissions is probably generations away.

What's Causing the Climate to Change?

The majority of climate scientists globally believe that humans are the cause of the climate changes of the past 150 years. They point out that all scientists accept that carbon dioxide enhances the greenhouse effect, that human activities have greatly increased the amount of carbon dioxide in the atmosphere, and that computer models of climate predict very well the changes the world has seen when anthropogenic effects are included in the models.

However, critics point out that gas bubbles in cores of glacial ice show that the earth's temperature and the levels of carbon dioxide in the atmosphere have been in lockstep for tens of thousands of years, long before anyone was burning fossil fuels, which suggests that natural climatic temperature cycles control carbon dioxide levels, not the other way around. Examples of the effects of these natural controls are well

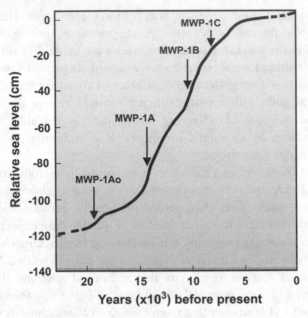

Figure 9.5
Generalized curve of sea level rise since the last ice age. MWP = meltwater pulse. MWP-1AO, 19,000 years ago; MWP-1A, 14,600 to 13,500 years ago; MWP-1B, 11,500 to 11,000 years ago; MWP-1C, ~8,200 to 7,600 years ago. (National Aeronautics and Space Administration)

documented and are not doubted by those who favor the effect of human influences (figure 9.5).

Those favoring human activities as the cause of global climate change believe that there are no astronomical factors strong enough to create the warming the globe has experienced in the past 150 years. Critics respond that there are indeed astronomical variables that are strong enough.[57] These include changes in the shape of the earth's orbit (eccentricity) and the earth's tilt and precession, all of which affect the amount of sunlight received on the earth's surface. The timescale of their variations is well known. Solar intensity (as indicated by sunspots) also varies periodically, and NASA believes that was a key factor in causing the "Little Ice Age" from the 1400s to the 1700s.

In 2003, an Israeli physicist and a Canadian geochemist proposed a new climate driver.[58] They envision slow movements of the solar system through the Milky Way galaxy as controlling the cosmic rays that bombard the earth's atmosphere. A reduction in cosmic radiation would lessen cloud cover and earth's reflectivity, warming the planet; the reverse would cause cooling. They say that the geological record of cosmic ray bombardment during the past 550 million years shows excellent agreement with climate fluctuations, trumping carbon dioxide.

Other scientists have noted a dramatic acceleration of the drift of magnetic poles of the earth since 1990, which affects global climate. They have also noted a decrease in magnetic intensity by as much as 10 percent in some regions, which allows increased cosmic radiation to impact the earth.[59] Other changes in the magnetosphere have increased the amount of cosmic radiation in polar areas, leading to heating of polar caps.

It may be that both anthropogenic and astronomical factors are involved in the climatic changes the world is experiencing, but no one knows how to determine the relative importance of each. Overlapping influences may be difficult to tease apart.

The view that human activities are not the major or only cause of climate change is not represented in the series of reports published by the IPCC, the latest one in 2007. In a 2003 poll conducted by German environmental researchers, two-thirds of more than 530 climate scientists from twenty-seven countries surveyed did not believe that "the current state of scientific knowledge is developed well enough to allow for a reasonable assessment of the effects of greenhouse gases."[60] About half of those polled stated that the science of climate change was not sufficiently settled to pass the issue over to policymakers.[61] The controversy remains unresolved.

10

Air Pollution: Lung Disease

The atmosphere almost looks like an eggshell on an egg, it's so very thin. We know we don't have much air; we need to protect what we have.

—Eileen Collins, NASA astronaut, on the view of Earth from space

When you can't breathe, nothing else matters.

—American Lung Association

The topic of air pollution cannot be completely separated from that of climate change. The carbon dioxide and methane gases that we spew into the air that are thought by most climatologists to be responsible for most climate change are pollutants, things not naturally in the air in such concentrations.

However, when air pollution is discussed, the emphasis is not on changes in global temperature and precipitation but on what is in the 3,400 gallons of air, 7 quarts per minute, that we inhale every day. The emphasis is on the maintenance of good health, in both humans and the other animals and plants we share the earth with. The amounts of carbon dioxide and methane in the air we breathe are not harmful to our bodies, but other things we have added to the air we inhale are—such things as soot, ozone, gaseous organic compounds, sulfur dioxide, lead, asbestos, rubber, arsenic, cadmium, mercury, and other interesting substances (figure 10.1).

The amounts of these substances we take in, along with our essential oxygen gas, have caused great increases in respiratory diseases such as bacterial infections, bronchitis, allergies, and asthma. They may also be responsible for many of the lung cancers not attributable to smoking. Data from Los Angeles indicate that air pollution thickens the wall of a person's carotid artery, a leading risk factor for heart attacks and strokes.[1] Each year, air pollution claims 70,000 lives in the United States; globally, an estimated 200,000 to 570,000 people die from it each year. The

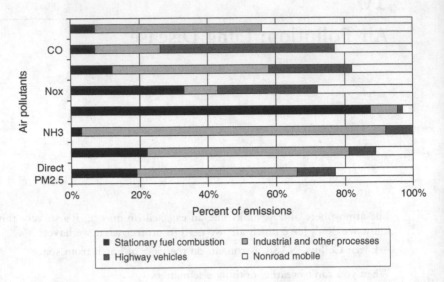

Figure 10.1
Distribution of emissions in the United States by source category for specific pollutants, 2007. (Environmental Protection Agency)

incidence of eye infections has also risen. Babies born in areas where air pollution is most severe have smaller lungs, lower birth weights, smaller heads, damaged DNA, lower IQs, and increased rates of birth defects. Women who live within a half-mile radius of areas of high emissions, such as major roads or factories, are two to four times more likely to give birth to children who develop cancer.

According to the American Academy of Pediatrics, children are more vulnerable than adults to many airborne contaminants. The cellular immaturity of children and the ongoing growth process account for this elevated risk. Since children breathe more rapidly and even inhale more pollutants per pound of body weight than do adults, even minor irritation caused by air pollution, which would produce only a slight response in an adult, can result in a dangerous level of swelling in the lining of a child's narrow airways. Increased exposure to air pollutants during childhood increases the risk of long-term damage to a child's lungs.

Worldwide, respiratory disease is second only to polluted water as the greatest killer of children. Air pollution causes elderly people to die prematurely from heart attacks because their weakened lungs cannot process enough oxygen. In New Delhi, 20 percent of the traffic police at busy intersections need regular medical attention for lung problems. Bus drivers in urban areas in the United States have increased cancer rates, chromosomal abnormalities, and DNA damage from breathing diesel fumes all day. According to the American Lung Association, Los Angeles has the worst air in the United States. Eight of the ten cities with the most polluted air are in California, with Houston at number seven and Charlotte, North Carolina at number ten.

Humans probably first experienced harm from air pollution when they built fires in poorly ventilated caves. Since that time, we have managed to pollute the air everywhere we have settled and even places we have yet to colonize, such as Antarctica and uninhabited parts of northern Canada. The pollution precedes our arrival; it is waiting patiently until we arrive. Probably the nation with the worst air pollution problem is China. People in the major cities walk around with surgical masks on their faces to filter out the particulate material in the air they must breathe. In a World Bank list of the world's most polluted places, China has twelve of the top eighteen cities, India has four, and Egypt (Cairo) and Indonesia (Jakarta) have one each.[2]

Acid rain produced from the sulfurous fumes emitted from coal-fired power plants and factories kills fish in northeastern lakes. Plants are similarly affected. Pollutant haze from soot and other aerosols reduces the penetration of sunlight and decreases photosynthesis. Commercial fruit size and weight decrease. Plant death in the field has increased because the weakened vegetation is more vulnerable to injury from diseases and pests. In summary, air pollution is harming our health, decreasing our longevity, and attacking our crops.

What's in the Air We Breathe?

Determinations of the amounts of the major pollutants in the air we breathe have been made for many decades and are reported annually by the American Lung Association and Environmental Protection Agency.[3] All of the pollutants are bad for human health, and the proportions of each vary yearly. Establishing the effect on human health of each is difficult because all are ingested simultaneously and their effects overlap. And there are no volunteers to conduct controlled experiments. But it is clear that progress on removing all of them is essential.

Soot

Soot is arguably the most damaging type of particle to humans of the air pollutants we eject from our fossil-fueled inventions. In government publications, soot is referred to as particulate matter (PM) and is classified by size. Particles less than 10 micrometers (four ten-thousandths of an inch) in diameter are designated as PM_{10}; those less than 2.5 micrometers are $PM_{2.5}$ (one ten-thousandth of an inch). Three-quarters of the particles less than 2.5 micrometers in diameter have diameters less than 0.1 micrometer (1/250,000 of an inch). Such tiny sizes are difficult to envision. In comparison, a strand of human hair is 100 micrometers in diameter and is about the smallest thing a human eye can distinguish; a coarse PM_{10} particle of 7 micrometers is the size of a red blood cell. When we discuss particulate matter in the air, we are talking about extremely small particles—particles that are invisible without using an optical microscope or an electron microscope.

Particulate matter is released from motor vehicles, power plants, and industrial processes in nearly equal amounts. We see it as the smoky exhaust from the tailpipes of cars, the dark material in the smokestacks of industry, the exhaust from coal-fired power plants, and the haze in city air that scatters sunlight.

In chapters 6 and 9, we considered the contribution of fossil fuels to climate change. They are thought by most climatologists to be public enemy number one in this regard, and their effect is growing as affluence and resulting car ownership increase in countries around the world. They are also high on the list of enemies when air pollution is considered (table 10.1). Cars emit about a third of America's air pollutants. Driving a car is probably a typical citizen's most polluting daily activity, in part because of the nature of hydrocarbons and in part because of the amount of driving Americans do. Americans drive 3 trillion miles each year. Assuming an average 20 miles per gallon, we burn 150 billion gallons of gasoline in a year. The love of Americans for their vehicles is evidenced by the fact that we devote 15 to 22 percent of our household expenditures to them, about twice as much as we spend for food.

More than 600 million cars are on the world's roads. The UN predicts that car ownership worldwide will increase to 1 billion to 1.5 billion by 2030 as populous developing nations such as China and India become more affluent. Auto sales in China are growing very fast, and the country has started to manufacture its own vehicles.

Another important source of soot in ports along American coastlines is international shipping, which has increased enormously during the past twenty years.[4] Shipping capacity has increased by 50 percent, and cargo ships have become one of the nation's leading sources of air pollution. Oceangoing ships burn the dirtiest grades of fuel. These low-grade hydrocarbons have sulfur levels 3,000 times that of gasoline. A single

Table 10.1
Contribution of motor vehicles to pollution

Air pollutant	Proportion from on-road motor vehicles	Note
Oxides of nitrogen (NO_x)	34%	Precursor to ground-level ozone (smog), which damages the respiratory system and injures plants
Volatile organic compounds (VOC)	34%	Precursor to ground-level ozone (smog), which damages the respiratory system and injures plants
Carbon monoxide (CO)	51%	Contributes to smog production; poisonous in high concentrations
Particulate matter (PM_{10})	10%	Does not include dust from paved and unpaved roads, the major source of particulate matter pollution (50% of the total)
Carbon dioxide (CO_2)	33%	Thought to be a primary contributor to global warming

Source: Federal Highway Administration (2002).

cargo ship coming into New York harbor can release in 1 hour as much pollution as 350,000 2004-model cars. Globally commercial ships release about 30 percent as much particulate pollution as all the world's cars.[5] Sixteen of these large ships emit as much sulfurous air pollution as all the cars in the world. There are more than 100,000 of these oceangoing transports plying the seas today.

On average, 43 percent of PM_{10} emissions at America's ports come from the diesel fuel burned by unregulated oceangoing ships and other marine vessels. Heavy diesel-burning trucks generate 31 percent of PM_{10} emissions, cargo-handling equipment 24 percent, trains 2 percent, and cars less than 1 percent. In the Los Angeles area, oceangoing ships, harbor tugs, and commercial boats such as passenger ferries emit many times more smog-forming pollutants than all the power plants in the southern California region combined. Growth forecasts are that overseas trade will triple by 2020, indicating that car and truck traffic will not be the only air quality problem in California as the century progresses.

A study in 1998 showed that the cancer risk for residents of Long Beach, to the immediate northeast of the ports of Los Angeles and Long

Beach, was twice as great as the risk for people in the surrounding area. Diesel particulates accounted for 70 percent of the risk.[6]

The effect PM has on our bodies depends to a large degree on its size and chemical composition. Those larger than 10 microns are kept out of our bodies by blockages such as nose hairs and the coughing mechanism in our throats. Particles smaller than 10 micrometers are trapped in our lungs, and those smaller than one-tenth of a micrometer are so minute they can pass through the lungs into the bloodstream, just like the oxygen molecules we need to survive. Soot contains up to forty cancer-causing chemicals.

Because the particles in PM are formed in so many ways, they can be formed of many substances. The larger ones are rich in soil dust, and all animals, including humans, have always lived with them. The smaller ones may be solids or liquids; some are solids suspended in liquids. PM smaller than 2.5 microns is so tiny that it stays suspended in the air for days or months before settling to the ground. This means that much of it comes from sources outside any particular region. Several states on the East Coast receive cryptodust from the Sahara Desert. The western United States receives cryptodust from the Gobi Desert in China. The relative percentages of homemade cryptodust and imported cryptodust are not known, but one would suspect that most is produced locally, meaning that we cannot blame others for the particulate pollution we inhale.

Short-term exposure to particle pollution can kill. Deaths can occur the same day that particle levels are high or one to two months later. In Britain, research has shown that people are twice as likely to die from respiratory disease when they are heavily exposed to soot emitted from vehicle exhausts. Particle pollution does not just cause people to die a few days earlier than they would otherwise. It also diminishes lung function, and it causes greater use of asthma medications and increased rates of school absenteeism, emergency room visits, and hospital admissions. Other adverse effects can be coughing, wheezing, cardiac arrhythmias, and heart attacks. Particle pollution from power plants in the United States leads to over 30,000 deaths each year, a number that can be compared to the 16,272 murders committed in 2008.[7]

Between 1980 and 2008, the concentration of particles nationwide decreased by 25 percent; from 2000 to 2008, the decrease was 8.3 percent.[8] These decreases have added several months to the life expectancy of the average person. Clearly there has been and continues to be significant progress in reducing the amount of particulates in the air, partly because of government regulations and partly because of improvements by the auto companies. The introduction of partially or completely electric vehicles should accelerate the downward trend.

In 2004, about one-third of Americans lived in an area with unhealthful short-term levels of $PM_{2.5}$ pollution.[9] Nearly 20 percent live in an area with unhealthful year-round levels. The worst American cities for particle pollution are Pittsburgh and Los Angeles. In December 2008, California enacted measures to force trucks to use filters to reduce diesel soot emissions by 85 percent, estimating that they would save about 600 lives per year.

Ozone

Ozone is an extremely reactive gas molecule composed of three oxygen atoms, in contrast to normal oxygen gas, which contains two oxygen atoms. It is formed by chemical reactions in the air near the ground from two gases that are generated from burning fossil fuels: nitrogen oxides and hydrocarbons. The hydrocarbons are commonly termed VOCs: volatile (easily evaporated) organic compounds. When the nitrogen compounds and VOCs come in contact with heat and sunlight, they combine to form ozone, the main component of the near-ground substance we call smog. Smog is thus a secondary pollutant; it does not come belching out of smokestacks or car exhausts. Smog is harmful to breathe because it attacks lung tissue. It is capable of causing inflammation in the lung at lower concentrations than any other gas. Those most affected are the elderly, children, people who work or exercise outdoors, and those with existing lung diseases such as emphysema and chronic bronchitis.

The current federal allowable amount of ground-level ozone is 75 parts per billion; one-third of the U.S. population lives in counties that exceed this standard.[10] However, we are making significant progress in smog reduction. Studies indicate that a decrease of 10 parts per billion ozone saves more than 3,700 lives annually and increases life expectancy by 2.7 years.[11]

Smog is a particularly serious problem for the 20 million asthma sufferers in the United States.[12] Every day, 40,000 people miss school or work because of asthma, 30,000 people have an asthma attack, 5,000 people visit the emergency room and account for one-quarter of all such visits, 1,000 people are admitted to the hospital, and 11 people die. In the smoggy San Joaquin Valley of California, 15 percent of the children have asthma, and Fresno, the valley's largest city, has the third highest rate of asthma in the country. On bad days, some schools hoist a red flag so parents can keep their children indoors. The prevalence of asthma has been increasing since the early 1980s, particularly among children, but the reasons are uncertain. Certainly ozone, smog, and particulate pollution can be suspected as complicit, but both of these villains have been declining in abundance. Complicating the analysis is that asthma has a genetic component. If one parent has asthma, chances are one in three that each child will have asthma. If both parents have asthma, it is 70 percent more likely that their children will have the disease.

Recent research has determined that ground-level ozone is associated with changes in male sperm quality.[13] It is estimated that at least 2.1 million couples in the United States have difficulty achieving pregnancy, with male infertility responsible for 40 to 50 percent of infertility cases. Hence, perhaps 1 million men may be able to attribute their difficulty in fertilizing their partner's eggs to ozone pollution. It is known that ozone and its reaction products can cross the blood-gas barrier and enter the bloodstream, and exposure to ozone can cause oxidative stress, which has been shown to disrupt testicular and sperm function. As with smoking, ozone exposure may trigger an inflammatory reaction in the male genital tract or the formation of circulating toxic species. Both events could cause a decline in sperm concentration. No other pollutant has been found to alter sperm quality or quantity.

Pregnant women exposed to the high concentrations of smog characteristic of large cities have triple the risk of having a child with heart malformations, the rate increasing from the normal 2 per 1,000 to 6 per 1,000.[14]

Sulfur

Two-thirds of the sulfur in America's air has been produced by human activities, 80 to 85 percent of it from burning coal to generate electricity. Coal contains 2 to 3 percent sulfur. The other 10 to 15 percent of atmospheric sulfur comes from oil refining, commercial ships, and the smelting of sulfide ores. The sulfur is released from smokestacks as sulfur dioxide gas (SO_2), which reacts with the oxygen and water vapor in the air to produce sulfuric acid (H_2SO_4), the major component of acid rain. Because measurements by the Environmental Protection Agency (EPA) are taken near the plant emitting the sulfur, smokestacks in factories are made to be very tall, putting the gas and resulting acid high in the atmosphere, where they are transported hundreds of miles before raining down.

Average acid rain is about ten times more acid than normal rain. It is an enhancement of the natural and slight acidity of rainwater,

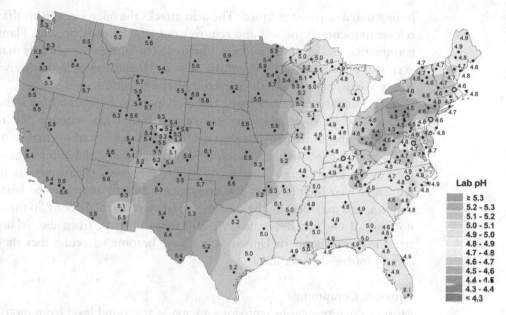

Figure 10.2
Variation in severity of acid rain. (Environmental Protection Agency, 2009)

analogous to the enhancement of the normal greenhouse effect thought to be caused by carbon dioxide and methane. The problem is worldwide but is particularly severe in coal-burning, heavily industrialized regions (figure 10.2). Northeastern North America has had an acid rain problem for many decades because of coal-burning industries in the American Midwest and persistent northeasterly winds. Lakes in the Northeast and in eastern Canada can be 50 to 100 times more acidic than unaffected lakes.

Our atmospheric sulfuric acid is carried into northeastern Canada, and in Nova Scotia some rivers are so acidic that salmon cannot live in them. Fortunately for both the American Northeast and eastern Canada, there has been significant progress in decreasing the release of sulfur from industrial smokestacks in the United States. Thanks to legislation enacted twenty years ago, sulfur emissions have dropped by more than 50 percent, but much further improvement is needed. A decline in the use of coal in favor of alternative sources of energy is required.

Both natural vegetation and crops are affected by acid rain. The waxy layer on the leaves that helps protect the plant from diseases is damaged. The acid soaks into the soil and damages plant roots, causing the plant to be stunted or perhaps killed. The acid attacks the microorganisms that release nutrients to the soil and removes nutrients already present. Plant germination and reproduction are inhibited. The cumulative effect is that even if the plant survives, it will be very weak and less able to survive strong winds, heavy rainfall or even a brief dry period.

Excess acidity causes mercury and aluminum to be leached from the soil, and when these elements are carried into lakes, aquatic life is affected. Natural mechanisms that neutralize the acidity of normal rainwater cannot cope with the excess acidity of acid rain, and many organisms in the lake cannot survive. As acidity increases, organisms die in sequence. First the plankton and crustaceans die off, and then the bass, crayfish, and mayflies. By the time the acidity has reached about 50 times normal, all the fish have died. At 200 times normal, frogs die. Many streams in northeastern United States have become so acidic that they can no longer support life.

Nitrogen Compounds

Most of the nitrogenous emissions are made at ground level from motor vehicles, in contrast to sulfurous emissions, which are released high into the air from smokestacks. Because nitrogen compounds are a major ingredient in the formation of ground-level smog, they are removed from the air very quickly, form little nitric acid (HNO_3), and contribute little to acid rain falling from thousands of feet above.

One-third of nitrogen emissions comes from motor vehicles. Before the advent of modern catalytic converters in cars in the mid-1990s, nitrogen emissions were much greater than they are today. The catalytic converter converts the nitrogenous compounds into nitrogen (the main gas in the air) and water. The converter is capable of reducing nitrogenous emissions by 95 percent but is effective only if the engine is warmed up. Many trips taken by walk-avoiding Americans are either too short or involve lots of stopping and starting, so the engine does not get hot enough for the catalytic converter to be effective.

Lead

Airborne lead is of diminished importance in the United States. Leaded paint has been banned since the 1970s, and the sale of leaded gasoline was totally banned in 1986. Lead in the air decreased by nearly 97 percent between 1980 and 2005, lakes and reservoir waters have seen decreases of as much as 70 percent, and human blood in the United States contains less than half as much lead as it did a few decades ago. Lead poisoning causes learning disabilities and antisocial behavior, among other maladies. Most of the larger car–owning nations of the world have either banned leaded gas or restricted its use.

Late in 2008, the EPA tightened the regulatory limit on airborne lead for the first time in thirty years, lowering the legal maximum to a tenth of what it had been on the grounds that it poses a more serious threat to young children than had been realized.[15] More than 300,000 American children show adverse effects from lead poisoning, and elevated lead can result in increased blood pressure and decreased kidney function in adults.

The vast majority of airborne lead, a neurotoxin that reduces children's IQ, now comes from lead smelters. The lead eventually falls to the ground, and most of a child's exposure comes from soil and indoor dust. However, there are fewer than 200 air lead monitoring stations nationwide, so it is not clear that airborne lead can be sufficiently monitored to make the new law enforceable. According to the EPA, there are 16,000 sources across the country emitting 1,300 tons of lead into the air each year, a sharp decrease from 74,000 tons emitted three decades ago.

Rubber

Car tires do not last forever, as all drivers are aware. But few people consider what happens to the rubber dust that has been ground off by friction between the rolling tires and the highway. The answer is that most of it flies upward and becomes an unwanted constituent in the air we breathe. Approximately 13 million cubic feet of rubber have been lost from wear on the 300 million tires that enter the scrap heap every year. Some of the rubber sticks to the highway and does not enter the air (skid marks), but the amount has never been determined. However, there can be no doubt that our lungs are partly coated with tiny rubber particles. A study in England in 2005 found that about a quarter of the particulate emissions from nonexhaust sources was from tire and brake wear.[16] In the United States, rubber is not a category in air pollution studies.

In addition to microscopic particles of rubber, tire wear releases large amounts of chemicals called polycyclic aromatic hydrocarbons into the air, and these chemicals are suspected of being carcinogenic. The volume of these chemicals released from tires as we drive is unknown.

Nanoparticles

Recent advances in technology have created a new type of air pollution—nannoparticles—microscopic particles that are present in more than 10,000 over-the-counter products. Nanoparticles are so small that a one-inch line contains 25 million of them.[17] The most commonly used nanoparticle is titanium dioxide. It is present in everything from medicine capsules and nutritional supplements, to cake icing and food additives, to skin creams, oils, and toothpaste. At least two million pounds of nanosized titanium dioxide are produced and used in the United States each year. The reason these particles are so dangerous is that, in laboratory experiments with mice, they damaged or destroyed the animals' DNA and chromosomes. It is assumed they can have the same effect on humans.

Because the particles are so small, they can enter the body by almost any pathway. They can be inhaled, ingested, and absorbed through the skin and eyes. They can invade the brain through the nose. They can get into the bloodstream, bone marrow, nerves, muscles, and lymph nodes. There are no federal regulations concerning the use of nanoparticles in consumer products.

Good News

It is clear that there has been major progress in the United States in cleansing the air. In the 1960s, 1970s, and 1990s, Congress enacted a series of Clean Air Acts that significantly strengthened regulation of air pollution. These acts set numerical limits on the concentrations of the major air pollutants and provided reporting and enforcement mechanisms. As a result, between 1980 and 2008, total emissions fell by 54 percent. Particulate emissions fell by 67 percent, sulfur dioxide by 58 percent, nitrogen dioxide by 41 percent, carbon monoxide by 56 percent, volatile organic compounds by 47 percent, and lead by 97 percent.[18] This has happened even as the country's population, economy, and vehicle traffic have exploded. For example, Los Angeles, the stereotypical smog capital of the nation, has gone from nearly 200 high-ozone days in the 1970s to fewer than 25 days a year now. Many areas of the Los Angeles basin are now smog free year round. According to the American Lung Association in 2007, 46 percent of the U.S. population lives in counties that have unhealthful levels of either ozone or particle pollution.

The World Health Organization states that 2.4 million people die each year from causes directly attributable to air pollution. Worldwide, more deaths per year are linked to air pollution than to automobile accidents, but those that stem from pollution are more subtle, often overlooked, and not worthy of news headlines.

Air Pollution and Ultraviolet Radiation

Two types of anthropogenic pollution in the air thousands of feet above our heads affect human lives: the effect on precipitation, which is of lesser importance, and the effect on the ozone layer, which is very serious.

Soot and Rainfall

Water vapor in the air normally condenses to liquid rain at elevations between 10,000 and 30,000 feet. Most of the condensation results from decreasing temperatures as elevation increases, an activity that is unaffected by human activities. The water droplets must grow to a certain size before they drop to the ground. Within the past ten years, it has been found that particles suspended in the air lower the amount of precipitation by preventing large droplets from forming.[19] The particles thus have a dampening effect on rainfall, which can have effects that extend far beyond suppressing local precipitation. For example, the incessant rainfall in the tropics produces much of the energy needed for global air circulation. Any change in rainfall there is certain to affect global climate, but no numbers are yet available. The quantitative effect of soot on tropical rainfall is an active field of current research.

The Ozone Layer

High above our heads around the globe at an elevation of 10 to 22 miles is a zone of air with a naturally occurring concentration of ozone about ten times as much as the amount in ground-level smog in Los Angeles. It is essential for the maintenance of life on earth. This zone has been known to meteorologists since 1913 but was brought to public notice only in 1985 when its thinning was recognized; it is known as the ozone layer. In public consciousness the word *ozone* is nearly always followed by the word *hole*, which meteorologists define as a depletion of ozone in the layer by at least one-third.

The amount of ozone in the 12-mile-thick zone in the stratosphere is determined by the balance between ultraviolet (UV) radiation from the sun and oxygen in the atmosphere. UV radiation splits the two-atom oxygen molecule into separate atoms, a small percentage of which recombines with available two-oxygen molecules to form the unstable and short-lived three-oxygen molecule called ozone. The balance between the rate of formation and the rate of decay of ozone results in a constant percentage of ozone in the layer. The amount of ozone is small—only about 1 molecule of ozone among 10 million molecules of normal oxygen.

One molecule in 10 million seems insignificant, but without it, life on earth would probably cease to exist. The ozone molecule absorbs 93 to 99 percent of the destructive and lethal UV radiation from the sun; only a few percent of the UV radiation emitted by the sun reaches the earth's surface. Ultraviolet radiation damages the DNA in skin cells, and the effect is cumulative over a person's lifetime. If not for the ozone layer, humans would have greatly increased rates of skin cancer, damaged immune systems, increased numbers of eye cataracts at earlier ages, genetic mutations, and lesser amounts of edible crops.

The importance of the ozone layer is clearly seen by considering the occurrence of skin cancer before the layer was damaged by chlorofluorocarbons (CFCs). The sun is the cause of at least 90 percent of skin cancers, so in a normal ozone layer, we would expect this malady to be more common at low latitudes, where the sun's intensity is higher, than at high latitudes. This is evident in the United States, where skin cancer is highly correlated with latitude (figure 10.3). It increases by 10 percent for each three degrees of latitude. For example, in Montana and Washington, annual deaths from skin cancer are 1.1 to 1.2 per 100,000 people; in Florida and Texas, it is twice as high.

The thinner the ozone layer is, the more UV radiation we receive and the greater is the likelihood of skin cancer. For each 1 percent loss of ozone molecules, skin cancer is expected to increase about 4 percent. If you lived near the South Pole, where ozone concentrations have decreased 75 percent, your skin cancer risk would increase by 300 percent.

Higher UV radiation also affects the reproduction and growth of the phytoplankton (one-celled plants) that live and float on the ocean surface and are at the base of the entire food chain in the ocean. Their abundance has declined over the past century. It is hard to imagine life in the sea without plankton. Increased UV radiation also decreases the survival rate of shrimp and fish larvae.

The Cause   The reason for the serious amount of loss of our protective ozone layer is the production of artificial chemicals called chlorofluoro-

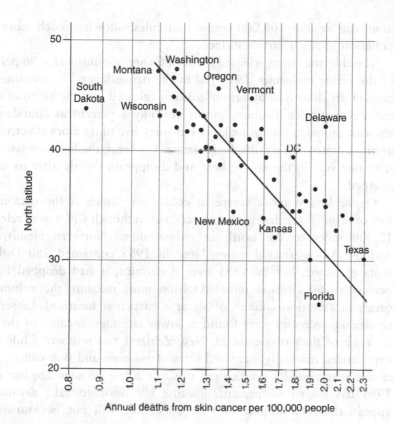

Figure 10.3
Annual deaths from skin cancer per 100,000 white males in the United States, compiled before thinning of the ozone layer. A latitudinal control is evident. (Department of Health and Human Services)

carbons (CFCs). They have been extensively used as refrigerants in refrigerators and air conditioners, for cleaning electronic circuit boards, in the manufacture of foams used for insulation, in insecticide and hair spray cans as propellants, and in rigid polystyrene plastic drinking cups.

Liquid CFCs are volatile chemicals and leak easily into the air. They rise in the atmosphere, and in a few years, they reach the ozone layer, where they remain for many decades, even under bombardment by UV radiation. Some types of CFCs can last in the ozone layer for more than a hundred years. They are destroyed very slowly by the UV rays, which separate the chlorine atoms from the rest of the CFC molecule. There are now five times more chlorine atoms in the upper atmosphere than before CFCs were invented. Chlorine attacks ozone, and each chlorine atom can destroy 100,000 ozone molecules, allowing much more UV radiation to reach earth's surface.

The chlorine atoms released from CFCs are responsible for 96 percent of the ozone molecules destroyed in the ozone layer. The remaining 4 percent are destroyed by bromine atoms released into the air from agricultural pesticides. Bromine atoms are only 1 percent as abundant as chlorine atoms in the ozone layer but forty-five times more effective per atom in destroying stratospheric ozone. However, the bromine lasts less than one year in the atmosphere and disappears shortly after its use is stopped.

Ozone loss is more severe in colder air, which is the reason the loss of ozone is greatest over Antarctica even though CFCs were released 12,000 miles to the north in industrialized Northern Hemisphere nations. An unaffected ozone layer in 1955 contained 380 Dobson units of ozone, but by 1995 over Antarctica, it had dropped by 76 percent to 90 Dobson units. (Dobson units measure the volume of ozone in a vertical column of air at a particular location). Lesser but significant decreases were found at lower latitudes. Because of the loss of much of their ozone shield, New Zealand and southern Chile have experienced markedly increased rates of sunburn and skin cancer. Skin cancer was virtually nonexistent in Chile a few decades ago but since 1994 has soared 66 percent. Because UV disorders take decades to appear, the total impact of the ozone hole will not be known for many years.

The Arctic is second to the Antarctic in terms of atmospheric chilliness, and so the loss of ozone there is no greater than 60 percent. In midlatitudes the loss is less than 10 percent.

The Cure   When the scientific community pointed out the decline and its likely effects to the leaders of the world's nations, there was legislative panic, a condition almost unheard of unless a shooting war is involved. In 1987 representatives of more than thirty industrial countries approved the Montreal Protocol to phase out the production and use of CFCs. In 1990 they tightened the restrictions and in 1992 tightened them still more. Most producing nations ceased production in 1996, and all nations are pledged to do so by 2010. The amount of chlorine in the ozone layer started decreasing in the mid-1990s and has been accompanied by a thickening of the layer. The minimum number of Dobson units of ozone in Antarctica was 92 in 1994 but increased to 112 in 2008. It is expected to be 2080 before full recovery occurs.

Because CFCs are used in a wide range of products, a substitute for them is essential. The replacement chemicals are hydrofluorocarbons (HFCs). They lack the chlorine atoms so deadly to the ozone layer but nevertheless are potent heat-trapping substances. Because they are released in tiny traces, they currently contribute less than 1 percent of the climate-warming effect of the carbon dioxide we produce. However, a recent analysis published by the National Academy of Sciences notes that the fast-paced growth in the use of HFCs in developing countries will soon make them a far bigger contributor to global warming than they now are.[20] Pound for pound, HFCs are hundreds to thousands of times more powerful ozone destroyers than carbon dioxide. By midcentury, emissions of HFCs could surge to cause global warming equivalent to the impact of between 28 and 45 percent of emissions of carbon dioxide.

Methyl Bromide

Methyl bromide is a broad-spectrum and very effective pesticide used to control pest insects, nematode worms, weeds, pathogens, and rodents.[21] Like CFCs, the chemical rises into the atmosphere and attacks the ozone layer. About 21,000 tons have been used annually in agriculture, primarily for soil fumigation (85 percent). Globally, about 72,000 tons are used annually, with use in North America the highest (38 percent), followed by Europe (28 percent), Asia and the Middle East (22 percent), and Africa and South America (12 percent).

Between 50 and 95 percent of the methyl bromide injected into the soil to sterilize it can eventually enter the atmosphere. In the United States, California strawberries and Florida tomatoes are the crops that use the most methyl bromide. Other crops that use this pesticide as a soil fumigant are tobacco, peppers, grapes, and nut and vine crops.

Production and importation of methyl bromide were scheduled to end in 2005 in the developed nations but have been delayed for lack of a suitable substitute. Developing countries have until 2015. There is no single alternative for all the functions of the chemical, and many different chemicals must be used to do the work it has been performing, which increases a farmer's costs and also the time spent in planting and growing the crop.

Air Pollution in Homes and Offices

Those concerned with public health have long understood the role of indoor environments in causing or exacerbating human diseases. Hippocrates was aware of the adverse effects of polluted indoor air in crowded cities and mines, and the biblical Israelites, a desert people, understood the dangers of living in damp housing (Leviticus 14:34–57).

Studies reveal that people in industrially developed countries such as the United States spend most of their time indoors, either at work or at home. The average American spends less than 1 hour per day outdoors.[22] Clearly Americans should be concerned about the possibility of pollution in their workplaces and homes, although most media attention is focused on outdoor air. Indoor air pollution is a quiet and neglected killer.

Because newer, energy-efficient office buildings tend to be tightly sealed, little outside air can enter, and indoor pollutants can reach high levels inside. Poor ventilation in office buildings causes about half of the indoor air pollution problems. The rest come from specific sources in offices, such as copying machines, electrical and telephone cables, mold and microbe-harboring air-conditioning systems and air ducts (which can lead to Legionnaire's disease), cleaning fluids, cigarette smoke, latex caulk and paint, pesticides, vinyl molding, linoleum tile, and building materials and furniture that emit air pollutants such as formaldehyde. For many people in the United States, the least healthy air they breathe all day is indoor air. The health problems in sealed buildings are called "sick building syndrome." An estimated one-fifth to one-third of all buildings in the United States are now considered "sick."[23]

A sick building not only can make workers ill but affects productivity as well. Researchers from the International Centre for Indoor Environment and Energy in Denmark have discovered that poor air quality inside office buildings reduces typing and reading speeds by up to 9 percent, the equivalent of losing 4.5 hours in a 48-hour working week.[24]

And it is not only office buildings that are sealed. Sealed homes are becoming more common because most homes now have air conditioning, and hence home owners keep their windows closed, at least during summer months. The air inside a typical home is on average two to five times more polluted than the air outside. In extreme cases, the air inside is 100 times more contaminated, largely because of household cleaners and pesticides.[25] In 1985 the EPA reported that toxic chemicals found in the air of almost every American home are three times more likely to cause some type of cancer than outdoor air pollutants are.

It is easy to recognize that using dung, crop residues, or wood for fuel indoors, as is typical for cooking and heating in homes in rural and less developed nations in Asia and sub-Saharan Africa, will generate large amounts of soot and perhaps other air pollutants as well. It is visible as it enters the air. Soot from cooking is not a major problem in developed and wealthy nations such as the United States, but even normal cooking using modern gas or electric stoves generates pollution. Experiments by the EPA have determined that deep-frying food for 15 minutes on a stove generates trillions of particles, nearly all with diameters less than 0.1 micrometer, the sizes that not only get into your smallest airways but are absorbed into your bloodstream. The deep frying increases the number of particles throughout the home by as much as tenfold.[26] Heating water for coffee produces about half as many particles. Exactly how these particles are generated and how toxic they are is unknown, and no one knows how to reduce their occurrence. We can be sure that no family is going to stop cooking.

Smoking

Tobacco is the world's largest single cause of preventable death. Five million people die every year from smoking, a number that will grow to perhaps 10 million in the next twenty years.[27] Cigarettes are well known to be the cause of emphysema and 80 percent of lung cancers, and they are implicated in cardiac and other health problems.[28] About 440,000 Americans die prematurely each year because of cigarette smoke. As a result of increased taxes on cigarettes, an increasing number of smoke-free areas in many cities, and intense publicity about the dangers of smoking, the number of smokers has been declining since 1981. Cigarette sales rose consistently from 2.5 billion in 1900 to a high of 631.5 billion in 1980, but have decreased every year since then to a low of 360 billion in 2007, according to the Department of Agriculture. In 2008, there were still 46 million smokers in the United States, 20.6 percent of the adult population. The average smoker smoked 8,295 cigarettes; at 20 cigarettes per pack, that is 415 packs, more than one pack a day.

Although 70 percent of smokers want to quit and 35 percent attempt to quit each year, less than 5 percent succeed.[29] The low rate of successful quitting and the high rate of relapse are related to the effect of nicotine addiction.

Prior to the past few years, tobacco growers enjoyed large government subsidies, awarded to them despite warning by the surgeon general, the government's chief medical official, fifty years ago and in succeeding years that cigarettes are a health hazard and cause lung cancer. Such was the power of the tobacco industry among national legislators. But its economic clout has been sinking. North Carolina is the nation's largest producer of tobacco, but the number of tobacco farms there decreased by nearly 80 percent between 1997 and 2007, according to the Department of Agriculture.

In 2009 the government passed legislation to restrict cigarette advertising further and placed control over production and marketing with the Food and Drug Administration (FDA). The FDA is responsible for approving the marketing and distribution of both food and drugs. Cigarettes are certainly not food, but because tobacco smoke is inhaled for its pharmacological effects, cigarettes are clearly drugs. Drugs require safety tests before they can be approved for sale. A product that kills a third of those who use it cannot pass such tests and must be banned.

Federal courts are stepping in where congressional leaders fear to tread. In May 2009, a three-judge panel of the U.S. Court of Appeals unanimously upheld a lower court ruling that declared the tobacco industry guilty of civil racketeering. The court affirmed that tobacco companies have been and continue to engage in a massive decades-long campaign to defraud the American public about the dangers of smoking. The lower court judge ruled that the federal government had failed to protect the public interest by bowing to political pressure. The decision was upheld on appeal. On February 19, 2010, the companies asked the Supreme Court to review the decision. As of mid-2010 the court had not ruled.

Second-Hand Smoke

Prior to the past two decades, it was believed that smokers were harming only themselves, but evidence since then has revealed that those who live or work with smokers are in danger as well. For everyone who is not a smoker but lives with one, the second-hand cigarette smoke they inhale daily is only slightly less dangerous than directly inhaling the gas from burning tobacco as the smoker does. According to the surgeon general, partners of smokers are at a 25 to 30 percent greater risk of coronary heart disease and 20 to 30 percent more likely to get lung cancer than partners of nonsmokers.[30] Second-hand smoke is also implicated in cervical cancer and hearing loss.

Recent research has noted the phenomenon of third-hand smoke, the invisible yet toxic gases and particles clinging to hair, skin, clothing and furnishings after smoking has ceased. The presence of these noxious substances is well recognized by nonsmokers when they enter a room where smoking has occurred. Those who come into contact with the smoker's residues receive chronic exposure to the chemicals and carcinogens left after the smoke has dissipated.

The harm to children from cigarettes starts before birth. Numerous studies have shown that women who smoke have a harder time getting pregnant and a greater risk of spontaneous abortion and having low-birth-weight babies. Children of mothers who smoke grow up to score 10 percent lower on standardized tests than other children.[31] This suggests that more than 33 million children are at risk of mental deficits from second-hand smoke. In addition, second-hand smoke is implicated in sudden infant death syndrome, respiratory problems, ear infections, and asthma attacks.

Radon Gas

Radon is a colorless, tasteless, odorless, radioactive gas produced by the natural disintegration of uranium in rocks and soils. It originates outdoors and is heavily diluted by the circulating air, and so poses no threat to those outdoors.

Inside houses in confined and poorly ventilated areas such as basements, radon may become concentrated, and long continued exposure to it may be dangerous. But few people live in basements in houses built on uranium-bearing rock or soil where cracked floors may be allowing radon to accumulate to dangerous amounts. And the average American moves ten to eleven times during a lifetime and so is unlikely to be exposed to high accumulations of radon for a significant length of time. Radon is a danger to only a trivial number of people.

Asbestos

Two diseases are associated with the inhalation of large amounts of asbestos: asbestosis and mesothelioma. Deaths from both are directly linked to long-term occupational exposure to asbestos in the preregulated workplace.

Asbestosis is a chronic lung disease caused by large amounts of asbestos that have been deposited in the lungs through inhalation. It causes scarring of the lung and decreased lung efficiency, putting stress on the heart, and it can result in heart failure. It is rare for asbestosis to develop in anyone who has not been exposed to large amounts of asbestos on a regular basis for at least ten years. Symptoms of the disease do not usually appear until fifteen to twenty years after initial exposure to asbestos. About 200 deaths per year are attributed to asbestosis, a bit more than double the number killed by lightning.

Mesothelioma is a cancer in the lung that has the same cause as asbestosis. It affects mostly men who worked in the construction industry in the 1970s and 1980s, before asbestos use was heavily regulated. Because the disease does not appear until ten to forty years after exposure, its incidence is still rising in 2009 but is expected to peak within the next few years. According to the Centers for Disease Control and Prevention in Atlanta, deaths from the disease climbed from 2,482 to 2,704 between 1999 and 2005. The annual death rate is about 14 per 1 million people. Mesothelioma is rapidly fatal once it appears.

Federal legislation requires that asbestos that is visibly deteriorating, such as in ceiling tiles or pipe wrappings, must be removed regardless of the number of fibers in the air. Hence, despite evidence that the removal process itself stirs up a higher level of asbestos fibers in the air than were present before removal, removal proceeds in all public buildings. Statistical experts say that the risk of a child's dying from an asbestos-related disease by attending a school containing asbestos is one-hundredth of the risk of that child being killed by lightning.[32] The chance of being hit by lightning is 1 in 750,000.

Black Lung Disease

About 40 percent of 140,000 active coal miners today work underground, an intrinsically unhealthy environment. Mining machines of various kinds rip into the coal, generating large amounts of coal dust. Inhalation of this dust has disastrous effects on human lungs, effects known collectively as black lung disease.

Black lung disease includes components of chronic bronchitis and emphysema, and it causes prolonged suffering and death. The accumulation and retention of coal dust in miners' lungs are directly correlated with their years of underground exposure, increasing from 13 percent of miners with less than ten years' exposure to 59 percent in those with more than forty years of exposure. Technological advances in mining equipment and the quality of ventilation, coupled with stricter federal regulation of underground mining operations, have reduced the number of miners needed below ground. Fewer of these miners develop black lung disease than was the case decades ago, but coal mining remains a dangerous occupation.

Multiple Chemical Sensitivity

Multiple chemical sensitivity refers to an unusually severe sensitivity or allergy-like reaction to many different kinds of pollutants including solvents, volatile organic compounds, petrol, diesel, smoke, and other chemicals and often includes problems with regard to pollen, house dust mites, and pet fur and dander. It is not a disease recognized by the American Medical Association or by many medical organizations that deal with allergies.

Nevertheless, some people are extraordinarily sensitive to the chemicals in ordinary household materials. Kathy Hemenway is one of these unfortunate people.[33] Her chemical sensitivities began when she was a child, when she noticed that perfume gave her a headache and her throat became sore when she was in fabric stores. Her sensitivity problems grew more severe in adulthood, so she started using natural, fragrance-free shampoos and soaps; avoiding air fresheners, fresh paint, pesticides and lawn-care chemicals; and becoming diligent about housekeeping, but only with natural cleaners such as baking soda and vinegar. She moved from homes with carpets to homes with hardwood floors.

But after an accidental exposure to nearby lawn chemicals at her home in Santa Cruz, California, Hemenway began to have trouble breathing and even more trouble sleeping. She grew agitated, jittery, and depressed and felt as if she were in a fog. She also became sensitive to many more substances than usual and had to use an oxygen tank to recover from even mild exposures, such as breathing exhaust fumes on the freeway.

She decided to get as far from civilization as possible and built a house in the remote high desert town of Snowflake, Arizona. The house has no paint, no carpets, no plywood, no particleboard, and no tarpaper. No pesticides were used on the building's foundation or on the land before the foundation was laid. The exterior of the house is made of masonry blocks, and most of the interior framing and roof are made of steel. The floors are glazed ceramic tile throughout the house, and in the bedroom the walls and ceiling are too. The house has radiant in-floor heating instead of forced-air heating to minimize blowing dust and eliminate the combustion by-products of forced-air heating.

Hemenway is one of many Americans who believe that sprays meant to freshen the air actually pollute it, that chemicals meant to beautify our yards in fact poison them, and that many of the products and materials that make modern life fast and convenient also make people sick.

Studies connect a host of suspect substances to many human illnesses, from headaches to immune disorders to cancer. But the complexities of human biochemistry and the vast number of chemicals everyone is exposed to make proving cause and effect very difficult.

Hemenway's story is one of many similar ones encountered by the Environmental Health Center in Dallas.[34] The people chronicled by the center are clearly more sensitive to the host of chemicals humans have introduced into the environment than are most other people. They are extreme examples of the effects these chemicals have on our bodies. The effects on most people are much more subtle and usually take longer to manifest themselves, and there is no way to eliminate them from the environment. They are in almost everything produced by the world's industries. They are part and parcel of our civilization. As Pogo, the opossum in the comic strip, said in 1971, "We have met the enemy and he is us."

11

Conclusion: Is There Hope?

It is because nations tend to stupidity and baseness that mankind moves so slowly; it is because individuals have a capacity for better things that it moves at all.

—George Gissing, *The Private Papers of Henry Ryecroft I*, 1903

Unless someone like you cares a whole awful lot, nothing is going to get better. It's not.

—Dr. Seuss, *The Lorax*, 1971

The key concept for everything that lives, plant or animal, is sustainability: satisfying daily needs without compromising the future of their descendants. An important distinction in this regard is the difference between needs and wants. Only a tiny percentage of Americans have real needs—those for adequate food, shelter, and clothing. However, they have an infinitely expanding and insatiable number of wants: new cars, fashionable clothes, expensive jewelry, trips to exotic locations, larger houses, and tastier food in pricier restaurants, to name a few. The futile attempt to satisfy these wants is the cause, either directly or indirectly, of most environmental problems: water, air, and soil pollution; climate change; energy shortages; garbage accumulation; and other, less noticeable things such as family breakups and an apparently increased rate of species extinctions.

So, How Are We Doing?

In an ideal world, environmental problems are easily and rapidly solved. Initially scientists uncover an unforeseen threat to humanity. They make their discovery known to the nation's leaders, the public contacts their representatives, and these policymakers implement incentives and regulations. The technology sector responds by devising and disseminating remedies. Within a brief period, socially responsible companies and industries adopt these remedies and change their ways to ameliorate the problems. The result is that the public's health and safety are protected.

In the real world, this is not what happens, basically because money is involved. The only ideology of business is to make money. In a capitalist economy, no one should be surprised that companies act unethically to make profits—they are beholden to the shareholders (you) to maximize those profits. Large industries have major investments in the way things are and fear they will be harmed financially by changes that are not of their making. And in capitalist economies, this generates resistance to change. We live in a world of crazy economics because we do not take account of true costs and externalities. Markets are good at fixing prices but incapable of recognizing costs. Damages to the public caused by industrial and agricultural activities, termed externalities, are not considered. As a result, public health and welfare usually take a back seat to industry profits.

To get away with this behavior, the industry typically emphasizes perceived uncertainties in the science that underlies the need for change. They mobilize their large corps of lobbyists in Washington, and perhaps in state capitals as well, and pressure is put on the nation's lawmakers to defeat proposed changes. These lobbying efforts can delay change for decades and perhaps indefinitely.

Corporations and other interest groups such as the fossil fuel industry, automobile industry, agricultural lobby, and many others spend money on items that they believe will result in increased profits. In this regard, it is significant that the amount spent on lobbying has increased every year during the past decade except 2009, an exception because of the deep recession that lowered spending. Spending on lobbying was $1.4 billion in 1999, $1.8 billion in 2002, $2.4 billion in 2005, and $3.3 billion in 2008. The money was distributed among 12,000 to 14,443 (2008) lobbyists. These expenditures are a major reason the federal government has, until the Obama administration, always favored coal and oil as energy sources rather than renewable and nonpolluting sources. It is also the reason that the efficiency of car engines is low and the reason Washington's agricultural policies have not favored organic farming rather than the health-destroying and pesticide-dowsing policies of factory farms.

Climate issues are a hot topic in Washington and serve as an example of the intensity of lobbying. There are many industrial groups that believe their profitability will be affected by pending climate legislation.

According to the Center for Public Integrity, 2,340 lobbyists dealing with climate issues were hired in 2008, and climate lobbyists in 2009 outnumber members of Congress by 4 to 1.

In addition to the enormous influence of moneyed lobbyists, many political appointees to decision-making positions at the federal level formerly worked for the companies they are charged with regulating. They are commonly suspected of placing the desires of their former employer above those of the public. In addition, there are numerous examples of federal research scientists being muzzled by political appointees who represent special interests at the expense of the public.[1] A recent publicized example was the restriction on the free speech of NASA scientists who disagreed with the government's position on climate change.[2]

Many examples of the effects of legislative lobbying can be cited. Perhaps the most outstanding is the production of tobacco. Cigarettes have been known to cause cancer and blood vessel disease for more than fifty years, but their production continues. Tobacco is an unregulated deadly drug that has been protected by its manufacturers with the cooperation of the nation's elected representatives. And until recently, the production was subsidized by the federal government. Tobacco's opponents have not been able to attract enough legislators to ban cigarette production, despite the fact that essentially all of them understand its harmful effects. Another example is the generation of carbon dioxide, the gas most responsible for climate change. The public, in poll after poll, has stated its desire that something be done about climate change. That gasoline and coal use are the main culprits are incontrovertible facts for scientists. Nevertheless, attempts to mandate better miles per gallon for automobiles and trucks have been successfully fought by the oil and automobile industries for many decades on the claim that they would be put out of business by the proposed changes, although this has been demonstrated otherwise by Japanese carmakers. The bankruptcies of General Motors and Chrysler are a clear case of imprudent production policies by these companies over the years. The contrast between their fate and the success of Japanese automakers could not be starker.

But there is hope for attitudinal change on the horizon. Many businesses are appreciating the facts that an environment where the climate is threatened, the seas are rising, clean drinking water is less available, and the migration of millions of people is a real possibility pose real threats and need to be considered in business decisions.

By any reasonable standard of environmental concern, the world in general and the United States in particular are doing poorly as conservationists. An international ranking of environmental performance puts the United States at the bottom of the Group of 8 industrialized nations and thirty-ninth among the 149 countries on the list.[3] The top ten countries, in this order, were Switzerland, Sweden, Norway, Finland, Austria, France, Latvia, Costa Rica, Colombia, and New Zealand.

The water we drink should, ideally, contain almost nothing but hydrogen and oxygen, the way it was when humans first entered the planetary scene a couple of hundred thousand years ago. We continue to pour poisons into our surface water supply and inject them below ground into our aquifers. The effect of this cocktail of noxious chemicals on human biochemistry is uncertain, but suspicion is growing that a significant part of current human metabolic maladies and hospitalizations results from them.[4] Not only are there poisonous artificial chemicals in the water we drink, but we have them in our homes in containers under the sink and in the furniture and carpets, and we slather them on ourselves and our children's bodies to ward off bugs and solar radiation.

The infrastructure that supports our way of life has been crumbling for decades, an unheralded fact that is only now appearing in the form of electrical blackouts in major cities. The lack of electrical transmission lines is delaying the integration of wind power into the national grid. The nation's underground constructions are in even worse shape. Underground water pipes break regularly, water losses are substantial, water shortages are spreading, and these problems are a continuing and worsening problem in America's cities.

The difficulty and long-term impossibility of living in the southern Mississippi River delta region and the adjacent Gulf Coast was forcefully brought home by Hurricanes Katrina and Rita in 2005, yet the clear message to abandon the area as unprotectable and unsustainable was lost on most Americans. Rising sea levels caused by global warming and only indirectly related to hurricanes are making small Pacific islands and low-lying parts of the world's coastlines progressively uninhabitable.

America's outlandishly large production of garbage is one of the few things for which the United States is not admired by the planet's 7 billion people who do not live there. Our per capita trash production compared to other industrialized nations is a graphic illustration of how wasteful our society is.

America's soil is increasingly lost to the ocean through stream runoff because of poor agricultural practices. These practices have also made our crops much less nourishing than they used to be, and the commercial food industry expands the variety of unhealthy artificial products they produce to fill the supermarkets. The unhealthy processed foods produced by American manufacturers are the main cause of the nation's obesity epidemic. The sugars, fats, and oils added to their products to increase taste and boost salability provide 42 percent of the average American's daily calories.

Beyond the world's shorelines, only a few large carnivorous fish are left in the sea, and it has proven difficult to resurrect their numbers.

Arguably the worst aspect of our lack of environmental concern is the misuse of fossil fuels. It is the main reason the United States placed low in the ranking of environmental performance already noted. We produce a large percentage of the world's industrial and agricultural products, and until the nation changes from using fossil fuels to using nonpolluting energy sources, we will continue to rank low on the list of environmentally responsible nations. The harm that the burning of coal and oil does to water and air is obvious and clear to all Americans, but the industry that produces coal and the oil refineries that produce refined petroleum products seem concerned only with preserving the unacceptable status quo, regardless of its cost to public health. They have fought the development of alternative energy sources for many decades and are only now being dragged kicking and screaming into the twenty-first century. It will be many decades, and perhaps the end of the century, before the use of hydrocarbons is phased out, as it must be eventually.

The use of nuclear power as a source of energy should be relegated to the dustbin of history. Its dangers are obvious, but it may take a few more disasters or some terrorist attacks on nuclear installations or on-site storage facilities by the world's fanatics for governments to shut down existing nuclear plants. The increased incidences of cancer and related maladies among nuclear plant workers, their families, and those who live near the plants do not seem to have penetrated public consciousness. Wind power, solar power, geothermal power, and other resources yet to be developed are less expensive than nuclear power, do not pollute, and are more than adequate to serve as energy sources for industrial civilization today and beyond.

Climate change is real and cannot be stopped during the lifetime of anyone now living, whether or not humans are partly or completely responsible for it. The world's emissions of carbon dioxide are still increasing, show no signs of abating in the foreseeable future, and have an atmospheric lifetime of at least hundreds of years. If the changing climate is partly due to astronomical factors, it is even more out of human control. The changes in temperature and precipitation that have already begun will have great regional economic consequences for agriculture and for people who live near low-lying coastlines. Because the prevention of climate change is impossible, accommodation is called for.

Air pollution in the United States has shown meaningful decreases of nearly all pollutants over the past few decades, but pressure must be kept on the motor vehicle and energy industries because they are responsible for most of the noxious materials in our air. The federal government and an increasing number of states appear determined to do this.

Everyone is harmed by environmental degradation. It is unfortunate that the environmental movement has become the turf of left-wing liberals. In poll after poll, Democrats emerge as more concerned about the environment than Republicans. Why has the view of environmentally concerned Republican President Theodore Roosevelt been forgotten by political conservatives? Why is conserving nature not viewed as worthwhile? After all, the normal functioning of the natural world stabilizes our climate, maintains the oxygen content of our air, fertilizes our soil, replenishes our water, and pollinates our crops. We are part of nature, so we should respect it and not modify it in a way that would damage us. That should be a commandment that everyone should follow.

Why Does Environmental Pollution Continue?

Because the most important impersonal goals to humans are jobs and money, and because most people in Western countries are reasonably satisfied with their living conditions, significant change is considered a threat and is resisted strenuously. Those who pollute are concerned that changes in the status quo will cost them money, or that they will become unemployed due to industrial changes, and they want such changes delayed or canceled. But as young people are increasingly aware, the days when a person could keep the same job for his or her entire working lifetime are gone. All workers must be prepared to alter their occupations whenever conditions change in order to survive in an industrial economy. It may not be anyone's first choice, but it is inevitable. It is the price of what is generally accepted as progress.

I believe most of the nation's pollution problems reflect the fact that many, if not most, Americans have too much money for the nation's good, a conclusion few Americans are willing to accept. The problem in the United States is affluence, not population growth. Extreme examples of expenditures may get the idea across: women spending thousands of dollars for dresses worn only one time to a movie premiere or Oscar awards ceremony; a man buying a two-seat car that costs $150,000 because it boosts his ego and accelerates from zero to 60 mph in under 5 seconds, a useless but macho attribute; or the purchase of a thirty-room mansion for a family of two or three people in an exclusive location. On a smaller scale that most Americans can relate to, we have men paying many hundreds of dollars for athletic shoes (called "sneakers" by an earlier generation) because they have an athletic celebrity's name on them; women paying hundreds of dollars for a fragrance they believe will make them more desirable; and individuals and families buying larger houses even as the size of the average household in the United States decreases.

Perhaps the clearest and most obvious sign of affluence is the uncountable number of weekly yard sales in the United States. Although there are events of this type in other countries, their number pales in comparison with those in this country. Only in America do people have so much stuff that they can give away in good condition mountains of it after a few months or years for pennies on the dollar.

Certainly the people who do these things have the right to do so; they earned the money and can dispose of it in any manner they choose. But such purchases reflect an ethical irresponsibility, aside from generating waste and pollution. Think of all the good and worthwhile things that could be accomplished with all the cash lavished on such pursuits.

Gimme More Stuff: The Growth Economy

During World War II, the federal government promoted slogans such as, "If you don't need it, DON'T BUY IT!" and others, such as "Use It Up, Wear It Out, Make It Do, or Do Without!" Today's urging is quite the opposite. A never-ending increase in consumption is a national goal endorsed by both governments and industry. Governments benefit from increased sales tax revenues. Industry promotes consumption with planned obsolescence and the production of items for one-time use. But the quest to satisfy wants is doomed by human nature to be unsatisfied, to say nothing of the limits of the earth's ability to provide the resources. Thanks to incessant advertising, most Americans have no problem justifying their excesses and consider it necessary to satisfy even the most extreme wants as a requirement for the survival of the nation's well-being. Dorothy L. Sayers, who invented the phrase, "It pays to advertise," expressed well the true meaning of the connection between advertising and consumption in her 1942 book *Why Work*: "A society in which consumption has to be artificially stimulated in order to keep production going is a society founded on trash and waste, and such a society is a house built upon sand."[5] There is an apparent belief that economic disaster will occur if restrictions are put on the number of varieties of cold breakfast cereals, cleaning products, drinks, and many other items. Such restrictions are considered an example of creeping socialism, a term that is anathema to nearly all Americans.

Choice is fine and desirable. No one should be forced to purchase an item because it is the only one available due to arbitrary restrictions placed on manufacturers. But is there no sensible limit? The tens of thousands of items in the average U.S. supermarket may well include more than fifty kinds, sizes, and brands of breakfast cereal. One hundred and fifty linear feet of shelves containing a cornucopia of essentially identical unhealthy cold breakfast cereals testifies to the waste (waist?) built into American society, a defect summed up by poet Gary Snyder in "Four 'Changes'": "Most of the production and consumption of modern societies is not necessary or conducive to spiritual and cultural growth, let alone survival. . . . Mankind has become a locust-like blight on the planet that will leave a bare cupboard for its own children—all the while in a kind of addict's dream of affluence, comfort, eternal progress. . . . *True affluence is not needing anything*" [italics added].[6] Snyder's viewpoint is, of course, economic heresy. Growth to most economists is as essential as the air we breathe. It is, they believe, the only force capable of lifting the poor out of poverty, feeding the world's growing population, meeting the costs of rising public spending, and stimulating technological development, not to mention increasingly expensive lifestyles. They see no limits to growth, ever. To most economists, shifting from vertical growth to horizontal development is seen as a naive and utopian idea with no chance against the dominant forces of economic capitalism.

Perhaps they are right; perhaps humans are inherently avaricious—grasping, greedy, and miserly. America's dominant Christian theology would agree that humans are inherently sinful but believes they can be redeemed. The concept of development over growth was first enunciated by John Stuart Mill, one of the founders of classical economics, in his 1848 treatise *Principles of Political Economy*. He predicted that once the work of economic growth was done, a "stationary" economy would emerge in which we could focus on human improvement: "There would be as much scope as ever for all kinds of mental culture, and moral and social progress . . . for improving the art of living and much more likelihood of it being improved, when minds cease to be engrossed by the art of getting on." So far Mill's expectation has not happened.

A growing number of experts are arguing that personal carbon virtue and collective environmentalism are futile as long as our economic system is built on the assumption of growth.[7] The science tells us that if we are serious about saving the earth, we must reshape our economy.

We are bombarded with advertisements to insulate our homes, turn down our thermostats, drive a little less, and so on. But you will not hear advice from either a Democratic or Republican government to "buy less stuff." Buying an energy-efficient TV is applauded; not buying one at all is at best odd. Unless there is an immediate crisis such as a cutoff of oil supplies, political leaders are unwilling to make the case for lowered consumption. Business leaders seem to lack the imagination to conceive of a prosperous America (or a prosperous world) where consumption is lessened. The only respected people who might carry the message are religious leaders. They can make the moral argument about humanity's responsibility to the environment.

The United States is the only nation where shopping is a recreational activity: I shop, therefore I am. Ecologist David Suzuki argues that consumerism has supplanted citizenship as the chief way that people participate in society. Martin Luther King, Jr., once noted that the root problem in American society was not racism or imperialism or militarism, but materialism. Interestingly for people of faith, some scholars now see consumption as the functional equivalent of a religion: millions live to consume and see shopping and owning as the activities that most give meaning to their lives. Consumerism can be seen as the first global religion. Consuming less may be the biggest thing people can do to reduce carbon emissions, but no one dares to mention it because if we did, it would threaten economic growth, the very thing that is causing the problem in the first place.

Of course, if we do not shop incessantly, factories will stop producing and workers will be laid off, they will not have money, and they will be forced to stop shopping. Tax income to the government will decrease and services will be cut. This is the logic of free-market capitalism: the economy must grow continuously or face collapse. With the environmental situation reaching the crisis point, it is time to stop pretending that mindlessly chasing economic growth is compatible with sustainability. And increased efficiency (using less per person) is not a substitute for conservation (using less in an absolute sense).

Consumerism in developed countries appears to have two speeds—fast and faster—and nobody has figured out how to slow this train before it creates an environmental and social wreck of historic proportions. The challenge is to redefine the meaning of "progress" and revamp economies and societies to work in harmony with the natural environment and serve all people. This means overturning some of the venerable pillars of twentieth-century progress:[8]

- The sole purpose of an economy is to generate wealth.
- Collecting material goods is the major goal of life.
- Waste is inevitable in a prosperous society.
- Mass poverty, while regrettable, is an unavoidable part of life.
- The environment is an afterthought to the economy.
- Species and ecosystems are valuable for economic and aesthetic reasons but do not have innate value.
- Ethics plays a negligible role in defining progress.

Overturning these venerable pillars is a formidable task. Without an intentional cultural shift that values sustainability over consumerism, no government pledges or technological advances will be enough to rescue humanity from unacceptably hazardous environmental and climate risks.

Does More Money Bring More Happiness?

Despite the obvious fact that nearly everyone wants to have more money, we are all familiar with the cliché that money cannot buy happiness, and this is verified by data from industrial economies. When growth in income is plotted against levels of happiness recorded in national polls in industrialized nations, there is no correlation. In the United States, for example, the average person's income more than doubled between 1957 and 2002 (with inflation factored out), yet the percentage of people reporting themselves to be "very happy" over that period remained flat, at 30 to 35 percent.[9] And despite people having increasing disposable income, between 1960 and 1990 the divorce rate doubled, the teen suicide rate tripled, the prison population quintupled, and the number of people suffering from depression soared.

This is the situation in well-off nations. But it is different in developing or undeveloped nations, where data reveal that happiness and income are indeed correlated, but only until income rises to about $13,000 per year. Less than 10 percent of households in the United States earn less than $13,000 a year. According to the Census Bureau, the median annual salary for an American man working full time in 2007 was $45,113. It was $35,102 for a woman. The median household income was $50,233.

Apparently money does deliver happiness for people who are trying to cover life's basic necessities. At higher incomes, the correlation vanishes: additional money has little or no relationship to happiness. When a person has enough money to cover basic necessities, happiness probably depends on interpersonal relationships.

If happiness is what people want, why not create economies that are geared to deliver it? Apparently growth economies do not do this. Is it possible to change the emphasis? Can the emphasis of industry and governments at all levels be changed from an emphasis on meaningless and environmentally destructive excessive consumption to an emphasis on well-being and consequent happiness?[10]

Why Do We Work?

I believe that a neglected but important aspect of the problem of growth versus the environment is that technological advances have made it unnecessary for everyone to work, but no one knows how to organize a society in which working is optional. So production is continually expanded to keep people employed, meaning that more useless "stuff" is always produced, much of it with built-in planned obsolescence (models of cars, clothing, other products).

There will always be those who work because they enjoy it—the people who do creative things such as painting, composing, researching, writing, teaching, or doing science to discover how things work. But those who do repetitive tasks that can lose their sparkle, such as helping factory machines put fenders on new cars that will replace existing but still functional older cars, manufacturing expensive clothing, checking olives on assembly lines to see that imperfect ones are rejected, or producing another brand of clothes-washing powder, would be happy to cease their daily work if there were an alternative way to supply their essential needs and a few of their wants. I am not aware of any futurists who consider this question.

The Law of Unintended Consequences

With amazing arrogance we presume omniscience and an understanding of the complexities of Nature, and with amazing impertinence, we firmly believe that we can better it. We have forgotten that we, ourselves, are just a part of

nature, an animal which seems to have taken the wrong turning, bent on total destruction.
—Daphne Sheldrick, *The Tsavo Story*, 1973

There are many examples of this human hubris and its failures in the environmental realm, as well as in other spheres of human concern. Among them is a particularly visible and striking example from Macquarie Island, a 50-mile spot of land in the Southern Ocean about halfway between Australia and Antarctica.[11] The island was discovered in 1810 and soon after, seafarers began visiting it to slaughter fur seals, elephant seals, and penguins. Among the passengers on the visiting ships were rats and mice. To counter them, the sailors in 1818 brought in cats and later rabbits to provide a food source for stranded seamen.

The cats fed on rabbits and native birds, exterminating two species of birds. The number of rabbits swelled to 130,000 by the 1970s and were stripping bare the island's vegetation. To counter this, scientists in 1968 introduced the myxomatosis virus, which is lethal to rabbits, and the European rabbit flea, which spreads it. By the early 1980s, the population of rabbits had fallen to about 20,000.

But the cats were hungry and began feeding on burrowing sea birds, threatening their existence. So researchers began shooting the cats, and by 2000, there were none left. With the cats gone, the rabbits began proliferating again despite the presence of the virus. By 2009, they had stripped as much as 40 percent of the island bare of vegetation. With much of the soil's stabilizing vegetation gone, landslides are increasing, one of which wiped out part of an important penguin colony.

Scientists' only course of action to prevent a possible ecosystem meltdown is to eradicate the remaining rabbits, mice, and rats. The cost estimate for this "solution" is at least $16 million and will take years.

The destruction of Macquarie Island is one of a large number of environmental examples of what is popularly known as the law of unintended consequences, which says, in effect, that knowledge is always incomplete and that unforeseen things are likely to happen after people start projects with good intentions. All purposeful actions will produce some unintended consequences.

Other environmental examples of the law of unintended consequences that are perhaps better known to Americans include the ethanol craze, initiated to decrease America's dependence on foreign oil, reduce tailpipe emissions from cars, and support agriculture. The major unintended effect, which certainly should have been foreseen, was a massive decrease in the world food supply as corn acreage expanded at the expense of other crops, and corn normally fed to chickens, cattle, and swine was diverted to ethanol production.[12]

Another example is the massive use of artificial nitrogen and phosphorous fertilizers in agriculture, which has had the unintended effect of producing a large and growing number of dead zones in shallow ocean waters around the world that are increasing in size over time as the use of fertilizers increases in developing countries.

A fourth calamity has been the deliberate introduction of the kudzu vine from Japan in 1876 into the southeastern United States. It was used in Asia for erosion control, to enrich soil, and it is high-quality forage for livestock. Unfortunately the Japanese did not bring any of kudzu's natural insect enemies along with the plant. Kudzu grows rapidly, up to 1 foot per day, and from 60 to 100 feet in one growing season. It has displaced native plants and has taken over large swaths of land in the southeastern United States that were formerly used for other purposes. The vine now hangs in curtains along highways all over the South and covers 7 million acres, smothering trees and shrubs in its march over the landscape (figure 11.1). Kudzu is invasive as far north as Connecticut.

Figure 11.1
Kudzu plants covering a house (center of photo) and surroundings. (Photo courtesy Jack Anthony, jjanthony.com/kudzu)

It is now considered a pest by the Department of Agriculture but difficult and expensive to eradicate.

A final and well-known example is the invention and widespread use of chlorofluorocarbons, which escape into the atmosphere and have severely damaged the ozone shield.

These unintended side effects can be more significant than intended effects. It is clear that trying to replace the way nature has operated for hundreds of millions of years is both futile and harmful and should be avoided whenever possible, which it usually is. "Nature knows best" is usually a sound working philosophy.

The clearest examples of the refusal to accept the law of unintended consequences are the repeated suggestions for geoengineering projects, also called climate engineering.[13] Rather than decreasing the use of fossil fuels, a group of distinguished environmental scientists suggested in 1965 spreading very small reflective particles over about 5 million square miles of ocean, so as to bounce about 1 percent more sunlight back into space. Within the past few years, another group has recommended fertilizing the oceans with iron to stimulate the growth of phytoplankton, which absorb carbon dioxide during photosynthesis. Another idea is to inject several million tons of sulfur dioxide into the stratosphere to form sulfuric acid, the component of tropospheric acid rain most environmentalists want to reduce (chapter 10). Clouds of sulfate particles would scatter sunlight before it gets to the lower atmosphere (the troposphere), making earth's surface cooler. A British physicist has suggested spraying microscopic droplets of seawater into the sky from a fleet of unmanned sailing vessels. The addition of the tiny droplets into clouds would cause an increased scattering of sunlight back into space.[14]

Many other geoengineering schemes to reduce climate change have been proposed, and they all have rather obvious chemical and ecological negative drawbacks, as well as effects we cannot foresee because of our permanently inadequate understanding of how the atmosphere-ocean dynamic and ecology will react to such unnatural interventions.[15] Geoengineering cannot solve the carbon dioxide problem, and it is foolish and dangerous in the extreme to even consider such schemes, akin to using gambling as a way to get out of debt and with much higher stakes.[16] But further research is inevitable because of human hubris. The idea of trying to reengineer the planet to avoid the need to reduce and eventually eliminate the use of fossil fuels would be laughable were it not so dangerous and impossible. It sometimes seems that humans are incurably stupid when their place in nature is concerned, a possibility immortalized in the film *The Age of Stupid* (2009) that deals in a scientifically valid manner with the coming effects of climate change.

What Can We Conclude from This Information?

The discussion thus far leads us to several conclusions:

1. The environment that humans depend on is being continually degraded.
2. The cause of the degradation is the desire for unnecessary and functionally useless material wealth.
3. Income above a certain level does not increase happiness.
4. The never-ending desire for more money cannot be satisfied.
5. The idea that continued growth is necessary for a nation's survival and a happy population is false.
6. A reorganization of the way economics operates is the only way to solve environmental problems.

The Environment and Spirituality

Many people try to be environmentally aware and minimize the size of their ecological footprint. Some may take shorter showers and install low-flush toilets to conserve water. Others concentrate their efforts on energy conservation, using energy-saving light bulbs and traveling by public transportation whenever it is available. Another group feels burdened by the excesses of the agricultural sector of society and buys organic food produced in a small foodshed. All of these efforts are commendable and worthwhile.

But it is questionable whether such individual efforts can be adopted nationwide or worldwide without a major overarching philosophical base accepted by most people. Such a base is provided by religious faith. The world's religions have many assets to lend to the effort to build sustainable progress, including moral authority, a long tradition of ethical teachings, and the political power that comes from having so many adherents. These are not limited by political viewpoint, age, gender, or religious denomination. Christians, Jews, and Muslims have the same philosophical base as far as the environment is concerned (table 11.1).[17] They all believe that the world and everything in it was created by God and it is blasphemous for humankind to destroy it. It is here for us not just to use but to care for. If life is sacred, so is the web that sustains it. People have a duty to be stewards of the earth.

Table 11.1
Selected religious perspectives on consumption

Faith	Perspective
Baha'i faith	"In all matters moderation is desirable. If a thing is carried to excess, it will prove a source of evil." (Baha'u'llah, Tablets of Baha'u'llah)
Buddhism	"Whoever in this world overcomes his selfish cravings, his sorrows fall away from him, like drops of water from a lotus flower." (Dhammapada, 336)
Christianity	"No one can be the slave of two masters. . . . You cannot be the slave both of God and money." (Matthew 6:24)
Confucianism	"Excess and deficiency are equally at fault." (Confucius, XI.15)
Daoism	"He who knows he has enough is rich." (Dao De Jing)
Hinduism	"That person who lives completely free from desires, without longing . . . attains peace." (Bhagavad-Gita, II.71)
Islam	"Eat and drink, but waste not by excess: He loves not the excessive." (Quran, 7.31)
Judaism	"Give me neither poverty nor riches." (Proverbs, 30:8)

Source: G. T. Gardner, *Inspiring Progress* (New York: Norton, 2006).

A recent poll found that 67 percent of Americans believe that harming the environment is a sin.[18] In 2008 and 2009, the Vatican officially agreed, listing pollution as an area of sinful behavior for today's believers and urged people to show more respect for the environment. The pope noted that there are signs that "creation is under threat" and said that government leaders have an obligation to work together for the "protection of the environment, and the safeguarding of resources and of the climate." He suggested that there should be more research into alternative energy and that industrialized countries should lower their energy consumption through technology or greater "ecological sensitivity" among individuals. "It is becoming more and more evident that the issue of environmental degradation challenges us to examine our lifestyle and the prevailing models of consumption and production, which are often unsustainable from a social, environmental and even economic point of view."[19]

That same day, January 2, 2010, leaders from the Southern Baptist Convention released a concurring statement: "There is undeniable evidence that the Earth—wildlife, water, land and air—can be damaged by human activity, and that people suffer as a result."

In July 2009 the Episcopal Convention endorsed the Earth Charter Initiative, a document written to promote the transition to sustainable ways of living and a global society founded on a shared ethical framework that includes respect and care for the community of life, ecological integrity, universal human rights, respect for diversity, economic justice, democracy, and a culture of peace. The House of Bishops at the convention urged the government to adopt "equitable subsidies for renewable energy (such as solar and wind turbine power, and research into new technologies) along with balancing its current subsidies for nonrenewable energy sources (oil, gas, and coal)." They also voted to support the adoption of a federal renewable energy standard that would require power plants to produce 20 percent of their electricity through renewable sources and committed the church to a 50 percent reduction in greenhouse gas emissions from the facilities it maintains by 2019.

At their 2003 convention, they had passed a resolution aimed at stopping mountaintop removal as a means of coal mining. They authorized the sending of action alerts to Episcopalians asking them to contact their elected officials about mountaintop removal. Imagine the possible effect if all church hierarchies in the United States were as aggressive as the Episcopalians in urging their adherents and the government to agitate for a sustainable society.

In October 2009, Ecumenical Patriarch Bartholemew, the leader of 250 million Orthodox Christians worldwide warned a gathering of scientists, policymakers, and other religious leaders in New Orleans that humanity's care for the environment is near a tipping point "where absolute limits to our survival are being reached." Noting that deforestation, water pollution, the collapse of fishing stocks, and other environmental crises indicate "we have lost our balance, externally and within," and urged his listeners to recognize their "sacred responsibility to the future."

That same month, the archbishop of Canterbury spoke to leaders of nine of the world's major faiths about the moral vision of the world's religions and their crucial role in tackling climate change. Environmental commitments of various kinds were made by several leaders.[20]

There is a growing movement among religious leaders who use their pulpits to stimulate environmental action. More than 10,000 congregations of Christian, Jewish, Islamic, Buddhist, and other faiths are working in thirty states as members of Interfaith Power and Light.[21] IPL is the brainchild of the Reverend Sally Bingham, a priest in the Episcopal Diocese of California. She says that most people want to do the right and moral thing, but that they do not always know what that is. She believes that this is the reason that religious leaders have an important role. "There are millions of people who don't listen to politicians and who are skeptical of science, but who will listen to their clergy," she notes.[22]

Faith communities own between 7 and 8 percent of the habitable land surface of the planet, run (or are involved in) half the world's schools, and control more than 7 percent of international financial investments.[23] They can be powerful political players in environmental protection and remediation.

Due in large part to the Enlightenment in the eighteenth century and the success of science in understanding the physics, chemistry, biology, and geology of the world we live in, the earth has come to be viewed as an impersonal machine, with only humans as meaningful creatures. Our culture seems to have lost touch with the view held by indigenous peoples that nature is alive and sacred and that we humans are an integral part of it all. Anyone who considers it to be so is often treated with ridicule or suspicion.

In the words of John Mohawk, Native American chief:

The natural world is our Bible. We don't have chapters and verses; we have trees and fish and animals. . . . The Indian sense of natural law is that nature informs us and it is our obligation to read nature as you would a book, to feel nature as you would a poem, to touch nature as you would yourself, to be part of that and step into its cycles as much as you can.[24]

Native Americans view environmental destruction as a sin.

The modern world's ethical framework appears to be slowly expanding and will eventually include the total environment. This framework affirms the right of all resources, including plants, animals, and earth materials, to continued existence and, at least in certain locations, continued existence in a natural state.

Expanding on John Stuart Mill's view of progress, Arnold Toynbee once noted that the measure of a civilization's growth is not its capacity to build empires or raise living standards, but the law of progressive simplification: the capacity to spend more and more time and energy on the nonmaterial side of life.[25] The world's religions have the capacity to challenge their institutions and adherents to make Toynbee's vision a reality.

The Ideal Environmental Life

Because of their powerful brains, humans have the ability to severely affect the functioning of the environment, often to the detriment not only of the organisms they share the planet with, but to themselves as well. Many of the actions humans have taken were done to increase their comfort, to satisfy "wants" rather than "needs." It will require a Herculean effort by the world's people to change this.

The use of fossil fuels must be stopped. Their production and refining pollute both the air and the water. In addition, the emissions from power plants and factories may be a partial or complete cause of climate change.

Factory farming and industrial agriculture are poisoning the soil, decreasing the nutritional value of the food produced, and poisoning the food as well. Organic agriculture can feed the world. Artificial chemicals are not necessary for successful farming.

The food processing industry seems interested only in increasing profits with no concern for the health effects on the people who eat their tasty but unhealthy products. There appear to be no ethical considerations in their corporate boardrooms. Anything that does not kill people immediately seems to be acceptable.

It is not possible for an individual to live an ideal environmental life in the modern world. It may never have been possible because of the apparently rapacious nature of the human species. Probably the best individuals can do is to be aware of their most damaging activities and try to hold them in check to the extent possible. Listed below are the most obvious things we should all try to do.

Food
- Buy locally and from a small foodshed
- Eat organic food
- Don't eat meat or dairy
- Eat more fish
- Avoid processed foods
- Buy fresh food in season
- Grow your own food
- Use reusable shopping bags

Energy
- Avoid using fossil fuels
- Use energy-saving illumination
- Walk or use public transportation
- Install solar panels in your home
- Limit use of air conditioning
- Buy energy-saving appliances
- Support alternative energy initiatives

- Don't buy a bigger house than you need for living
- Recycle as much as possible
- Use clothes until they decay
- Buy clothes made of natural fabrics
- Plant trees to absorb carbon dioxide

Water
- Install low-flush toilets
- Take short showers
- Drink tap water, not bottled water
- Do not have a large grass lawn

Notes

Chapter 1

1. J. F. Kenny, N. L. Barber, S. S. Hutson, K. S. Linsey, J. K. Lovelace, and M. A. Maupin, "Estimated Use of Water in the United States in 2005," U.S. Geological Survey Circular 1344, 2009.

2. Circle of Blue, "U.S. Faces Era of Water Scarcity," July 9, 2008. Available at http://circleofblue.org/waternews/2008/world/us-faces-era-of-water-scarcity/; L. Song, "Rethinking Water Management" Earth, January 1, 2010, 36–45.

3. E. H. Clarke II, "Water Prices Rising Worldwide," Earth Policy Institute, March 7, 2007. Available at http://earth-policy.org/Updates/2007/Update64 printable.htm.

4. Environmental Protection Agency, "Outdoor Water Use in the United States," 2008.

5. Circle of Blue, "U.S. Faces Era of Water Scarcity."

6. S. Kelly, "Land Use History of North America: Colorado Plateau," n.d. Available at http://cpluhna.nau.edu/places/grand_canyon3.htm

7. A. Powers and B. Boxall, "Colorado River Water Deal Is Reached," Los Angeles Times, December 14, 2007, A16. Available at http://articles.latimes.com/2007/dec/14/nation/na-colorado14.

8. J. Mattera, "The Leaps and Leaks of the 2008 Great Lakes Compact," October 3, 2008. Available at http://toledocitypaper.com/view_article.php?id=2178; F. Barringer, "Growth Stirs a Battle to Draw More Water from the Great Lakes." New York Times, August 12, 2005, A12.

9. Draft for discussion at SOLEC 2008, "State of the Great Lakes," 2008.

10. Environmental Protection Agency, "State of the Great Lakes," 2005. Available at http://epa.gov/glnpo/solec/sogl2007/7056_WaterWithdrawals.pdf.

11. U.S. National Atlas, "Water Use in the United States, 2008", 2009. Available at http://nationalatlas.gov/articles/water/a_wateruse.html; N. D. Kristof, "Cancer from the Kitchen?" New York Times, December 6, 2009.

12. K. F. Dennehy, "High Plains Regional Ground-Water Study." U.S. Geological Survey Fact Sheet FS-091-00, 2000.

13. "Liquidating Our Assets," *World-Watch*, September–October 1997, 39.

14. Kenny et al., "Estimated Use of Water in the United States in 2005."

15. E. Royte, " A Tall, Cool Drink of ... Sewage?" *New York Times Magazine*, August 10, 2008, 10–12.

16. K. Kloosterman, "EWA Squeezes Drinking Water from Thin Air," *Israel 21c*, October 30, 2008. Available at http://israel21c.org/.../ewa-squeezes-drinking-water-from-thin-air

17. CopperWiki, "Virtual Water," n.d. Available at http://copperwiki.org/index.php/Virtual_Water.

18. W. L. Anderson, "The Economics of Water in the West," Ludwig von Mises Institute, July 2, 2004. Available at http://mises.org/story/1557#.

19. D. Bacher, "Report Reveals Enormous Cost of Agricultural Water Subsidies," *Fish Sniffer*, January 2, 2005. Available at http://fishsniffer.com/dbachere/050102subsidies.html.

20. P. O'Driscoll, "Americans Use Less Water," *USA Today*, March 11, 2004.

21. P. Torcellini, N. Long, and R. Judkoff, "Consumptive Water User for U.S. Power Production," National Renewable Energy Laboratory, 2003.

22. A. Vickers, "Lawn Binge," *Boston Globe*, June 10, 2007.

23. Environmental Protection Agency, "Water and Wastewater Pricing," n.d. Available at http://epa.gov/cgi-bin/epaprintonly.cgi.

24. H. Stallworth, "Water and Wastewater Pricing," Environmental Protection Agency, 2000.

25. M. Margolis, "Drinking Water Goes Digital," *Newsweek*, August 27, 2001, 7.

26. "Water Industry Enters the Smart Meters Arena," *Energy World*, February 2008, 20.

27. H. Stallworth, "Water and Wastewater Pricing." EPA Report 832-F-03-027, 2003.

28. D. Brinn, "Israeli System Turns Contaminated Water into Drinking Water—Instantly," *Israel 21c*, March 18, 2007; J. Yabroff, "Water for the World," *Newsweek*, June 18, 2007, 20.

29. V. Smil, "Water News: Bad, Good and Virtual," *American Scientist*, September–October 2008, 399–407.

30. J. Yauck, "Taking a Fresh Look at Desalination," *Geotimes*, October 2006, 42–43.

31. "Perspectives," U.S. Geological Survey Fact Sheet 075–03, 2003.

32. C. Duhigg, "That Tap Water Is Legal But May Be Unhealthy," *New York Times*, December 17, 2009.

33. Gallup Poll, April 20, 2007.

34. Grinning Planet, "Water Pollution Facts," 2005. grinningplanet.com/2005/07.../water-pollution-facts-article.htm.

35. C. Duhigg, "Health Ills Abound as Farm Runoff Fouls Wells," *New York Times*, September 18, 2009.

36. M. Dorfman and K. S. Rosselot, "Testing the Waters: A Guide to Water Quality at Vacation Beaches, 18th edition" National Resources Defense Council, 2008; Grinning Planet, "Water Pollution Facts."

37. R. Showstack, "Survey of Emerging Contaminants in U.S. Indicates Need for Further Research," *EOS*, American Geophysical Union, March 26, 2002, 146.

38. K. S. Betts, "First Ecosystems Analysis Reveals Data Gaps," *Environmental Science and Technology*, November 1, 2002, 404A.

39. C. Lock, "A Portrait of Pollution," *Science News*, May 22, 2004, 325–326.

40. S. Goodman, "Mix of Common Farm Pesticides Deadly to Salmon—Study," March 3, 2009. Available at http://eenews.net/public/Greenwire/2009/03/03/17.

41. Ralph Nader Research Institute, "The Health Effects of Drinking Water Contamination." Available at http://belcraft.com/DrinkingWaterContamination .htm, 2005.

42. M. Cone, "Autism Epidemic Not Caused by Shifts in Diagnoses; Environmental Factors Likely," *Environmental Health News*, January 9, 2009. Available at http://environmentalhealthnews.org/ehs/news/autism-and -environment; S. Higgs, "Autism and the Indiana Environment Blog," November 15, 2009. Available at http://bloomingtonalternative.com/node/10225.

43. "Chemicals Are Destroying Our Brains," *Ecologist*, October, 2004, 8.

44. "Mothers and Daughters Linked by Pollution," *Ecologist*, July–August 2006, 12.

45. "Unborn Babies Polluted with Industrial Chemicals," *Environment*, September–October 2005, 9; "Toxic Transfer," *New Scientist*, September 10, 2005, 6.

46. Centers for Disease Control and Prevention, "National Report on Human Exposure to Environmental Chemicals," 2010. Available at http://cdc.gov/ exposurereport/.

47. K. Doheny, "Household Chemicals May Show Up in Blood. Study by Environmental Group Shows Toxic Chemicals End Up in Blood Samples," WebMD, May 5, 2009. Available at webmd.com/news/20090501/household -chemicals-may-show-up-in-blood.

48. K. Evans, "Society's Experiment: There Are a Hundred Plus Poisons in Your Blood," July 9, 2009. Available at http://naturalnews.com/z026584_chemicals _cancer_blood.html.

49. Grinning Planet, "Water Pollution Facts," 2005.

50. "Chesapeake Bay Watch," *Washington Post*, November 8, 2008, A16; J. Achenbach, "A 'Dead Zone' in the Gulf of Mexico," *Washington Post*, July 31, 2008, A2; H. Blatt, *America's Food* (Cambridge, MA: MIT Press, 2008), 31–32, 163, 165–166.

51. M. Hawthorne, "CDC Releases Disputed Report on Great Lakes," *Chicago Tribune*, January 14, 2009.

52. "The Hudson: A River Runs Through an Environmental Controversy," *Environmental Health Perspectives*, April 2002, A184-A1187; M. Hill, "GE Dredging of PCBs from Upper Hudson River Begins," New York Times, May 15, 2009. _ McKinley, Jr., "Heading to Texas, Hudson's Toxic Mud Stirs Town," *New York Times*, May 31, 2009, 14.

53. D. A. Fahrenthold, "Bay Is a Threat to Humans, Too," *Washington Post*, July 7, 2009, E4.

54. H. Blatt, *America's Food* (Cambridge, Mass.: MIT Press, 2008), 37–38.

55. N. D. Kristof, "It's Time to Learn from Frogs," *New York Times*, June 28, 2009, 9.

56. G. Lean, "Official: Men Really Are the Weaker Sex," *Independent* (London), December 7, 2008; J. Glausiusz, "Is Pollution Weeding Out Male Babies?" *Discover*, January 2008, 36–37. P. W. McRandle, "Hormone-Altering Chemicals in Everyday Products," *WorldWatch*, March–April 2007, 5; N. D. Kristof, "Chemicals and Our Health," *New York Times*, July 16, 2009, A27.

57. G. Lyons, "Official: Men Really Are the Weaker Sex". *The Independent* (England), December 7, 2008.

58. The Endocrine Society, "Endocrine-Disrupting Chemicals," 2009.

59. "World Oil Pollution: Causes, Prevention and Cleanup," Ocean Link, n.d. Available at http://oceanlink.island.info/ocean_matters/oil.html.

60. C. Duhigg, "EPA Vows Better Effort on Water," *New York Times*, October 16, 2009.; C. Duhigg, "Millions in U.S. Drink Dirty Water, Records Show," *New York Times*, December 8, 2009.

Chapter 2

1. D. Leonhardt, Piling Up Monuments of Waste," *New York Times*, November 19, 2008, B1, B6.

2. S. Burns, "Infrastructure Needs—and Gets—Help," *Earth*, May 2009, 86.

3. A. Cohen, "Public Works: When 'Big Government' Plays Its Role," *New York Times*, November 13, 2007, A28.

4. American Water Works Association, "New or Repaired Water Mains," (Washington, D.C.: Environmental Protection Agency, 2002). Available at http://epa.gov/OGWDW/disinfection/tcr/pdfs/whitepaper_tcr_watermains.pdf.

5. Ibid.

6. M. Lavelle, "Water Woes," *U.S. News & World Report*, June 4, 2007, 44.

7. E. Gies, Water Wars: Is Water a Human Right or a Commodity?" *Worldwatch*, March-April, 2009, 22–27; *Environmental Science and Technology*, September 1, 2001, 360A.

8. R. D. Morris, "Pipe Dreams," *New York Times*, October 3, 2007, A25.

9. S. Twedt, "A Sea of Drinking Water Lost Between Treatment, Tap," *Pittsburgh Post-Gazette*, July 15, 2002.

10. American Water Works Association, "New or Repaired Water Mains."

11. M. Zapotosky and A. C. Davis, "Water Main Breaks Hit Area," *Washington Post*, January 18, 2009, C1.

12. Twedt, "A Sea of Drinking Water Lost Between Treatment, Tap."

13. Lavelle, "Water Woes," 37.

14. J. Tibbetts, "Down, Dirty, and Out of Date," *Environmental Health Perspectives*, July 2005, A466.

15. Lavelle, "Water Woes," 44.

16. Ibid., 45.

17. P. D. Thacker, "Americans Are Swimming in Sewage," *Environmental Science and Technology*, December 15, 2004, 483A.

18. J. Tibbetts, "Down, Dirty, and Out of Date," A466.

19. Ibid.

20. C. Duhigg, "Sewers at Capacity, Waste Poisons Waterways," *New York Times*, November 23, 2009.

21. P. D. Thacker, "Americans Are Swimming in Sewage," 483A.

22. Whitman, "The Sickening Sewer Crisis," *U.S. News & World Report*, June 12, 2000, 17.

23. Ibid.; J. Down, M. Mendoza, and J. Pritchard, "Studies Find Factories Release Pharmaceuticals," Associated Press, April 20, 2009. Available at http://google .com/hostednews/ap/article/ALeqM5liPvot6FmM_Fi44QmYyLaqZOmHW...

24. H. Blatt, *America's Food* (Cambridge, Mass.: MIT Press, 2008), 35, 37–38.

25. M. Bluejay, "How Much Electricity Costs and How They Charge You," 2008. Available at http://michaelbluejay.com/electricity/cost.html.

26. R. N. Anderson, "The Distributed Storage-Generation 'Smart' Electric Grid of the Future," in *Workshop Proceedings, The 10–50 Solution: Technologies and Policies for a Low-Carbon Future*, Pew Center on Global Climate Change and the National Commission on Energy Policy, 2004.

27. P. Behr, "Electricity Overseer Says Grid Must Grow," *Washington Post*, October 17, 2007, D2.

28. U.S. Department of Energy, "Gridworks: Overview of the Electric Grid," Available at http://energetics.com/gridworks/grid.html.

29. Ibid.

30. T. Gellings and K. E. Yeager, "Transforming the Electric Infrastructure," *Physics Today*, December 2004, 45–51.

31. Department of Energy, "The Smart Grid: An Introduction," 2008; P. Fairley, "Building an Interstate Highway System for Energy," June 10, 2009. Available at http://discovermagazine.com/2009/jun/10-building-interstate-highway-system -for-energy/articl.

32. R. Gramlich and M. Goggin, "Green Power Superhighways," American Wind Energy Association/Solar Energy Industries Association, February 2009; E. Wood, "Green Superhighway: Overhauling the Grid to Accommodate Renewables," April 23, 2009. Available at http://renewableenergyworld.com/rea/news/print/article/2009/04/green-superhighway-ov.

33. H. Knight, "Edison's Revenge," *New Scientist*, October 11, 2008, 19.

34. B. V. V. Washington, "Why America Has So Many Potholes," *Time*, June 24, 2001.

35. D. Stout and M. L. Wald, "Highway Fund Shortfall May Halt Road Projects," *New York Times*, September 6, 2008, A10.

36. J. W. Schoen, "U.S. Highways Badly in Need of Repair," MSNBC.com, 2007. Available at http://msnbc.msn.com/id/20095291/print/1/displaymode/1098/; C. L. Jenkins, "One-Third of Roads Rated in Bad Shape," *Washington Post*, May 9, 2009, B1.

37. Ibid.

38. Association of American Railroads, "Truck Size and Weight Limits," 2008. Available at http://aar.org/GovernmentAffairs/~/media/AAR/PositionPapers/281.ashx.

39. Federal Highway Administration, "Bridge Inventory-Total and Deficient, 1996 to 2007, and by State," 2007.

40. D. P. Billington, "One Bridge Doesn't Fit All," *New York Times*, August 18, 2007, A13.

41. "Report: Repairing U.S. Bridges Would Cost $140 Billion," 2008. Available at http://cnn.com/2008/US/07/28/bridge.report/index.html.

42. F. Aherns, "A Switch on the Tracks: Railroads Roar Ahead," *Washington Post*, April 21, 2008, A1.

43. General Accounting Office, "Railroad Bridges and Tunnels: Federal Role in Providing Oversight," 2007. Available at http://gao.gov/htext/d07770.html.

44. D. B. Buemi, "Amtrak, Solar Cost Comparisons and the 2-4x Myth," 2007. Available at http://renewableenergyaccess.com/rea/news/printstory;jsessionid=CB2E1DE4F20DB.

45. D. Eggen, "High-Speed Rail Drives Obama's Transportation Agenda," *Washington Post*, March 8, 2009, A1.

46. "A Little Hope for Amtrak," *New York Times*, November 1, 2007, A26.

47. Harris Interactive, "Americans Would Like to See a Larger Share of Passengers and Freight Going by Rail in the Future," Poll Number 14, February 8, 2006.

48. J. P. M. Syvitski, C. J. Vorosmarty, A. J. Kettner, and P. Green, "Impact of Humans on the Flux of Terrestrial Sediment to the Global Coastal Ocean," *Science*, April 15, 2005, 376–380.

49. J. G. Workman, "Deadbeat Dams," Property and Environment Research Center, 2006. Available at http://perc.org/articles/article849.php?view=print.

50. American Society of Civil Engineers, "2009 Report on America's Infrastructure," 2009. Available at http://asce.org/reportcard/2005/index2005.cfm.

51. L. Roth, testimony of the American Society of Civil Engineers to the House Committee on Transportation and Infrastructure, May 8, 2007.

52. J. Leslie, "Before the Flood," *New York Times,* January 22, 2007, A19.

53. Ibid.

54. Library of Congress, "Dam Removal: Issues, Considerations, and Controversies," National Council for Science and the Environment, 2006. Available at http://policyarchive.org/handle/10207/2865.

55. J. C. Marks, "Down Go the Dams," *Scientific American*, March 2007, 66–71.

56. Leslie, "Before the Flood," A19.

57. American Society of Civil Engineers, "2009 Report on America's Infrastructure."

58. "U.S. Air Traffic Facilities Are Showing Their Age," *International Herald Tribune*, December 19, 2008, 20.

59. M. J. Sniffen, "High Risk of Runway Collision Plagues U.S.," *Aviation*, December 6, 2007. Available at http://aviation.com/safety/071206-ap-us-runway-collision-danger.html.

60. American Society of Civil Engineers, "2009 Report for America's Infrastructure, Aviation," 2005. Available at http://asce.org/reportcard/2005/page.cfm?id=21&printer=1.

61. A. Prud'homme, "There Will Be Floods," *New York Times*, February 27, 2008, A25.

Chapter 3

1. P. Peduzzi, "Is Climate Change Increasing the Frequency of Hazardous Events?" *Environment Times*, 2004. Available at http://nationmaster.com/encyclopedia/Extreme-weather.

2. H. Black, "Unnatural Disaster: Human Factors in the Mississippi Floods," *Environmental Health Perspectives*, 116 (2008); A390–A393.

3. M. Twain, "Life on the Mississippi," 1883.

4. G. P. Johnson, R. R. Holmes, Jr., and L. A. Waite, "The Great Flood of 1993 on the Upper Mississippi River 10 years Later," U.S. Geological Survey Fact Sheet 2004–3024, 2004.

5. T. Karl, "The Rising Cost of Global Warming: I. Global Warming and Extreme Weather," 1998. Available at http://pirg.org/reports/enviro/cost98/page3.htm; Climate Solutions, "Floods and Mudslides: An Increase in Intense Rainfalls Would Mean More Flooding and Mudslides," n.d. Available at http://climate solutions.org/pages/globalWmg6.html; T. R. Karl, J. M. Melillo, and T. C. Peterson, *Global Climate Change Impacts in the United States* (Cambridge: Cambridge University Press, 2009).

6. "Levee." n.d. Available at http://absoluteastronomy.com/topics/Levee.

7. P. Whoriskey and J. Warrick, "Floodwall Overtopping May Not Be to Blame," *Washington Post*, October 8, 2005, A11; J. K. Bourne, Jr., "A City's Faulty Armor," *National Geographic online*, May 2007 (also in August issue).

8. K. E. Trenberth, "Warmer Oceans, Stronger Hurricanes," *Scientific American*, July 2007, 45–51; National Oceanographic and Atmospheric Administration, "Atlantic Hurricane Climatology and Overview," n.d.. Available at http://ncdc.noaa.gov/oa/climate/research/hurricane-climatology.html.

9. C. Dean, "Will Warming Lead to a Rise in Hurricanes?" *New York Times*, May 29, 2007, F1, F4.

10. C. Dean, "Surprises in a New Tally of Areas Vulnerable to Hurricanes," *New York Times*, October 10, 2006, F4.

11. "Majority Unprepared for Hurricanes, Poll Says," *International Herald Tribune*, June 1, 2007, 5.

12. "Issues Associated with Catastrophic Coastal Storms," Woodrow Wilson National Fellowship Foundation, 1997. Available at http://woodrow.org/teachers/esi/1997/53/csestdy.htm.

13. K. M. Crossett, T. J. Culliton, P. C. Wiley, and T. R. Goodspeed, "Population Trends along the Coastal United States: 1980–2008," National Oceanographic and Atmospheric Administration, 2004.

14. J. A. Groen and A. E. Polivka, "Hurricane Katrina Refugees" Who They Are, and How They Are Faring," *Monthly Labor Review*, March 2008, 32–51; C. E. Colten, R. W. Kates, and S. B. Laska, "Three Years After Katrina," *Environment*, September-October 2008, 36–47; "Checking In with New Orleans" *Washington Post*, October 16, 2009.

15. Colten, Kates, and. Laska, "Three Years after Katrina," 36–47; S. S. Hsu, "A Warning about Disaster Housing," *Washington Post*, July 8, 2009.

16. S. S. Hsu, "A Warning About Disaster Housing", *Washington Post*, July 8, 2009.

17. Louisiana Environmental Action Network, "Hurricane Katrina & Rita Environmental Data," 2006. Available at http://leanweb.org.

18. Karl, Melillo, and Peterson, *Global Climate Change Impacts in the United States*.

19. P. Whoriskey, "LA Plan to Reclaim Land Would Divert the Mississippi," *Washington Post*, May 1, 2007, A3.

20. M. Tuhus, "Treading Water," *E Magazine*, January–February 2008, 20–21.

21. "Here Comes the Mud," *Bulletin of the American Meteorological Society*, March 2006, 164.

22. "Poll: Abandon Flooded Areas of New Orleans," September 9, 2005. Available at http://chron.com/disp/story.mpl/topstory2/3346460.html.

23. H. Lambourne, "New Orleans 'Risks Extinction,'" February 3, 2006. Available at http://news.bbc.co.uk/2/hi/science/nature/4673586.stm.

24. "Levees Not Enough to Protect New Orleans," *Earth Magazine,* August 2009, 7.

25. J. K. Bourne, Jr., "New Orleans: A Perilous Future," *National Geographic,* August 2007, 32–67.

26. Ibid.

27. P. Whoriskey, "Florida Appears to be Losing Its Sunny Magnetism," *Washington Post,* January 9, 2008, A3.

28. M. Gall, K. A. Borden, and S. L. Cutter, "When Do Losses Count?" *Bulletin of the American Meteorological Society,* June 2009, 799–809.

29. Environmental Defense, "Global Warming Disrupts Insurance Coverage," n.d. Available at http://environmentaldefense.org/page.cfm?tagID=14044; K. Breslau, "The Insurance Climate Change," *Newsweek,* January 29, 2007, p 36–38; A. Goodnough, "Florida Acts to Lower Home Insurance Cost," *New York Times,* January 23, 2007, A12.

30. J. Garreau, "A Dream Blown Away," *Washington Post,* December 2, 2006, C1.

31. P. Whoriskey, "Florida's Big Hurricane Gamble," *Washington Post,* February 20, 2007, A2.

32. "Promises, Promises," *Bulletin of the American Meteorological Society,* November, 2008, 1627.

33. S. Gupta, "Bigger Hurricanes Bring More Tornadoes," *Earth,* December 2009, 18.

Chapter 4

1. Environmental Protection Agency, "Municipal Solid Waste Generation, Recycling, and Disposal in the United States: Facts and Figures for 2007," 2008.

2. A. Streeter, "From Kitchen Scraps to Biogas—Stockholm Embraces Garbage Disposals," 2008. Available at http://treehugger.com/files/2008/09/kitchen-scraps -to-biogas.php.

3. PC disposal.com, "Computer Recycling," n.d. Available at http://pcdisposal .com/computer-recycling-article.htm.

4. Tiny Tots Diaper Service, "Disposable Diapers Consume the Landfills," 2007. Available at http://tinytots.com/ds/barges.html.

5. Wikipedia, "Diaper," n.d. Available at http://wikipedia.org/wiki/Diapers.

6. P. O'Mara, "A Tale of Two Diapers," *Mothering,* n.d. Available at http:// mothering.com/guest_editors/quiet_place/138.html.

7. Ibid.; J. Berry, "The Big Issue: Plastic Bags," November 9, 2009. Available at Earth911.com.

8. "No More Plastic Bags," *New York Times,* September 30, 2008, A26; American Chemistry Council, "U.S. Recycling of Plastic Bags and Film Reaches Record High in 2007," February 5, 2009; T. F. Lindeman, "Reusable Shop-

ping Bags Gain in Popularity as a Way to Help Environment," *Pittsburgh Post-Gazette*, May 27, 2007; E. Rosenthal, "Irish Help to Banish a Plastic Nuisance," *International Herald Tribune*, February 1, 2008, 1, 8.

9. Environmental Protection Agency, "Municipal Solid Waste Generation."

10. Zero Waste America, "Landfills: Hazardous to the Environment," n.d. Available at http://zerowasteamerica.org/Landfills.htm.

11. Environmental Research Foundation, "The Basics of Landfills: How They Are Constructed and Why They Fail," n.d. Available at http://zerowasteamerica .org/BasicsOfLandfills.htm.

12. S. Hannaford, "Solid Waste Oligopoly," February 5, 2008. Available at http://oligopolywatch.com/2008/02/05.html.

13. Prometheus Energy, "Landfill Gas," 2007. Available at http://prometheus -energy.com/whatwedo/landfillgas.php.

14. R. Goldstein, "2006 Update: The State of Landfill Gas Utilization Projects," Environmental Protection Agency, 2007.

15. Landfill Methane Outreach Program, "Landfill Gas Energy Projects and Candidate Landfills," 2008; B. Guzzone and C. Leatherwood, "Landfill Gas Use Trends in the United States," *Biocycle*, September 2007, 57–58; Environmental Protection Agency, "An Overview of Landfill Gas Energy Opportunities in the U.S." June 2008.

16. E. Royte, *Garbage Land: On the Secret Trail of Trash* (Back Bay, MA: Back Bay Books, 2006).

17. E. Douglas, "Dig Up the Dump," *New Scientist*, October 4, 2008, 34–35.

18. Ibid.

19. "Landfill Link to Birth Defects Strengthened," *Environmental Science and Technology*, November 1, 2001.

20. "Superfund Scrutiny," *Environmental Science and Technology*," July 1, 2007, 4488; Environmental Protection Agency, "Hazardous Waste Sites on the National Priority List by State and Outlying Area: 2007."

21. K. N. Probst and D. Sherman, "Success for Superfund: A New Approach for Keeping Score," Resources for the Future, April 2004.

22. Office of Housing and Urban Development, "Brownfields Frequently Asked Questions," 2009. Available at http://hud.gov/offices/cpd/economicdevelopment/ programs/bedi/bfieldsfaq.cfm; American Society of Civil Engineers, "Report Card for America's Infrastructure: Hazardous Waste," 2005.

23. Ibid.

24. T. McNicol, "The Ultimate Garbage Disposal," *Discover*, May 2007, 20; M. Behar, "The Prophet of Garbage," *Popular Science*, March 2005, 56–90; N. Blackburn, "Israeli Startup Develops Ambitious Plant to Rid the World of Waste," *Israel21c*, December 2, 2007. Available at http://israel21c.org/.../Israel -startup-develops-ambitious-plant-to-rid-the world of waste^

25. S. Deneen, "How to Recycle Practically Anything," *E Magazine*, May-June 2006, 26–31, 63.

26. Ibid.

27. Ocean Conservancy, Trash Travels, 2010.

28. M. Verespej, "Study Ranks Plastics' Share of Marine Debris Problem," March 12, 2009. Available at http://plasticsnews.com/headlines2.html?cat=18id =1236889278; L. Hoshaw, "Afloat in the Ocean, Expanding Islands of Trash," *New York Times*, November 10, 2009. Available at http://nytimes.com/2009/ 11/10/science/10patch.html?th=&emc=th&pagewanted=print

Chapter 5

1. H. Blatt, *America's Food: What You Don't Know About What You Eat* (Cambridge, MA: MIT Press, 2008).

2. D. R. Montgomery, "Soil Erosion and Agricultural Sustainability," *Proceedings National Academy of Sciences*, August 14, 2007, 13268–13272.

3. Natural Resources Conservation Service, *National Resources Inventory, 2003 Annual NRI* (Washington, D.C.: Department of Agriculture, 2003).

4. J. Woodward and I. Foster, "Erosion and Suspended Sediment Transfer in River Catchments," *Geography* 82:4 (1997), 353–376.

5. Seafriends, "Soil Erosion and Conservation," 2000. Available at http:// seafriends.org.nz/enviro/soil/erosion.htm.

6. J. C. Knox, "Agricultural Influence on Landscape Sensitivity in the Upper Mississippi Valley," *Catena* 42 (2001), 193–224.

7. H. Blatt, *Our Geologic Environment* (Upper Saddle River, N.J.: Prentice Hall, 1997), 67.

8. Dalai Lama XIV, *Little Book of Inner Peace* (London: Element Books, 2002), 144.

9. H. Blatt, *America's Food*, 2005, 46–48.

10. U.S. Department of Agriculture, "Soil Erosion in Agricultural Systems," n.d. Available at http://msu.edu/user/dunnjef1/rd491/soile.htm.

11. R. Pryor, "No-Till Farming in Dryland Cropping Systems," University of Nebraska, Lincoln Institute of Agriculture and Natural Resources, January 7, 2009; N. Bakalar, "Unsustainable Soil Use Can Cause Civilizations to Collapse," *Discover*, January 2008, 54.

12. "Soil Erosion in Agricultural Systems."

13. K. Kloosterman, "An 'Essential' Green Oil to Fight Unwanted Pests," *Israel 21c*, December 8, 2008.

14. environmental working group, foods you'll want to buy organic, n. D. foodnews.org/highpesticidefoods.php.

15. "Organic Agriculture: Implementing Ecology-Based Practices," *Organic Trade Association Newsletter*, October–November 2001, 3.

16. "Greener Greens," *Consumer Reports*, January 1998, 14–15.

17. M. Day, "Dipping into Danger," *New Scientist*, February 21, 1998, 5.

18. C. Cox and M. Surgan, Unidentified Inert Ingredients in Pesticides: Implications for Human and Environmental Health, *Environmental Health Perspectives*, December 2006,1803–1806.

19. Pesticide Data Program, *Annual Summary Calendar Year 2005* (Washington, D.C.: Department of Agriculture, 2006). Available at http://ams.usda.gov/science/pdp.

20. Organic Consumers Association, "U.S. Organic Food Sales Up 22%, Hit $17 Billion in 2006," Organic Consumers Association, 2007; C. Dimitri and L. Oberholtzer, "Marketing U.S. Organic Foods: Recent Trends from Farms to Consumers," USDA Economic Information Bulletin 58 (Washington, D.C.: U.S. Department of Agriculture, 2009); T. A. Smith and B-H. Lin, "Consumers Willing to Pay a Premium for Organic Produce," *Amber Waves,* March 2009; J. Schmit, "Organic Food Sales Feel the Bite from Sluggish Economy," *USA Today,* August 19, 2008.

21. D. Pimental, P. Hepperly, J. Hanson, D. Douds, and R. Seidel, "Environmental, Energetic, and Economic Comparisons of Organic and Conventional Farming Systems," *Bioscience* 55 (2005), 573–582; B. Halweil, "Can Organic Farming Feed Us All?" *Worldwatch*, May–June 2006, 18–24; E. Hamer and M. Anslow, "Ten Reasons Why Organic Can Feed the World," *Ecologist*, March 1, 2008. Available at http://theecologist.org//10_reasons_why_organic_can_feed_the_world.html

22. K. Roseboro, Organic Coexistence Reseach Paper Skews Facts to Support Dubious Conclusion, 2004. Available at cropchoice.com.

23. Greenpeace, GM Contamination Register ReportT, 2007.

24. W. Neuman and A. Pollack, "rise of the superweeds." New york times, may 3, 2010.

25. M. Moss, "Food Companies Try, But Can't Guarantee Safety," *New York Times*, May 15, 2009.

26. L. Saad, "Seven in Ten Americans Reacted to a Food Scare in the Past Year," August 1, 2007. Available at http://galluppoll.com/content/?ci=28264.

27. G. Harris, "Ill from Food? Investigations Vary by State," *New York Times*, April 20, 2009, A1, A17.

28. "Globalization and Food Safety," *Environmental Health Perspectives*, August 2008, A33; J. M. Fonseca and S. Ravishankar, "Safer Salads," *American Scientist*, November–December 2007, 494–501.

29. F. Gale and J. C. Busby, "Imports from China and Food Safety Issues," Department of Agriculture Economic Information Bulletin 52 (Washington, D.C.: U.S. Department of Agriculture, 2009).

30. A. Barrionuevo, "Food Imports Often Escape Scrutiny," *New York Times*, May 1, 2007, C1, C6.

31. Worldwatch Institute, "Genetically Modified Crops Only 9 Percent of Primary Global Crop Production" (Washington, D.C.: Worldwatch Institute, 2008).

32. Institute for Responsible Technology, "Genetically Modified Foods Pose Huge Health Risk," May 20, 2009; A. Turpen, "Three Approved GMO Crops Linked to Organ Damage, New Study Shows," January 13, 2010. Available at http://naturalnews.com/z027931_GMO_crops_organ_damage.html.

33. Worldwatch Institute, "Genetically Modified Crops."

34. D. Gutierrez, "Doctors Warn About Dangers of Genetically Modified Food," February 25, 2010. Available at http://natural news.com/z028245_gm_food _side_effects.html.

35. U.S. Department of Agriculture, Economic Research Service, "Agricultural Productivity in the United States" (Washington, D.C.: Department of Agriculture, 2008).

36. B. Halweill, "Still No Free Lunch: Nutrient Levels in U.S. Food Supply Eroded by Pursuit of High Yields" (Boulder, CO: Organic Center, 2007).

37. T. Parker-Pope, "Vitamin Pills: A False Hope?" *New York Times*, February 17, 2009.

38. Ibid.

39. J. A. Lockwood, *Six-Legged Soldiers: Using Insects as Weapons of War* (New York: Oxford University Press, 2008).

40. P. Muir, "Why Not?" Oregon State University, n.d.. Available at http://oregonstate.edu/~muirp/whynot.htm.

41. D. A. Pfeiffer, "Eating Fossil Fuels," 2004. Available at http://fromthewilder ness.com/free/.../100303_eating_oil.html

42. E. Engelhaupt, "Do Food Miles Matter?" *Environmental Science and Technology*, May 15, 2008, 3482; C. L. Weber and H. S. Matthews, "Food-Miles and the Relative Climate Impacts of Food Choices in the United States," *Environmental Science and Technology*, May 15, 2008, pp. 3508–3513; B. Trivedi, "Dinner's Dirty Secret," *New Scientist*, September 13, 2008, 28–32; S. De Weerdt, "Is Local Food Better?" *Worldwatch*, May–June 2009, 6–10.

43. J. Siikamaki, "Climate Change and U.S. Agriculture: Examining the Connections," *Environment*, July–August 2008, 37–49.

44. Blatt, *America's Food*, 134–135.

45. A. Moritz, "Eating Meat Kills More People Than Previously Thought," *Natural News*, March 30, 2009. Available at http://natural news.com/025957 _meat_cancer_disease.html.

46. American Dietetic Association, "Vegetarian Diets," 2003. Available at http://eatright.org/cps/rde/xchg/ada/hs.xsl/advocasy_933_enu_html.htm.

47. Vegetarian Quotes, n. d. Available at http://indianchild.com/vegeterianism _quotes.htm.

48. R. A. Myers and B. Worm, "Rapid Worldwide Depletion of Predatory Fish Communities," *Nature*, May 15, 2003, 280–283.

49. "World Fish Restoration a Global Challenge, UN Report Warns," 2005. Available at http://.organicconsumers.org/politics/fishrestore30705.cfm

50. J. Adler, B. Campbell, V. Karpouzi, K. Kaschner, and D. Pauly, "Forage Fish: From Ecosystems to Markets," *Annual Review of Environment and Resources* 33 (2008): 153–166.

51. P. Chek, "You Are What You Eat—Processed Foods," 2002. Available at http://chekinstitute.com/printfriendly.cfm?select=42.

52. Ibid.; C. Willyard, "Estrogen linked to Fish Immunity Problems" *Earth*, September 2009, 25.

53. Food Commission, *The Food Commission Guide to Food Additives* (London: Food Commission, 2004).

54. "Livestock Floor the Environment with Estrogen," *Environmental Science and Technology*, July 1, 2004, 241A.

55. S. Jobling et al., "Wild Intersex Roach (*Rutilus rutilus*) Have Reduced Fertility." *Biology and Reproduction*, August 2002, 515–524; "Hormones in the Water Devastate Wild Fish," *New Scientist*, May 26, 2007, 16.

56. "Intersex Fish Found Near California Coast"; "Want a Sex Change? Buy Some Suncream!" *Ecologist*, April 6, 2006, 12.

57. Cornell University, "Milk And Human Health," n. d.

58. M. Pollan, "Unhappy Meals," *New York Times Magazine*, January 28, 2007, 38–45; D. Barber, "Change We Can Stomach," *New York Times*, May 11, 2008, 13.

59. M. Adams, "Fraudulent "'Smart Choices' Food Labeling Program Crumbles as Food Manufacturers Flee Scrutiny." Available at http://natural news.com/z027380_food_labeling_nutrition.html.

60. M. Adams, "Fraudulent 'Smart Choices' Food Labeling Program Crumbles As Food Manufacturers Flee Scrutiny," *Natural News*, 2009. Available at http://naturalnews.com/z027380_food_labeling_nutrition.html.

61. L. Layton, "FDA Seeks Better Nutrition Labeling," *Washington Post*, October 21, 2009.

62. M. Pollan, *In Defense of Food: An Eater's Manifesto* (New York: Penguin, 2008).

63. G. Harris, "Ill from Food? Investigations Vary by State," *New York Times*, April 20, 2009, A1, A17.

Chapter 6

1. *Boston Globe*, June 24, 2007.

2. British Petroleum, *Statistical Review of World Energy*, 2008.

3. M. R. Copulos, testimony to the Foreign Relations Committee, U.S. Senate, March 30, 2006; M. R. Copulos, "The Hidden Cost of Oil," 2007. http://setamericafree.org/saf_hidden costofoil010507.pdf

4. Friends of the Earth, "Big Oil, Bigger Giveaways," July 2008.

5. J. Sawin, "Charting a New Energy Future," in *State of the World 2003*, ed. L. Starke (New York, Norton, 2003), 85–109.

6. "Steady as She Goes," *Economist*, April 22, 2006, 5–6.

7. American Petroleum Institute, "The Truth about Oil and Gasoline: An API Primer," 2008.

8. British Petroleum, "Statistical Review of World Energy," 2008.

9. Copulos, testimony to the Foreign Relations Committee.

10. A. E. Cha, "China's Automobiles Are One More Demand on World Oil Supply," *Washington Post*, August 28, 2008, A1.

11. Copulos, testimony to the Foreign Relations Committee.

12. D. Woynillowicz, "Tar Sands Fever," *Worldwatch*, September–October 2007, 8–13.

13. C. Hatch and M. Price, "Canada's Toxic Tar Sands: The Most Destructive Project on Earth," Environmental Defense (Canada), 2008.

14. R. M. Pollastro, L. N. R. Roberts, T. A. Cook, and M. D. Lewan, "Assessment of Undiscovered Technically Recoverable Oil and Gas Resources of The Bakken Formation, Williston Basin, Montana and North Dakota, 2008" U. S. Geological Survey Open-File Report 2008-1353, 2009.

15. J. Birger, "Oil Shale May Finally Have Its Moment." 2007. Available at http:// royaldutchshellpia.com/2007/10/31/fortune-magazine-oil-shale-may-finally -have-its-moment/ "Oil-Shale Extraction Technology Has a New Owner," Society of Petroleum Engineers, February 1, 2008. Available at http://spe.org/.../oil -shale-extraction-technology-has-a-new-owner.

16. Energy Information Administration, "Natural Gas Navigator," 2008.

17. Federal Energy Regulatory Commission, "North American LNG Terminals," 2009.

18. Energy Information Administration, "Unconventional Natural Gas Production Rapidly Growing Share of U.S. Total Production," 2004. Available at http:// theoildrum.com/files/Conventional_Unconventional.png.

19. J. Hurdle, "U.S. Gas Drilling Boom Stirs Water Worries," February 24, 2009. Available at http://reuters.com/articlePrint?articleId=USTRE51O0FW 20090225.

20. J. Hurdle, "Gas Drillers Battle Pennsylvania Pollution Concerns," Reuters, May 3, 2009. Available at http://reuters.com/articlePrint?articleId=USTRE542 2TG20090504; J. Mouawad and C. Krauss, "Dark Side of a Natural Gas Boom," *New York Times*, December 8, 2009; J. Hurdle, "Pennsylvania Says Natgas Drilling Risks Inevitable," Reuters, March 20, 2009. Available at http:// reuters.com/articleidUSTRE52J6AP20090320; C. Weaver, "Environmental Concerns Rise in Northeastern Pennsylvania as Natural Gas Drilling Spreads," January 2, 2010. Available at www1.voanews.com/environmental-concerns-rise –in-northeastern-pennsylvania-as natural-gas-drilling-spreads-80502507.html; D. Hopey, "Marcellus Shale Gas Drilling Put Under Microscope," Pittsburgh Post-Gazette, June 13, 2010. Available at http://post-gazette.com/pg/10164/ 1065304-455.stm.

21. Hurdle, Gas Drillers Battle Pennsylvania Pollution Concerns, Available at Http://Reuters.Com/Articleprint?Articleid=Ustre5422tg20090504.

22. C. Ruppel, "Trapping Methane Hydrates for Unconventional Natural Gas," *Elements*, June 2007, 193–199.

23. Ibid.

24. A. Goho, " Energy on Ice," *Science News*, June 25, 2005, 410–412; J. Pelley, "Gas Hydrates on the Front Burner," *Environmental Science and Technology*, October 15, 2008, 7550–7551; J. Eilperin, "Points to Major Source of Natural Gas in Alaska," *Washington Post*, November 12, 2008, A6.

25. H. Blatt, *America's Environmental Report Card* (Cambridge, MA: MIT Press, 2005), 97.

26. Energy Information Administration, "Coal Production in the United States—An Historical Overview," 2006 Available at http://tonto.eia.doe.gov/ftproot/coal/coal_production_review.pdf.

27. "US to Shift away from Petroleum, towards Coal by 2030." *Energy World*, February 2007, 5.

28. "Goldman Environmental Prize Winners," *Worldwatch*, July–August 2009, 26.

29. B. Block, U. S. Activist Battles West Virginia Coal Industry, Available at http://worldchanging.com/archives/009824.html.

30. Environment News Service, "Climate Scientist James Hansen Arrested In Mountaintop Removal Protest," 2009. Available at http://Ens-Newswire.Com/Ens/Jun2009/2009-06-23-01.Asp.

31. Energy Information Administration, "Annual Energy Review: Coal Production, Selected Years, 1949–2007," 2007.

32. NationMaster, "Energy Statistics by Country," 2008. Available at http://nationmaster.com/graph/ene_ele_con-energy-electricity-consumption.

33. S. Mufson and B. Harden, "Coal Can't Fill World's Burning Appetite," *Washington Post*, March 20, 2008, A1; C. Krauss, An Export in Solid Supply," *New York Times*, March 19, 2008, C1, C7.

34. E. Shuster, "Tracking New Coal-Fired Power Plants," National Energy Technology Laboratory, January 5, 2009.

35. EcoBridge, "Causes of Global Warming," n.d.

36. Energy Information Administration, "Fuel Costs for Electricity Generation," 2009.

37. M. Clayton, "U.S. Coal Power Boom Suddenly Wanes," *Christian Science Monitor*, March 4, 2008.

38. Sierra Club, "Dirty Coal Power," 2007.

39. C. Abrams, "America's Biggest Polluters: Carbon Dioxide Emissions from Power Plants in 2007" (Boston: Environment America Research and Policy Center, November 2009).

40. C. Duhigg, "Cleansing the Air at the Expense of Waterways," *New York Times*, October 13, 2009.

41. T. Yulsman, "Coal Waste Dumps: Ticking Toxic Time Bombs," Center for Environmental Journalism, University of Colorado, December 26, 2008; Earth• Justice, "Coal Ash Pollution Contaminates Groundwater, Increases Cancer Risks," September 4, 2007; "Tons of Coal Ash Piling Up across U.S., Analysis Says," *Washington Post*, January 10, 2009, A2.

42. "Tennessee's Coal Ash Sludge Brought to Georgia," *Atlanta Journal-Constitution*, June 5, 2009; J. Eilperin, "Disposal of Coal Ash Rises as Environmental Issue," *Washington Post*, January 16, 2009, A4.

43. T. Devitt, "Mercury Miasma," University of Wisconsin, 2004. Available at http://whyfiles.org/201mercury/images/combustion_mercury.gif.

44. "Coal's Heavy Cost," *International Herald Tribune*, August 24, 2007; Environmental News Service, "Mercury Emissions Up at Coal-Burning Power Plants," November 2008. Available at http://ens-newswire.com/ens/nov2008/2008-11 -21-092.html; D. C. Evers and C. T. Driscoll Jr., "The Danger Downwind," *New York Times*, April 26, 2007, A25.

45. Environmental Defense Fund, "Coal-Fired Power Plants Are Big Contributors to Sooty Particle Pollution in Eastern States," April 24, 2007.

46. A. Gabbard, "Coal Combustion: Nuclear Resource or Danger?" *ORNL Oak Ridge National Laboratory Review* 26 (1993), 14-22.

47. J. P. McBride, R. E. Moore, J. P. Witherspoon, and R. E. Blanco, "Radiological Impact of Airborne Effluents of Coal and Nuclear Plants," *Science*, December 8, 1978.

48. W. Sweet, "Better Planet," *Discover*, August 2007; J. Russell, "Coal Use Rises Dramatically Despite Impacts on Climate and Health," *Worldwatch*, March–April 2008, 24.

49. D. P. Schrag, "Preparing to Capture Carbon," *Science*, February 9, 2007, 812–813; P. Viebahn, M. Fischedick, and D. Vallentin, "Carbon Capture and Storage," in *State of the World 2009* (New York: Norton, 2009), 99–102; S. Mufson, "Coal's Future Wagered on Carbon Capture," *Washington Post*, August 11, 2009, A8.

50. R. Edwards, "Clean-Coal Debut in Germany," ABC News, September 22, 2008. Available at http://abcnews.go.com/print?id=5844357.

51. Worldwatch Institute, "Coal Use Rises Dramatically Despite Impacts on Climate and Health," March–April 2008, 24.

52. L. R. Brown, "U.S. Moving toward Ban on New Coal-Fired Power Plants," Earth Policy Institute, February 14, 2008.

53. C. J. Hanley, "Scientists See Coal as Key Challenge" October 22, 2007, Available at http://blnz.com/news/2007/10/22/Scientists_coal_challenge_0580 .html; P. Krugman, "Empire of Carbon," *New York Times*, May 15, 2009.

54. M. Casey, "World's Coal Dependency Hits Environment," Associated Press, October, 5, 2007.

55. K. Bradsher and D. Barboza, "Pollution from Chinese Coal Casts a Global Shadow," *New York Times*, June 11, 2006, 1, 12–13. A15.

56. BBC News, 2009. Available at http://Newsvote.BBC.Co.Uk/Mpapps/ Pagetools/Print/News.Bbc.Co.Uk/2/Hi/Americas/7784969.Stm?A...

57. C. Flavin, "Low-Carbon Energy: A Roadmap," 33, in *Worldwatch Institute Report* 178, ed. L. Mastny (Washington, D. C.: Worldwatch Institute, 2008).

58. B. Block, "Study Finds Rich U.S. Energy-Efficiency Potential," Worldwatch Institute, July 31, 2009. Available at worldwatch.org/node/6212?emc=el&m=27 9787&l=5&v=d0647aa050.

59. "How to Cool the World," *Economist*, May 12, 2007, 59–60.

60. M. Parfit, "Freedom!" *National Geographic*, August 2005.

61. Worldwatch Institute, "CFL Sales Skyrocket," 2008.

62. E. Rosenthal and F. Barringer, "Green Promise Seen in Switch to LED," *New York Times*, May 30, 2009, A1, A3.

63. K. Galbreath, "Study Sees Energy Efficiency of Homes and Offices as Arena for Savings," *International Herald Tribune*, July 31, 2009, 15; F. Barringer, "White Roofs Catch on as Energy Cost Cutters," *New York Times*, July 30, 2009, A1, A17.

64. Friends of the Earth, "Public Transportation, Gas Prices, and Climate," 2007. Available at http://action.foe.org/content.jsp?...2007...transportation ...-United States

65. C. Krauss, "Gas Prices Send Surge of Riders to Mass Transit," *New York Times*, May 10, 2008, A1, A15.

66. Environmental Defense Fund, "Transportation by the Numbers," July 30, 2008. edf.org/article.cfm?contentID=8161; "Transit Trance," *Worldwatch*, May–June 2009, 5.

67. L. H. Sun, "New Ridership Record Shows U.S. Still Lured to Mass Transit," *Washington Post*, December 8, 2008, A8; "Invest in Mass Transit," *Washington Post*, December 9, 2008, A18.

Chapter 7

1. Energy Information Administration, "Annual Energy Outlook 2009 Early Release," December 17, 2008; M. W. Wald, "The Power of Renewables," *Scientific American*, March 2009, 56–61.

2. I. Bowles, "Home-Grown Power," *New York Times*, March 7, 2009, A21.

3. E. Kintisch, "Alternative Energy," *Discover*, January 2007, 21–23; T. Doggett, "U.S. Renewable Energy Faces Weak Economy, Old Grid," Reuters, February 23, 2009. Available at http://reuters.com/articlePrint?articleId=USTRE51M5R7 20090223; Energy Information Administration, "Annual Energy Outlook 2009," December 17, 2008.

4. J. Sawin, "Renewables 2007 Global Status Report," 2008. Available at worldwatch.org/node/5630; K. Bradsher, "Green Power Takes Root in the Chinese Desert," *New York Times*, July 3, 2009.

5. Z. Li, "China's Renewable Energy Law Takes Effect," Renewable Energy Access, January 19, 2006. Available at http://renewableenergyaccess.com/rea/news/story?id=41932; E. Martinot and L. Junfeng, "Powering China's Development: The Role of Renewable Energy," 2008. Available at http://renewablee nergyworld.com/rea/news/print?id=51586; M. Standaert, "China Poised to Go All Out with Clean Tech," *San Francisco Chronicle*, May 10, 2009. C. Zeppe-zauer and C. Camabuci, "A New Revolution: China Hikes Wind and Solar Power Targets," October 9, 2009. Available at http://renewableenergyworld. com/rea/news/print/article/2009/10/a-new-revolution-chin...; S. Mufson, "China Steps Up, Slowly But Surely," *Washington Post*, October 24, 2009.

6. M. Renner, "Working for People and the Environment," Worldwatch Institute, 2008; J. Sawin, "American Energy: The Renewable Path to Energy Security," Worldwatch Institute/Center for American Progress, 2006, 10; B. Thurner, "U.S. Renewable Energy Firms Cash In," UPI, 2006. Available at http://upi .com/InternationalIntelligence/view.php?StoryID=20051229-020019-5852r; D. R. Holst, F. Kahrl, M. Khanna, and J. Baka, "Clean Energy and Climate Policy for U.S. Growth and Job Creation," October 25, 2009. Available at http://are .berkeley.edu/~dwrh/CERES_Web/Docs/ES_DRHFK091025.pdf.

7. R. D. Perlack, L. L. Wright, A. F. Turhollow, R. L. Graham, B. J. Stokes, and D. C. Erbach, "Biomass as Feedstock for a Bioenergy and Bioproducts Industry: The Technical Feasibility of a Billion-Ton Annual Supply," U.S. Department of Energy/U.S. Department of Agriculture, 2005.

8. H. Blatt, *America's Food: What You Don't Know About What You Eat* (Cambridge, MA. MIT Press, 2008), 138.

9. B. A. Gloy, " Biogas: What Options for Slurry Power in the US?" October 21, 2008. Available at http://renewableenergyworld.com/rea/news/print?id =53901; H. Fountain, "Down on the Farm, an Endless Cycle of Waste," *New York Times*, December 29, 2009.

10. K. Silverstein, "The Appeal of Animal Waste," n.d.. Available at http:// renewableenergyaccess.com/rea/news/printstory;jsessionid=4B5CEC38DD...

11. N. H. Niman, "A Load of Manure," *New York Times*, March 4, 2006, A13.

12. A. Simms, "The Rough Guide to Economic Nonsense," *Economist*, July–August 2004, 8; "The Solar Subsidy Crutch or an Uneven Playing Field?", 2006 Available at http://renewableenergyaccess.com/rea/news/story?id-44723.

13. J. Earley and A. McKeown, "Smart Choices for Biofuels," Sierra Club/Worldwatch Institute, 2009; P. C. Wescott, "U.S. Ethanol Expansion Driving Changes Throughout the Agricultural Sector," *Amber Waves*, September 2007.

14. M. L. Wald, "Corn Power Put to the Test," *New York Times*, February 7, 2006, F3.

15. T. Alexander and L. Gordon, "What's Stopping Us? The Hurdles to Commercializing Cellulosic Ethanol," March 5, 2009; J. Decker, "Going against the

Grain: Ethanol from Lignocellulosics," *Renewable Energy World Magazine*, November–December 2008. Available at http://renewableenergyworld.com/rea/news/article/2009/01/going-against-the-grain-ethano....

16. J. Surowiecki, "Deal Sweeteners," *New Yorker*, November 27, 2006, 92.

17. R. Zalesky, "Integrating Biofuels into the Fuel Supply," *Geotimes*, March 2007, 31–33.

18. M. Johnston and T. Holloway, "A Global Comparison of National Biodiesel Production Potentials," *Environmental Science and Technology*, October 24, 2007, 7967–7973.

19. E. Gies, "Biodiesel Basics," *E Magazine*, March–April 2006, 54.

20. C. W. Schmidt, "Biodiesel," *Environmental Health Perspectives*, February 2007, A86-A91; D. C. Holzman, "The Carbon Footprint of Biofuels," *Environmental Health Perspectives*, June 2008, A246-A252.

21. E. Rosenthal, "Two Studies Find Benefits Outweighed by Environmental Costs," *International Herald Tribune*, February 8, 2008, 1, 5.

22. "Duke Study Critiques Corn Ethanol," March 6, 2009. Available at http://renewableenergyworld.com/rea/news/print/article/2009/03/duke-study-critiques-co...

23. A. Cutler, "Bioelectricity Promises More 'Miles per Acre' Than Ethanol," Available at http://rea.com/rea/news/article/2009/05/bioelectricity

24. D. Graham-Rowe, "Hydro's Dirty Secret Revealed," *New Scientist*, February 26, 2005, 8.

25. "Tapping the Power of the Sea," *Economist*, April 28, 2007, 67.

26. M. W. Johnston, "Harnessing the Tides: Marine Power Update 2009," February 20, 2009. Available at http://renewableenergyworld.com/rea/news/article/2009/02/harnessing-the-tides-marine-power-update-2009?cmpid=WNL...

27. D. Elliott, "Marine Renewables: Opening the Tidal Current Option," *Energy World*, February 2004, 10–13.

28. A. Kumar, "Uranium Lode in Va. Is Feared, Coveted," *Washington Post*, January 2, 2008, B1.

29. World Nuclear Association, "Supply of Uranium," June 2008.

30. Energy Information Administration, "Nuclear Electric Power," *Annual Energy Review 2007*; Nuclear Energy Institute, "Nuclear Facts," 2007.

31. M. W. Wald, "Nuclear Power May Be in Early Stages of a Revival," *New York Times*, October 24, 2008, B3.

32. M. Freemantle, "Nuclear Power for the Future," *Chemical and Engineering News*, September 13, 2004, 31–35.

33. C. T. Whitman and P. Moore, "Nuclear Power Will Drive the Future," *International Herald Tribune*, May 16, 2006, 10; Nuclear Energy Institute, "Nuclear Facts."

34. J. Tester and H. P. Meissner "The Future of Geothermal Energy," MIT, 2007.

35. "US Geothermal Capacity Could Top 10 GW," October 2, 2009. Available at http://renewableenergyworld.com/rea/news/print/article/2009/10/us-geothermal-capacity-...

36. Geothermal Energy Association, "About Geothermal Energy," 2007. Available at http://geo-energy.org/about GE/currentUse.asp.

37. "Power-Laden Winds Sweep North America," *Science News*, July 16, 2005, 36.

38. D. Appleyard, "Wind Installations Continue to Break Records across the Globe," February 4, 2009. Available at http://renewableenergyworld.com/rea/news/print/article/2009/02/wind-installations-co....

39. M. Bolinger and R. Wiser, "Surpassing Expectations: State of the US Wind Power Market,", 2008. Available at http://renewableenergyworld.com/rea/news/print?id=53498.

40. Department of Energy, "20% Wind Energy by 2030," July 2008.

41. E. Salerno and J. Isaacs, "The Economic Reach of Wind," April 17, 2009. Available at http://renewableenergyworld.com/rea/news/print/article/2009/04/the-economic-reach-of-...

42. D. Gutierrez, "EarthTronics Wind Turbine Generates Electricity Even in Low Wind Speeds," November 23, 2009. Available at http://naturalnews.com/z027550_wind_turbine_renewable_energy.html.

43. T. Bryant, "Mid-Atlantic Offshore Wind Potential: 330 GW," Renewable Energy Access, February 7, 2007. Available at http://renewableenergyaccess.com/rea/news/printstory;jsessionid=404F17EE790EC22....

44. G. Edge, "Delivering Offshore Wind Power in Europe," European Wind Energy Association, 2007.

45. C. D. Barry, "Lessons Learned by Offshore Oil Industry Boost Offshore Wind Energy," *Renewable Energy World*, June 13, 2008. Available at http://renewableenergyworld.com/rea/news/print?id=52760; E. de Vries, "Float On: Floating Offshore Wind Opens Up the Deep," Renewable Energy World, April 24, 2008. Available at http://renewableenergyworld.com/rea/news/print?id=52031.

46. "Offshore Wind Power Set to Expand," *Worldwatch*, January–February 2009, 4.

47. "PV Costs to Decrease 40% by 2010," May 23, 2007. Available at http://renewableenergyaccess.com/rea/news/printstory;jsessionid=7D3E3466AD6862

48. "Sandia and Stirling Set Solar-to-Grid Conversion Efficiency Record," *Renewable Energy Weekly*, February 20, 2008; "From 40.7 to 42.8% Solar Cell Efficiency," July 30, 2007. Available at http://renewableenergyaccess.com/rea/news/printstory;jsessionid=040847333A&C...

49. D. Schneider, "Solar Energy's Red Queen," *American Scientist*, January–February 2008, 24.

50. "Powering Up," *Economist*, September 10, 2006, 91–92.

51. D. Appleyard, "Utility-Scale Thin-Film: Three New Plants in Germany Total Almost 50 MW," March 11, 2009. Available at http://renewableenergyworld .com/rea/news/print/article/2009/03/utility-scale-thin-film-th...

52. R. Baxter, "Energy Storage: How Solar Can Always Meet Demand," October 15, 2008. Available at http://renewableenergyworld.com/rea/news/ print?id=53861.

53. "Trapping Sunlight," *Economist*.com, September 13, 2007. Available at http://economist.com/world/na/PrinterFriendly.cfm?story_id=9804148;   D.   R. Mills and R. G. Morgan, "A Solar-Powered Economy: How Solar Thermal Can Replace Coal, Gas and Oil," July 3, 2008. Available at http://renewableenergy world.com/rea/news/print?id=52693.

54. D. P. Buemi, "Amtrak, Solar Cost Comparisons & the 2-4x Myth," 2007. Available at http://renewableenergyworld.com/rea/news/article/2007/05/amtrak -solar-cost-comparisons...

55. C. Morris and N. Hopkins, "Why the Sunny, Windy United States, Is So Far Behind Calm, Cloudy Germany in Renewable Energy Generation," *World-watch*, May–June 2008, 20–25.

56. J. Burgermeister, "Germany: The World's First Major Renewable Energy Economy," *Renewable Energy World*, April 3, 2009. Available at http:// renewableenergyworld.com/rea/news/article/2009/04/germany-the-worlds -first-majo....

57. R. Fried, "Sun Shines," *E Magazine*, March–April 2007, 46.

58. Ibid.

59. Northeast Sustainable Energy Association, "Fuel Cells," n.d. Available at http://nesea.org/energy/info/fuelcells.html; "Fuel Cell Basics: Applications," n.d. Available at http://fuelcells.org/basics/apps.html; M. K. Heiman and B. D. Solomon, "Fueling U.S. Transportation: The Hydrogen Economy and its Alternatives," *Environment*, October 2007, 10–25.

60. M. Sardella, "The Hydrogen Hallucination," November 10, 2003. Available at http://solaraccess.com/ news/story?storyid=5497&p=1; S. Satyapal, J. Petrivic, and G. Thomas, "Gassing Up with Hydrogen," *Scientific American*, April 2007, 81–87.

61. "The Car That Makes Its Own Fuel," Isracast.com, February 3, 2006.

62. C. Simpson, "Solar Power Is Here and There, But Can It Be Everywhere?" *Renewable Energy World*, April 7, 2009. Available at http://renewableenergy world.com/rea/news/article/2009/04/solar-power-is-here-and-there-...

63. "Obama's Renewable Energy Speech," May 28, 2009. Available at http://solarfeeds.com/the-green-market-blog/7306-obamas-renewable-energy -revolution-s....

Chapter 8

1. European Nuclear Society, "Nuclear Power Plants, World-Wide," 2009; A. Faiola, "Nuclear Power Regains Support," *Washington Post*, November 24, 2009.

2. Keystone Center, "Nuclear Power Joint Fact-Finding," June 2007.

3. M. Lavelle, "The Nuclear Option," *U.S. News & World Report*, October 10, 2007, 32–33; N. Straub and P. Behr, "Energy Regulatory Chief Says New Coal, Nuclear Plants May Be Unnecessary," *New York Times*, April 22, 2009.

4. J. Riccio, "Nuclear Power Crawling Forward," Worldwatch Institute, June 18, 2008.

5. K. Charman, "Brave Nuclear World, Part 1," *Worldwatch*, May–June 2006, 30.

6. G. Bourget, "The True Cost of Nuclear Power," Mendocino Environmental Center, May 3, 2001. Available at mecgrassroots.org/NEWSL/ISS38/38.07 CostNuclear.html.

7. T. Friedman, "The Power of Green," *New York Times Magazine*, April 15, 2007, 41–51.

8. Charman, "Brave Nuclear World? Part 1."

9. T. Blakeslee, "Nuclear Power: The Safe and Easy Way," October 14, 2008. Available at http://renewableenergyworld.com/rea/news/print?id=53803.

10. A. E. Cha, "China Embraces Nuclear Future," *Washington Post*, May 29, 2007, D1.

11. Keystone Center, "Nuclear Power Joint Fact-Finding."

12. Public Citizen, "Nuclear Power and Global Warming," 2007.

13. H. Edwards, "Radioactive Blunders Double in a Decade," *New Scientist*, February 9, 2002, 6.

14. G. Wehrfritz and A. Webb, "Breach Of Faith" *Newsweek*, September 30, 2002, 41–42

15. K. Grossman, "The Push To Revive Nuclear Power," 2002. Available at http://greens.org/S-R/28/28-21.html.

16. Keystone Center, "Nuclear Power Joint Fact-Finding"; M. L. Wald, "Nuclear Plant, Closed after Corrosion, Will Reopen," *New York Times*, March 9, 2004, A16.

17. "Nuclear Meltdown Narrowly Averted," *Ecologist*, November 2006, 9.

18. G. Wehrfritz and A. Webb, "Breach of Faith," *Newsweek*, September 30, 2002, 41–43.

19. Keystone Center, "Nuclear Power Joint Fact-Finding."

20. H. Blatt, *America's Environmental Report Card: Are We Making the Grade?* (Cambridge, MA: MIT Press, 2005), 195; J. L. Mangano, J. Sherman, C. Chang, A. Dave, E. Feinberg, and M. Frim,, Elevated Childhood Cancer Incidence Proximate to U.S. Nuclear Power Plants," *Archives of Environmental Health*, February 2003, 74–82; "Nuclear Power Plants Shut," *Ecologist*, July–August 2002, 8; W. Hoffmann, C. Terschueren, and D. B. Richardson, "Childhood Leukemia in the Vicinity of the Geesthacht Nuclear Establishments near Hamburg, Germany," *Environmental Health Perspectives*, June 2007, 947; T. Riley, "The Nuke Next Door," *E Magazine*, May–June 2004, 14–16; *Ecologist*, September, 2006, 13, and June 2007, 9.

21. J. J. Mangano, J. M. Gould, E. J. Sternglass, J. D. Sherman, J. Brown, and W. Mcdonnell, "Infant Death and Childhood Cancer Reductions After Nuclear Plant Closings in The United States," *Archives Of Environmental Health*, January–February, 2002, 23–31.

22. Ibid.

23. "State of the Planet," *Ecologist*, September 2002, 9; D. B. Richardson and S. Wing, "Radiation and Mortality of workers at Oak Ridge National Laboratory: Positive Associations for Doses Received at Older Ages," *Environmental Health Perspectives*, August 1999, 649; R. Edwards, "Fathering Cancer," *New Scientist*, June 22, 2002, 22.

24. Nuclear Energy Institute, "Nuclear Facts," n.d. Available at http://nei.org/doc.asp?catnum=2&catid=106&docid=&format=print

25. The Keystone Center, "Nuclear Power Joint Fact-Finding," 2007.

26. L. Stiffler, "53 Million Gallons in Danger of Leaking," March 20, 2008. Available at http://seattlepi.com/local/355909_tanks21.html.

27. K. Kloosterman, "Israeli Discovery Converts Dangerous Radioactive Waste into Clean Energy," *Israel 21c*, March 18, 2007.

28. C. Maag, "Nuclear Site Nears End of Its Conversion to a Park," *New York Times*, September 20, 2006, A20.

29. "Nuclear Trains," *New Scientist*, February 8, 2006, 6.

30. J. I. Dawson and R. G. Darst, "Russia's Proposal for a Global Nuclear Waste Repository: Safe, Secure, and Environmentally Just?" *Environment*, May 2005, 11–21.

31. N. Straub and P. Behr, "Energy Regulatory Chief Says New Coal, Nuclear Plants May Be Unnecessary," *New York Times*, April 22, 2009.

32. H. Scheer, "Nuclear Energy Belongs in the Technology Museum," November 22, 2004. Available at http://renewableenergyaccess.com/rea/news/story?id=19012.

Chapter 9

1. C. Brownlee, "Inherit the Warmer Wind," *Science News*, December 2, 2006, 262–264.

2. J. Roach, "Global Warming Unstoppable for 100 Years, Study Says," *National Geographic News*, March 17, 2005.

3. E. Youngsteadt, "When Its Environment Changes, so Does a Sparrow's Tune," May 15, 2009. Available at http://sciencenow.sciencemag.org/cgi/content/full/2009/515/1.

4. Environmental Protection Agency, "Climate Change—Health and Environmental Effects: Ecosystems and Biodiversity," n.d. Available at http://epa.gov/climatechange/effects/eco.html.

5. T. R. Karl, J. M. Melillo, and T. C. Peterson (eds.), *Global Climate Change Impacts in the United States* (Cambridge: Cambridge University Press, 2009).

6. K. P. Green, "Is the Polar Bear Endangered, or Just Conveniently Charismatic?" (Washington, D.C.: American Enterprise Institute, May 2008).

7. Ibid.

8. D. A. Fahrenthold, "Obama Team Retains Bush Polar Bear Policy," *San Francisco Chronicle*, May 9, 2009, A4.

9. "NRDC Calls on the Obama administration to Rescue the Polar Bear," 2009. Available at http://rushprnews.com/2009/02/05/nrdc-calls-on-the-obama -administration-to-rescue-the-polar-bear/

10. World Wildlife Fund, "Polar Bear," n.d. Available at http://worldwildlife .org/polarbears/; A. C. Revkin, "More Polar Bear Populations in Decline," *New York Times Dot Earth*, July 6, 2009.

11. A. V. Revkin, "More on the Polar Bear's Fate," *New York Times,* March 24, 2009.

12. "Diving Polar Bear Confirms Inuit Observations," *New Scientist*, September 27, 2007, 16.

13. D. A. Fahrenthold, "Saving Species No Longer a Beauty Contest," *Washington Post*, June 29, 2009, A1.

14. W. F. Jasper, "Heat or Cold: Which Is More Deadly?" December 23, 2008. Available at http://thenewamerican.com/tech-mainmenu-30/environment/621 ?tmpl=component&print

15. Ibid.

16. B. Lomborg, "Chill Out," *Washington Post*, October 7, 2007, B1.

17. B. Orlove, "Glacier Retreat: Reviewing the Limits of Human Adaptation to Climate Change," *Environment,* May–June 2009, 22–34.

18. P. Damassa, "Polar Warming and Its Global Consequences," *Earth Trends*, February 2007; Karl, Melillo, and Peterson, *Global Climate Change Impacts in the United States,*.

19. A. C. Revkin, "In Greenland, Ice and Instability," *New York Times*, January 8, 2008 F1, F4.

20. J. Hansen, "Climate Change: On the Edge," February 17, 2006. Available at http://independent.co.uk/environment/climate-change-on-the-edge-466818.html.

21. P. Brown, "Ice Caps Melting Fast: Say Goodbye to the Big Apple," AlterNet, October 10, 2007. Available at http://alternet.org/module/printversion/64735.

22. Ibid.

23. "Climate Change Could Displace 25 Million by 2010," *Hindu*, June 10, 2009.

24. D. Struck, "In Arctic Ice, Lessons on Effects of Warming," *Washington Post*, June 9, 2007, A11.

25. J. Eilperin, "Antarctic Ice Sheet Is Melting Rapidly," *Washington Post*, March 3, 2006, A1.

26. R. Bilton, "Antarctica's Cold Awakening," BBC News, February 26, 2009. Available at http://bbc.co.uk/2/hi/science/nature/7909305.stm…

27. E. Engeler, "Antarctic Glaciers Melting Faster Than Thought," February 26, 2009. Available at http://usatoday.com/weather/.../2009-02-25-warming_N .htm.

28. F. Demarthon, "Spotlight on the Poles," CNRS International Magazine, April 2007, 18–23.

29. C. Dean, "As Alaska Glaciers Melt, It's Land That's Rising," New York Times, May 18, 2009, A1, A11; B. P. Kelly, T. Ainsworth, D. A. Boyce, Jr., E. Hood, P. Murphy, and J. Powell, "Climate Change: Predicted Impacts on Juneau," Scientific Panel on Climate Change City and Borough of Juneau, 2007.

30. Ecologist, "Ocean CO_2 'Sponge' Effect Slowing Down," October, 2007; J. Eilperin, "As Emissions Increase, Carbon 'Sinks' Get Clogged," Washington Post, December 3, 2009.

31. Woods Hole Oceanographic Institution, "In Carbon Dioxide-Rich Environment, Some Ocean Dwellers Increase Shell Production," December 3, 2009. Available at http://sciencedaily.com/releases/2009/12/091201182622.htm.

32. R. B. Gagosian, "Abrupt Climate Change: Should We e Worried?" World Economic Forum, Davos, Switzerland, January 27, 2003; National Academy of Sciences, National Research Council Committee on Abrupt Climate Change (Washington, D.C.: National Academy Press, 2002); R. A. Kerr, "Sea Change in the Atlantic," Science, January 2, 2004, 35.

33. J. Hansen, "Twenty Years Later: Tipping Points Near on Global Warming." Available at http://huffingtonpost.com/dr-james-hansen/twenty-years-later-tippin _b_108766.html.

34. F. Pearce, "But Here's What They Didn't Tell Us," New Scientist, February 10, 2007, 7–8.

35. S. Escalon, "Oceans and Climate: Stability Under Threat," CNRS International Magazine Quarterly October 2006, 21.

36. University Corporation for Atmospheric Research, "Most of Arctic's Near-Surface Permafrost May Thaw by 2100," December 19, 2005. Available at http://ucar.edu/news/releases/2005/permafrost.shtml.

37. I. Sample, "Warming Hits 'Tipping Point,'" Guardian [London], August 11, 2005.

38. S. Connor, "Exclusive: The Methane Time Bomb," Independent [UK], September 23, 2008; M. Inman, "Methane Bubbling Up from Undersea Permafrost?" National Geographic News, December 19, 2008.

39. F. Pearce, "Climate Warming as Siberia Melts," New Scientist, August 13, 2005, 12.

40. F. Pearce, "Dark Future Looms for Arctic Tundra," New Scientist, January 21, 2006, 15.

41. Environmental Protection Agency, "Inventory of U, S. Greenhouse Gas Emissions and Sinks: 1990–2007," April 2009.

42. Intergovernmental Panel on Climate Change, 4th Assessment Report, Table 2.14, 2007; D. Archer, M. Eby, V. Brovkin, A. Ridgwell, L. Cao, U. Mikolaje-

wicz, K. Caldeira, . Matsumoto, G. Munhoven, A. Montenegro, and K. Tokos, "Atmospheric Lifetime of Fossil Fuel Carbon Dioxide," *Annual Review of Earth and Planetary Sciences*, 2009, 117–134.

43. Environmental Protection Agency, "Inventory of U.S. Greenhouse Gas Emissions and Sinks."

44. R. Goodland and J. Anhang, "Livestock and Climate Change," *Worldwatch*, November–December 2009, 10–19; F. MacKay, "Looking for a Solution to Cows' Climate Problem," *New York Times*, November 17, 2009.

45. R. C. Kaufmann, "Using the Market to Address Climate Change," 103–106, in *State of the World 2009*, ed. L. Starke (New York: Norton, 2009); J. Hansen, "Cap and Fade," *New York Times*, December 7, 2009.

46. A. Kollmuss, "Carbon Offsets Oil," *Worldwatch*, July–August 2007, 9–14.

47. "Forests and the Planet," *New York Times*, May 29, 2009, A24; B. Heinrich, "Clear-Cutting the Truth about Trees," *New York Times*, December 20, 2009.

48. D. Howden, "Deforestation: The Hidden Cause of Global Warming," *Independent* [England], May, 14, 2007.

49. "Fear Of Methane Emissions From Oceans," *Energy World*, November, 2009, 4.

50. J. Hansen and L. Nazarenko, "Soot Climate Forcing Via Snow and Ice Albedos," *Proceedings of the National Academy of Sciences* 101 (2004): 423–428.

51. "Black and White: Soot on Ice." Available at http://nasa.gov/vision/earth/environment/arctic_soot.html.

52. R. Engelman, "Sealing the Deal to Save the Climate," 170–171, in *State of the World 2009*, ed. L. Starke (New York) Norton, 2009).

53. D. Kreutzer, "The Economic Impact Of Cap And Trade". Heritage Foundation, 2009.

54. Ibid.

55. M. Casey and S. Borenstein, "Google, NASA Team Up to Save the World," *Jerusalem Post*, December 18, 2009.

56. J. Hopkinson, "Environment Study Looks at Future of Coastlines," 2009 Available at http://delmarvanow.com/apps/pbcs.dll/article?AID=20091105/WCT01/911050330&tem...

57. E. Ahronovitz, "Who's Afraid of Global Warming?" February 13, 2009, 19–22; W. J. Broad, "In Ancient Fossils, Seeds of a New Debate on Warming," *New York Times*, November 7, 2006, F1, F4; "Global Warming: Sun Takes Some Heat," *Environment*, October 2004, 7; Environmental Protection Agency, "Past Climate Change," n.d. Available at http://epa.gov/climatechange/science/pastcc.html; G. Stanhill, "A Perspective on Global Warming, Dimming, and Brightening," EOS [American Geophysical Union], January 30, 2007, 58.

58. N. Shaviv and J. Veizer, "Celestial Driver Of Paleozoic Climate?" *GSA Today*, July, 2003, 4–10.

59. Geochange, 2010, Available at : http://geo-change.org/

60. financialpost.com/story.html?id=597d0677-2a05-47b4-b34f-b84068db11f48p=4.

61. R. T. Patterson, "Read the Sunspots," *Financial Post*, June 20, 2007.

Chapter 10

1. "Smog Clogs Arteries," *Science News*, December 11, 2004, 372–373; American Lung Association, "Key Facts about Air Pollution," 2004.

2. S. Slanina, "Air Pollution in China," May 13, 2008. Available at http://eoearth.org/article/Air_pollution_in_China.

3. American Lung Association, "State of the Air Report, 2008; Environmental Protection Agency, "Latest Findings on National Air Quality: Status and Trends through 2008, 2009."

4. D. Bailey et al., "Harboring Pollution: The Dirty Truth about U.S. Ports," Natural Resources Defense Council, 2004; K. Lydersen and J. Eilperin, "EPA Proposal to Cut Great Lakes Ship Emissions Stirs the Waters," *Washington Post*, October 23, 2009.

5. "Sick Ships," *Worldwatch*, May–June 2009, 5.

6. F. Barringer, "California Air Is Cleaner, But Troubles Remain," *New York Times*, August 3, 2005, A1, A16.

7. Sierra Club, "Dirty Coal Power," n.d.

8. Environmental Protection Agency, "Latest Findings on National Air Quality," 2009; "Drop in U.S. Air Pollution Linked to Longer Lifespans," CNNhealth.com, January 21, 2009.

9. N. Bakalar, "Standards: Even Approved Amount of Ozone Is Found Harmful," *New York Times*, February 28, 2006, F6.

10. M. Janofsky, "Many Counties Failing Fine-Particle Air Rules," *New York Times*, December 18, 2004, A12.

11. American Lung Association, "Particle Pollution," in *State of the Air Report 2008*; B. Weinhold, "Ozone Nation: EPA Standard Panned by the People," *Environmental Health Perspectives*, July 2008. Available at http://ehponline.org/members/2008/116-7/spheres.html; C. Potera, "Cleaner Air, Longer Life," *Environmental Health Perspectives*, April 2009. Available at http://ehponline.org/docs/2009/117-4/forum.html.

12. Asthma and Allergy Foundation of America, "Asthma Facts and Figures," 2009.

13. T. Tillett, "Sperm Alert," *Environmental Health Perspectives*, March 2006, A177.

14. J. Selim, "Fetuses Take Air Pollution to Heart," *Discover*, April 2002, 12.

15. J. Eilperin, "EPA Places Stricter Regulations on Airborne Lead," *Washington Post*, October 17, 2008, A5.

16. P. Thomas, "Tyre Dust: Where the Rubber Meets the Road," *Ecologist*, November 2005, 14–18.

17. A. Schneider, "Amid Nanotech's Dazzling Promise, Health Risks Grow," March 24, 2010. Available at aolnews.com/nanotech/article/amid-nanotechs -dazzling-promise-health- risks-grow/19401235.

18. Environmental Protection Agency, "Air Quality Trends," 2009. Available at http://epa.gov/airtrends/aqtrends.html#comparison.

19. J. Siegel, "More Air Pollution Means Less Rain in Hilly Areas," *Jerusalem Post*, March 9, 2007, 7.

20. A. C. Revkin, "Ozone Solution Poses a Growing Climate Threat," *New York Times Dot Earth*, June 22, 2009; D. A. Fahrenthold, "Chemicals That Eased One Woe Worsen Another," *Washington Post*, July 20, 2009, A7.

21. P. D. Thacker, "No Silver Bullet to Replace Methyl Bromide," *Environmental Science and Technology*, January 1, 2005, 13A.

22. J. M. Samer, M. S. Marburg, and J. U. Spengler, "Health Effects and Sources of Indoor Air Pollution," American Lung Association, 1988.

23. T. Socha, "Air Pollution Causes and Effects," September 11, 2007. Available at http://healthandenergy.com/air_pollution causes htm

24. "Pollution Bad for Productivity," *Ecologist*, October 2004, 10.

25. "Spring Forward Environmentally," Worldwatch Institute, March 23, 2007; Socha, "Air Pollution Causes and Effects."

26. B. Booth and K. Betts, "Cooking Spews Out Ultrafine Particles," *Environmental Science and Technology*, April 15, 2004, 141A–142A.

27. R. N. Proctor, "The Golden Weed, America's Most Deadly Drug," *Science*, May 4, 2007, 692–693.

28. MedHeadlines, "Smoking on (Very Slow) Decline in US," November 18, 2008.

29. American Cancer Society, "Cigarette Smoking," 2003.

30. M. Kaufman, "U.S. Details Dangers of Secondhand Smoking," *Washington Post*, June 28, 2006, A1.

31. B. Harder, "Living in a Fog," *Science News*, January 15, 2005, 38.

32. M. E. Gunter, "Asbestos As a Metaphor for Teaching Risk Perception," *Journal Of Geological Education*, 42:17–24, 1994.

33. K. Ravn, "In Search of a Nontoxic Home," *Los Angeles Times*, July 27, 2009.

34. S. H. Hviid, "Personal Stories from Environmentally Ill Patients around the Environmental Health Center in Dallas," 1999. Available at http://ehcd.com/ websteen/ehcd_patient_stories.htm.

Chapter 11

1. "Kennedy Slams Bush Policy On Environment At Sundance," *The Daily Herald*, September 18, 2005. Available At http://harktheherald.com/modules

.php?op=modload&name=news&file=article&sid=64514; H. Cooper and J. M. Broder, "BP's Ties to Agency Are Long and Complex," *New York Times*, May 25, 2010; K. Kindy "Analysis Finds Uneasy Mix in Auto Industry and Regulation," *Washington Post*, March 9, 2010.

2. "James Hansen Talks About Climate Change," *Worldwatch*, July–August, 2008, 28.

3. F. Barringer, "U.S. Given Poor Marks on the Environment," *New York Times*, January 23, 2008, A8.

4. S. L. Baker, "Stop Eating Processed and Fried Foods and You'll Restore the Body's Natural Defenses, Study Finds," November 26, 2009. Available at http://naturalnews.com/z027589_processed_foods_health.html.

5. H. Oldmeadow, ed., *Why Work, World Wisdom* (New York: Penguin, 2005), 217–228.

6. G. Snyder, "Four Changes," 1970.

7. Nine excellent two-page articles by different authors centering around the idea that Western economies are not sustainable and need to be modified to center on development rather than unfettered growth are published in *New Scientist*, October 18, 2008, 40–54.

8. G. T. Gardner, *Inspiring Progress* (New York: Norton, 2006), 21.

9. Ibid., 107.

10. Ibid., 115–127.

11. "At a Glance Environment," *Dallas Morning News*, January 17, 2009.

12. J. Ziegler, "U. N. Official Says Crops As Biofuel 'Crime Against Humanity'," October 27, 2007. Available at http://helenair.com/lifestyles/article_4859a6d5-a734-586d-a075-

13. S. O'Connell, "Can Geo-engineering Rebuild the Planet?" February 16, 2009. Available at http://telegraph.co.uk/earth/4641586/Can-geo-engineering-rebuild-the-planet.html; "Governments 'Eager' to Divert Climate Funding to Geoengineering," *Ecologist*, December 17, 2009.

14. F. Schwab, "All Aboard For Geoengineering?" *Earth*, January, 2010, 88; "Taking A Bite Out Of Geoengineering," *Bulletin of the American Meteorological Society*, July, 2009, 911; J. Pelley, "Potential of Geoengineering Highly Uncertain," *Environmental Science & Technology*" November 15, 2009, 8472–8473; S.O'Connell, "Can Geo-Engineering Rebuild The Planet?" Available at http://telegraph.co.uk/earth/4641586/can-geo-engineering-rebuild the planet .html; R. Kunzig, "Geoengineering: How to Cool Earth—At a Price," *Scientific American*, October 20, 2008.

15. R. Kunzig, "Geoengineering: How to Cool the Earth—at a Price," *Scientific American*, October 20, 2008.; K. Caldeira, "Taking a Bite out of Geoengineering," *Bulletin of the American Meteorological Society*, July 2009, 911; H. Fountain, "More Carbon Dioxide May Create a Racket in the Seas," *New York Times*, December 29, 2009.

16. Ecological Society of America, "Geoengineering to Mitigate Global Warming May Cause Other Environmental Harm," August 7, 2009. Available at: http://sciencedaily.com/releases/2009/08/090806080142.htm; J. Tierney, "The Earth Is Warming? Adjust the Thermostat," *New York Times*, August 11, 2008.

17. R. Oliver, "All about: Religion and the Environment," 2008. Available at http://cnn.com/2008/WORLD/asiapcf/01/27/eco.about.religion;L. Sevier, "Life, Religion and Everything," *Ecologist*, September 2007, 38–42.

18. "Faith in Action: Communities of Faith Bring Hope for the Planet," Sierra Club, 2008.

19. "Pope Urges Respect For The Environment,"UPI, January 2, 2010. Available at http://upi.com/Top_News/International/2010/;01/02/Pope-urges-respect-for-environment/U...

20. "Impact Of Religions Will Have 'Deeper Roots' Than Copenhagen," November 2, 2009.Available at http://guardian.co.uk/environment/2009/nov/02/impact-religion-copenhagen

21. D. Browning, "Religion and Climate Change," Environmental Defense Fund, January 8, 2010.

22. D. Browning, "Religion and Climate Change," January 8, 2010.

23. M. Anslow, "Impact of Religions Will Have 'Deeper Roots' Than Copenhagen," Available at http://guardian.co.uk/environment/2009/nov/02/impact-religion-copenhagen_

24. L. Sevier, "Life, Religion and Everything," *Ecologist*, September 2007, 40.

25. "Soulful Living At Work," Health Education Alliance For Life and Longevity, n.d.. Available at http://heall.com/soul/spiritatwork.html.

Additional Readings

Introduction

Myers, S. S. *Global Environmental Change: The Threat to Human Health.* Washington, DC: *Worldwatch*, 2009.

Chapter 1: Water

Alley, W. M. "Tracking U.S. Groundwater." *Environment*, April 2006, 11–25.

Bergkamp, G., and C. W. Sadoff. "Water in a Sustainable Economy," 107–122. In *State of the World 2008*, ed. L. Starke. New York: Norton, 2008.

Circle of Blue. *U.S. Faces Era of Water Scarcity: Profligate Use Hurts in Unexpected Places.* July 9, 2008. Available at csrwire.com/PressReleasePrint.php ?=12592.

Duncan, D. E. "The Pollution Within." *National Geographic*, October 2006, 116–143.

Environmental Working Group. *A National Assessment of Tap Water Quality.* Washington, DC: Environmental Working Group. 2005.

George, R. *The Big Necessity: The Unmentionable World of Human Waste and Why It Matters.* New York: Holt, 2008.

Gies, E. "Water Wars: Is Water a Human Right or a Commodity?" *Worldwatch*, March–April 2009, 22–27.

Gilliom, R. I. "Pesticides in U.S. Streams and Groundwater." *Environmental Science and Technology*, May 15, 2007, 3409–3414.

Jones, T. "Great Lakes Key Front in Water Wars: Western, Southern States Covet Midwest Resource." *Chicago Tribune*, October 28, 2007.

Lean, G. "Official: Men Really Are the Weaker Sex." *Independent* (London), December 7, 2008.

Song, L. "Rethinking Water Management." *Earth*, January 2010, 36–45.

Yauck, J. "Taking a Fresh Look at Desalination." *Geotimes*, October 2006, 42–43.

Chapter 2: Infrastructure

American Society of Civil Engineers. *2009 Report on America's Infrastructure.* Reston, VA: American Society of Civil Engineers, 2009. Available at asce.org/reportcard/2009/index2009.cfm.

Bourne, J. K., Jr. "New Orleans: A Perilous Future." *National Geographic,* August 2007, 32–67.

Doyle, M. W., and D. G. Havlick. "Infrastructure and Environment." *Annual Review of Environment and Resources* 34 (2009): 349–373.

Gellings, C. W., and K. E. Yeager. "Transforming the Electric Infrastructure." *Physics Today,* December 2004, 45–51.

Gramlich, R., and M. Goggin. "Green Power Superhighways." Washington, DC: American Wind Energy Association/Solar Energy Industries Association, February 2009.

Herbert, B. "What the Future May Hold." *New York Times,* November 17, 2009.

Lavelle, M. "Water Woes." *U.S. News and World Report,* June 4, 2007, 37–53.

Lekic, S. "Runway Incursions a Top Concern at Overcrowded Airports." *Aviation,* November 9, 2007.

Mark, J. C. "Down Go the Dams." *Scientific American,* March 2007, 66–71.

McLaughlin, E. C. "Experts: Leadership, Money Keys to Building Bridges." 2007. Available at CNN.com/US cnn.com/2007/US/08/02/bridge.infrastructure/index.html.

Powers, K. "Aging Infrastructure: Dam Safety." Washington, DC: Congressional Research Service, 2005.

Roth, L. Testimony of the American Society of Civil Engineers to the House Committee on Transportation and Infrastructure, May 8, 2007.

Schoen, J. W. "U.S. Highways Badly in Need of Repair." MSNBC.com, August 3, 2007. Available at msnbc.msn.com/id/20095291/print/1/displaymode/1098/.

Schwartz, S. I. "Catch Me, I'm Falling." *New York Times,* August 13, 2007, A19.

Snoeyink, V. L., et al. *Drinking Water Distribution Systems: Assessing and Reducing Risks.* Washington, DC: National Academies Press, 2006.

Tibbetts, J. "Down, Dirty, and Out of Date." *Environmental Health Perspectives,* July 2005, A465–A467.

U.S. Environmental Protection Agency. *Sustaining Our Nation's Water Infrastructure.* Washington, DC: U.S. Environmental Protection Agency, 2006.

Chapter 3: Floods

Bourne, J. K., Jr. "New Orleans: A Perilous Future." *National Geographic,* August 2007, 32–67.

Colten, C. E., R. W. Kates, and S. B. Laska. "Three Years After Katrina." *Environment*, September–October 2008, 36–47.

Crossett, K. M., T. J. Culliton, P. C. Wiley, and T. R. Goodspeed. "Population Trends Along the Coastal United States: 1980–2008." Washington, DC: National Oceanographic and Atmospheric Administration, 2004.

Johnson, G. P., R. R. Holmes, Jr., and L. A. Waite. "The Great Flood of 1993 on the Upper Mississippi River—10 Years Later." Fact sheet 2004–3024. Reston, VA: U.S. Geological Survey, 2004.

"New Orleans and the Delta." *American Experience*. PBS. September 2, 2005. Available at http://pbs.org/now/science/neworleans.html.

O'Connor, J. E., and J. E. Costa. "Large Floods in the United States: Where They Happen and Why." Circular 1245. Reston, VA: U.S. Geological Survey, 2003.

O'Connor, J. E., and J. E. Costa. "The World's Largest Floods, Past and Present: Their Causes and Magnitudes." Circular 1254. Reston, VA: U.S. Geological Survey 2005.

Pielke, R. A., Jr. "Nine Fallacies of Floods." *Climatic Change* 42 (1999): 413–438.

Saulny, S. "Development Rises on St. Louis Area Flood Plains." *New York Times*, May 15, 2007, A13.

Trenberth, K. E. "Warmer Oceans, Stronger Hurricanes." *Scientific American*, July 2007, 45–51.

Whoriskey, P. "La. Plan to Reclaim Land Would Divert the Mississippi." *Washington Post*, May 1, 2007, A3.

Chapter 4: Garbage

Deneen, S. "How to Recycle Practically Anything." *E Magazine*, May–June 2006, 26–31, 63.

Douglas, E. "Better by Design." *New Scientist*, January 6, 2007, 31–35.

Gugliotta, G. "Retiring Tires: A Heated Debate on Using Them as Fuel." *Discover*, February 2008, 26–27.

Gurskis, J. "How Green Are Your Grocery Bags?" Fox News, March 27, 2009.

Ocean Conservancy. "A Rising Tide of Ocean Debris." Washington, DC: Ocean Conservancy, 2009.

Robbins, E. "Flushing Forests." *Worldwatch* May–June 2010, 6–11.

Royte, E. *Garbage Land: On the Secret Trail of Trash*. Boston: Back Bay Books, 2006.

Thomas, P. "The Lethal Consequences of Breathing Fire." *Ecologist*, September 2006, 44–47. U.S. Environmental Protection Agency. *Municipal Solid Waste Generation, Recycling, and Disposal in the United States: Facts and Figures for 2007*. Washington, DC: Washington, D.C.: U.S. Environmental Protection Agency, 2008.

Chapter 5: Soil, Crops, and Food

Adams, M., "Nutrition Can Save America," November 28, 2009. Available at http://naturalnews.com/report_Nutrition_Health_America_0.html.

Blatt, H., *America's Food: What You Don't Know About What You Eat* Cambridge, MA: MIT Press, 2008.

Foer, J. F. *Eating Animals.* New York: Little, Brown, 2009.

Greene, C., E. Slattery, and W. D. McBride. "America's Organic Farmers Face Issues and Opportunities." *Amber Waves,* June, 2010.

Halweill, B. "Still No Free Lunch: Nutrient Levels in U.S. Food Supply Eroded by Pursuit of High Yields." Boulder, CO: Organic Center, 2007.

Halweill, B. "Meat Production Continues to Rise." Washington, DC: Worldwatch Institute, August 20, 2008.

Kallman, M. "Genetically Modified Crops and the Future of World Agriculture." *Earth Trends,* May 2008.

McKeown, A. "Fish Farming Continues to Grow as World Fisheries Stagnate." Washington, DC: Worldwatch Institute, December 17, 2008.

Pollan, M. "Unhappy Meals." *New York Times Magazine,* January 28, 2007, 38–45.

Pollan, M. "Farmer in Chief," *New York Times Magazine,* October 9, 2008, 64–71.

Schoonover, H., and Muller, M., "Food Without Thought: How U.S. Farm Policy Contributes to Obesity." Minneapolis, MN: Institute for Agriculture and Trade Policy, 2006.

U.S. Department of Agriculture, *Food and Drug Atlas,* 2010.

U.S. Food and Drug Administration, "The Food Defect Action Levels: Levels of Natural or Unavoidable Defects in Foods That Present No Health Hazards for Humans." Washington, DC: U.S. Food and Drug Administration, 2005.

Chapter 6: Fossil Fuels

Collett, T. S. "Gas Hydrates as a Future Energy Resource." *Geotimes,* November 2004, 24–27.

Combs, S. "General U.S. Government Energy Web Sites: Window on State Government, State of Texas." N.d. Available at http://window.state.tx.us/specialrpt/energy/furtherinfo/.

Energy Information Administration. "Annual Energy Review 2007." Washington, DC: Energy Information Administration, 2008.

Hatch, C., and M. Price. "Canada's Toxic Tar Sands: The Most Destructive Project on Earth." Toronto: Environmental Defence (Canada), 2008.

Lynch, P. "The True Cost of Fossil Fuels," May 12, 2008. Available at http://renewableenergyworld.com/rea/news/recolumnists/story?id=52359.

Madsen, T., and R. Sargent. "Making Sense of the 'Coal Rush.'" Boston: U.S. PIRG Education Fund, National Association of State PIRGs, 2006.

Sample, I. "Final Warning." *New Scientist*, June 28, 2008, 32–37.

Sever, M. "Coal's Staying Power." *Geotimes*, September 2006, 18–19.

Schon, S. C., and A. A, Small III. "Climate Change and the Potential of Coal Gasification." *Geotimes*, September 2006, 21–23.

Five articles by various authors, "Peak Oil Forum." *Worldwatch*, January–February 2006, 9–24: T. Prugh, Peak Oil Forum, 9; K. Aleklett, Oil: A Bumpy Road Ahead, 10–12; R. Cavaney, Global Oil Production About to Peak? A Recurring Myth, 13–15; C. Flavin, Over the Peak, 16–18; R. K. Kaufmann, Planning for the Peak in World Oil Production, 19–21; V. Smil, Peak Oil: A Catastrophist Cult and Complex Realities, 22–24.

U.S. Environmental Protection Agency. "EPA's Report on the Environment; Highlights of National Trends." Washington, DC: U.S. Environmental Protection Agency, 2008.

Chapter 7: Alternative Energy Sources

General

Ballesteros, A. R., J. Coequyt, M. Furtado, J. Inventor, W. Krewitt, D. Mittler, O. Schafer, S. Simon, S. Teske, and A. Zervos. "Future Investment." Washington, DC: Greenpeace, July 2007.

Flavin, C. "Building a Low-Carbon Economy," 75–90. In *2008 State of the World*, ed. L. Starke. New York: Norton, 2008.

Flavin, C. "Low-Carbon Energy: A Roadmap." Institute Report 178. Washington, DC: *Worldwatch*, 2008.

Goldemberg, J. "Energy Choices Toward a Sustainable Future," *Environment*, December 2007, 6–17.

Greenpeace. "Energy Revolution." Washington, DC: Greenpeace, 2008.

Jacobson, M. Z. "Review of Solutions to Global Warming, Air Pollution, and Energy Security." *Energy and Environmental Science* 2 (2009): 148–173.

McCulloch, R. "Clean Energy, Bright Future." Washington, DC: Environment America Research and Policy Center, 2009.

Sawin, J. L., and W. R. Moomaw. "An Enduring Energy Future," 130–150. In *2009 State of the World*, ed. L. Starke. New York: Norton, 2009.

Sawin, J. L., and W. R. Moomaw. "Renewable Revolution: Low-Carbon Energy by 2030." Washington, D.C.: *Worldwatch*, 2009.

Weiss, C., and W. B. Bonvillian. "Stimulating a Revolution in Sustainable Energy Technology." *Environment*, July–August 2009, 10–20.

Biofuels

Coyle, W. "The Future of Biofuels: A Global Perspective." *Amber Waves*, November 2007.

Earley, J., and A. McKeown. "Smart Choices for Biofuels." Washington, DC: Sierra Club/Worldwatch Institute, 2009.

Holzman, D. C. "The Carbon Footprint of Biofuels." *Environmental Health Perspectives*, June 2008, A246–A252.

Naylor, R. L., A. J. Liska, M. B. Burke, W. P. Falcon, J. C. Gaskell, S. D. Rozelle, and K. G. Cassman "The Ripple Effect: Biofuels, Food Security, and the Environment." *Environment*, November 2007, 30–43.

Thomas, P. "Biofuels." *Ecologist*, March 2007, 27–45.

Energy from Moving Water

Jeffries, E. "Ocean Motion Power." *Worldwatch*, July–August 2008, 22–27.

Johnston, M. W. "Harnessing the Tides: Marine Power Update 2009." February 2009. Available at renewableenergyworld.com/rea/news/article/2009/02/harnessing-the-tides-marine-power-update-2009?cmpid=WNL/

Energy from Nuclear Fission

Brooks, M. "Is It All Over for Nuclear Power?" *New Scientist*, April 22, 2006, 33–37.

Hultman, N. E., J. G. Koomey, and D. M. Kammen. "What History Can Teach Us about the Future Costs of U.S. Nuclear Power." *Environmental Science and Technology*, April 1, 2007, 2088–2093.

Energy from the Earth's Interior

Green, B. D., and R. G. Nix. "Geothermal—The Energy under Our Feet." Washington, DC: National Renewable Energy Laboratory, November 2006.

Lacey, S. "Technological Innovation Driving Renewed Interest in Geothermal Energy." Renewable Energy World.com, October 9, 2008. Available at http://renewableenergyworld.com/rea/news/story?id=53805.

Tester, J. "The Future of Geothermal Energy." Cambridge, MA: MIT, 2007.

Energy from the Wind

Bolinger, M., and R. Wiser. "Surpassing Expectations: State of the US Wind Power Market." September 4, 2008.

U.S. Department of Energy. "20% Wind Energy by 2030." Washington, DC: U.S. Department of Energy, July 2008.

U.S. Department of Energy. "Annual Report on U.S. Wind Power Installation, Cost, and Performance Trends: 2007." Washington, DC: U.S. Department of Energy, 2008.

Energy from the Sun

Mills, D. R., and R. G. Morgan. "A Solar-Powered Economy: How Solar Thermal Can Replace Coal, Gas and Oil." July 3, 2008. Available at http://renewableenergyworld.com/rea/news/print?id=52693.

Zweibel, K., J. Mason, and V. Fthenakis. "A Solar Grand Plan." *Scientific American*, January 2008, 64–73.

Fuel Cells

Satyapal, S., J. Petrivic, and G. Thomas. "Gassing up with Hydrogen." *Scientific American*, April 2007, 80–87.

Service, R. F. "The Hydrogen Backlash." *Science*, August 13, 2004, 958–961.

"Clean Vehicles: Fuel Cell Vehicles." Cambridge, MA: Union of Concerned Scientists, n.d.

Chapter 8: The Nuclear Energy Controversy

Charman, K. "Brave Nuclear World? Part 1." *Worldwatch*, May–June, 2006, 26–31; Part 2, July–August 2006, 12–20.

Dawson, J. I., and R. G. Darst. "Russia's Proposal for a Global Nuclear Waste Repository: Safe, Secure, and Environmentally Just?" *Environment*, May 2005, 11–21.

Hultman, N. E., J. G. Koomey, and D. M. Kammen. "What History Can Teach Us about the Future Costs of U.S. Nuclear Power." *Environmental Science and Technology*, April 1, 2007, 2088–2093.

Keystone Center. "Nuclear Power Joint Fact-Finding," Keystone, CO: Keystone Center, June 2007.

Motavalli, J. "A Nuclear Phoenix?" *E Magazine*, July–August 2007, 26–36.

Ramana, M. V. "Nuclear Power: Economic, Safety, Health, and Environmental Issues of Near-Term Technologies." *Annual Review of Environment and Resources* 34 (2009): 127–152.

"Staff Responses to Frequently Asked Questions Concerning Decommissioning of Nuclear Power Reactors." Available at eu.decom.be/faqs/faqf4.htm.

Chapter 9: Climate Change

Blaustein, R. J. "Recarbonizing the Earth." *Worldwatch*, January–February 2010, 24–28.

Burger, N., N. Ecla, T. Light, and M. Toman. "In Search of Effective and Viable Policies to Reduce Greenhouse Gases." *Environment*, May–June 2009, 8–18.

Flavin, C. "Low Carbon Energy: A Roadmap." In Report 178, ed. L. Mastny. Washington, DC: Worldwatch Institute, 2008.

Gardner, G. T., and P. C. Stern, "The Short List." *Environment*, September–October 2008, 12–24.

Intergovernmental Panel on Climate ChangeClimate Change 2007: Synthesis Report, 2007.

Levinson, D. H., and J. H. Lawrimore (eds.). "State of the Climate in 2007." *Bulletin of the American Meteorological Society, Supplement*, July 2008.

Monaghan, A. "Antarctica and Climate Change." *Worldwatch*, January–February 2009, 6–12.

Nisbet, M. C. "Communicating Climate Change." *Environment*, March–April 2009, 12–23.

Orlove, B. "Glacier Retreat: Reviewing the Limits of Human Adaptation to Climate Change." *Environment*, May–June 2009, 22–34.

Pearce, F. *With Speed and Violence: Why Scientists Fear Tipping Points in Climate Change.* Boston: Beacon Press, 2007.

Randall, D. "What on Earth? The Concerned Citizen's Guide to Global Warming" *e Independent [UK]*, December 6, 2009. Available at //independent.co.uk/environment/climate-change/what-on-earth-the-concerned-citizen's-guide-to-global-warming-1835069.htm.

Scherr, S. J., and S. Sthapit. *Mitigating Climate Change through Food and Land Use.* Washington, DC: Worldwatch Institute, 2008.

Siikamäki, J. "Climate Change and U.S. Agriculture: Examining the Connections." *Environment*, July–August 2008, 36–49.

Starke, L. (ed.). *2009 State of the World.* New York: Norton, 2009.

Trumper, K., M. Bertzky, B. Dickson, G. van der Heijden, M. Jenkins, and R. Manning. *The Natural Fix? The Role of Ecosystems in Climate Mitigation.* New York: United Nations Environmental Programme, 2009.

U.S. Environmental Protection Agency. "Inventory of U.S. Greenhouse Gas Emissions and Sinks: 1990-2007." Washington, DC: U.S. Environmental Protection Agency, April 2009.

Chapter 10: Air Pollution

American Lung Association. "State of the Air Report 2008." Washington, DC: American Lung Association, 2008.

Kahl, J. S., et al. "Have U.S. Surface Waters Responded to the 1990 Clean Air Amendments?" *Environmental Science and Technology*, December 15, 2004, 484A–490A.

U.S. Environmental Protection Agency. "National Air Quality Status and Trends Through 2007." Washington, DC: U.S. Environmental Protection Agency, 2008.

U.S. Environmental Protection Agency. "The Inside Story: A Guide to Indoor Air Quality." Washington, DC: U.S. Environmental Protection Agency, April 1, 2009.

Chapter 11: Conclusion

Adams, M. "Why the Free Market Doesn't Work: Consumption vs. Conservation." *Natural News,* April 23, 2009.

Diamond, J. "What's Your Consumption Factor?" *New York Times,* January 2, 2008, A17.

Diamond, J. "Will Big Business Save the Earth?" *New York Times,* December 6, 2009.

Eyles, J., and N. Consitt. "What's at Risk: Environmental Influences on Human Health." *Environment,* October 2004, 24–39.

Gardner, G. T. *Inspiring Progress.* New York: Norton, 2006.

Hayward, S. F. *Index of Leading Economic Indicators 2008.* 13th ed. San Francisco, Pacific Research Institute, 2008.

Jackson, T. *Prosperity without Growth? The Transition to a Sustainable Economy.* London: UK Sustainable Development Commission, 2009.

Mazur, L., and L. Miles. *Conversations with Green Gurus.* Chichester, United Kingdom: John Wiley, 2009.

Schweickart, D. "A New Capitalism—or a New World?" *Worldwatch,* September–October 2009, 12–19.

Sierra Club. "Faith in Action: Communities of Faith Bring Hope for the Planet." Washington, DC: Sierra Club, 2008.

Starke, L. (ed.). *Vital Signs 2009.* Washington, D.C.: Worldwatch Institute, 2009.

U.S. Environmental Protection Agency. "EPA's Report on the Environment; Highlights of National Trends." Washington, DC: U.S. Environmental Protection Agency, 2008.

Index

Printed in the United States
by Baker & Taylor Publisher Services